An Introduction to The Principles of Medical Imaging

An Introduction to
The Principles of Medical Imaging

Chris Guy
Imperial College, London

Dominic ffytche
Institute of Psychiatry, London

Imperial College Press

Published by

Imperial College Press
57 Shelton Street
Covent Garden
London WC2H 9HE

Distributed by

World Scientific Publishing Co. Pte. Ltd.
P O Box 128, Farrer Road, Singapore 912805
USA office: Suite 1B, 1060 Main Street, River Edge, NJ 07661
UK office: 57 Shelton Street, Covent Garden, London WC2H 9HE

British Library Cataloguing-in-Publication Data
A catalogue record for this book is available from the British Library.

AN INTRODUCTION TO THE PRINCIPLES OF MEDICAL IMAGING

ISBN 1-86094-138-9

This book is printed on acid and chlorine free paper.

Printed in Singapore by FuIsland Offset Printing

PREFACE

We were originally motivated to write *An Introduction to the Principles of Medical Imaging* simply because one of us, CNG, could not find one text, at an appropriate level, to support an optional undergraduate course for physicists. Shortly after we started the book, we realised that both medical students in general and beginning medical research scientists could also benefit from the conceptual approach that we had chosen for this book. Thus from the outset we have been keeping in mind the needs of two apparently different parishes of readers. Although we have had to take care over the different levels of assumed knowledge for the two parishes, our teaching experience has shown that they share one very important characteristic; neither group by choice uses the symbolic language of applied mathematics in their own descriptions of the physical world. Thus it is inappropriate to introduce medical imaging with too heavy a reliance on purely mathematical descriptions. Instead we have chosen to use words and pictures wherever possible. In many ways a more conceptual, even qualitative approach to the subject is actually more appropriate since neither human biology nor medicine are exact sciences. The instrumentation and technologies used in medical imaging are certainly susceptible to exact mathematical analysis, but individual human body is not. In health, human beings come in a wide variety of shapes, sizes and constitutions so that, when a disease process is to be diagnosed in one individual, this has to be accomplished against a background of a wide range of normality. This requires an informed qualitative judgement on the part of the clinician. When imaging forms part of an investigation its results often provide another partial clue to the solution of a complex puzzle, rather than a single definitive answer.

The increasing use of quantitative measurement and computers in clinical medicine is gradually changing the traditional education of medical students. In addition the use of modern imaging methods, as an aid to diagnosis, is blurring the traditional relationship between clinical and technical staff, shaped historically by the century long use of x-ray radiography. Roughly speaking the technologists, medical physicists, radiographers were, historically, just employed to maintain the x-ray sets, take the exposures ordered by the radiologist, develop the films and hand over the pictorial results on film. Traditionally the medic's job really began when she received the films and this terminated the involvement of the technologists, barring instrument malfunction A clear and strict division of legal and ethical responsibility in medicine is of course absolutely necessary and will remain unchanged in the future. We suspect however that the related division in scientific understanding will not survive long into the next century. Medics increasingly need a strong understanding of the technical principles and especially the limitations of the instruments that they use; the traditional preserve of the hospital technologist. Furthermore modern imaging is gradually moving away from the production of qualitative static "photographs" of anatomy towards the generation and use of graphs and maps of numbers, extracted from digitised images and measurements. The diagnostic process can now call upon more than one type of investigation whose results have to be combined into a unified picture of the problem. This influx of new and qualitatively different data has created a problem of how best to present the clinician with the results of combined imaging investigations. One obvious approach, is to dream up ways of combining the results in terms of multi-coloured tokens. This however is simply replacing one qualitative picture with another and assumes that the clinician will not or cannot be concerned with how a particular combination of tokens has been assembled. An automated approach can only be partially successful and is rather limited in comparison with the power of the informed human intellect.

Throughout the book we emphasise the general principle at the expense of detailed specific technical detail. We do this because the principles are likely to be more enduring than any particular current practical implementation of a technology. As far as possible we have minimised the amount of mathematics used, in accord with Bragg's dictum that all good science is explicable to a school child. There is however one important general area where we have found this to be impossible. The central theme of the book is tomographic reconstruction and this relies on a mathematical result from Fourier analysis called the Central Slice Theorem. We have chosen to tackle this subject more or less head on rather than skirt around it, because it enriches our discussion of image making throughout the book and enables us to treat the vast subject of MRI in more depth than would otherwise have been possible. In fact the use of Fourier methods has become so imbedded in signal and image processing in general that we see no harm in the medical user of data analysis packages being given more insight into this underlying central mathematical idea.

To serve our two parishes , two extra ingredients had to be added to the book, one for each parish. The medics need an overview of basic atomic and nuclear physics and how mathematics has become the engine of the exact sciences. The physicist needs to know the clinical context in which technological tools are used. Our second chapter is primarily aimed at the medic, our final chapter is for the physicist. Since most readers will not proceed from here to the end of the book in a sequential fashion we have provided a route map with chapter dependencies for the skimmers and dippers. Dedicated sequential readers, skimmers and dippers alike will benefit from attempting the questions and short problems at the end of each chapter. None of the problems involve the use of any mathematics beyond school A level, some will be benefit from the use of a programmable calculator or home computer.

The Organisation of the Book

Chapter 1 + Appendices A ,B

Chapter 1 together with the appendices describe the principles of tomographic reconstruction using words and pictures. All the later chapters dealing with the separate technologies rely on this central chapter, since after data has been collected, all tomographic imaging schemes reconstruct their particular pictures using almost the same basic scheme. Readers with a medical background will, it is hoped, benefit from the introduction to Fourier analysis and the concept of K space given in Appendix A. A more rigorous derivation of the central slice theorem, intended primarily for the physicist, is provided in appendix B. We encourage the medical reader to struggle with these appendices, since as we have already said, Fourier methods underpin a great deal of standard image and data processing.

Chapter 2

Chapter 2 is written primarily for non-scientists. It starts with a discussion of how physics and engineering construct approximate but successful mathematical models of the physical world and emphasises the underlying unity in methods and concepts employed. Imaging technologies using x-rays, γ–rays and magnetic resonance all depend on a detailed quantitative description of the atomic and nuclear structure of individual atoms and the ways in which light (photons) and energetic particles, such as electrons, can exchange energy with these objects. Sections 2.2 to 2.5 provide a qualitative background to atomic and nuclear physics. Section 2.6 is devoted to the enumeration and description of the ways in which atoms and photons interact. These sections lay the foundation for understanding x-ray contrast mechanisms used in chapter 4 and the scattering of γ–rays, used in diagnostic nuclear medicine, described in chapter 5. The final section on magnetic resonance is a preparation for the more detailed discussion of MRI in chapter 6. This introductory section is intended to emphasise the fact that MRI depends ultimately on intrinsic nuclear properties (the nuclear magnetic moment) and that the resonance is a generic process, very similar to the absorption and emission of visible light by atoms.

Chapter 3

Much of this chapter will be new to both parishes. It provides the standard definitions used in ionising radiation measurement and radiological protection and a brief overview of the two vast subjects of radiation biology and radiation physics. The chapter emphasises the modern view, that ionising radiation is intrinsically dangerous and harmful to biological tissue but that the risk of permanent damage can and is now kept within perfectly acceptable bounds by careful monitoring and dose control.

Chapters 4 to 6

In turn these chapters describe the principles underlying the modern imaging methods of x-ray radiography and CT (4), SPECT and PET (5) and MRI (6) All of the tomographic applications use nearly the same mathematical methods to reconstruct cross sectional images of human anatomy which are described in chapter 1. Each individual technology uses the physics of electromagnetic waves and particles, described in chapter 2, to obtain the data required for image reconstruction. We have been intentionally brief as far as detailed engineering is concerned, concentrating on underlying physical principles In each chapter we describe the contrast mechanisms, the factors affecting spatial and temporal resolution and the more general engineering constraints.

Chapter 6 on MRI is by far the largest and perhaps the most difficult chapter of the book for both parishes. The size is intentional, the difficulty is probably inevitable. NMR and MRI are so versatile in their application that, over the past fifty years, a very wide range of applications of several different manoeuvres has been developed. The chapter deals first with the basic physics of NMR, before embarking on MRI. Within its rather shorter lifetime, MRI has thrown up a wide range of different ways of combining contrast flavours, nearly all of which can be obtained from standard clinical machines.

Chapter 7

Diagnostic ultrasound imaging, the subject of chapter 7, is not a tomographic technique and thus falls outside the general theme of the book. Our brief description of ultrasound is included for completeness, since ultrasound is second only to x-ray projection radiography in its frequency of clinical use.

Chapter 8

This chapter is intended mainly for the physicist and engineer. Our aim is to illustrate just how imaging, providing some pieces of evidence, fits into the very human activity of medical diagnosis. The final section, on neuroimaging, illustrates one of many medical research areas using modern imaging and the one which just happens to be close to our own interests. The human brain is sometimes referred to as the last frontier of science. The last twenty five years have witnessed a second revolution in medical imaging but this has been largely confined, in its tremendous success, to human anatomy. The last frontier is likely to be the scene of a third revolution in the imaging, the creation of movies of human function. Thus our last section points to the future illustrating how far we have to go and the limitations of the tools that are presently in use.

Appendices

Appendix A provides a reminder of basic wave concepts such frequency and phase and then introduces the ideas underlying Fourier analysis and its use in image and signal processing. Appendix B gives a reasonably rigorous account of the central slice theorem and filtered back projection and describes the simpler filtered modifications that are commonly used in SPECT and PET. Appendix C is devoted to an account of simplified Bloch equations of NMR and an introduction to the idea of the rotating frame of reference commonly used in NMR and MRI to describe the properties of resonance signal.

Acknowledgements

We are indebted to colleagues who have toiled through the earlier drafts of this book, pointing out errors and making valuable suggestions. In particular we would like to thank Dr Gonzalo Alarcon for reading all of it and Dr McRobbie for politely correcting some of our more quaint misconceptions about realities of real modern medical imaging practice. CNG would also like to thank the several years of Imperial students who bore the brunt of his medical imaging teaching apprenticeship. An introduction to medical imaging would hardly be complete without the inclusion of examples of real images. For this we are grateful to Ms Judy Godber of Seimens UK, Mrs Margaret Dakin of St Thomas' Clinical PET centre, Dr Steven Williams of the Institute of Psychology and Dr Donald McRobbie of Charing Cross Hospital. The cover cartoon is reproduced from "Mog's Mumps" a delightful children's book by Helen Nicoll and Jan Pienkowski. We are grateful to them both for allowing us to use it as our cover. Our thanks are also due to the staff of World Scientific Publishing, in particular Dr.Richard Lim, Dr John Navas and Sook Cheng Lim for their help with formatting and their tolerance of long delayed returns of the draft.

A Glossary of Units and Common Terms

Scale : Powers of ten used throughout Science

pico (10^{-12}) p	tera (10^{12}) T
nano (10^{-9}) n	giga (10^{9}) G
micro (10^{-6}) μ	mega (10^{6}) M
milli (10^{-3}) m	kilo (10^{3}) k

SI Units

Length	metre	- m
Mass	kilogram	- kg
Time	second	- s
Force	newton	- N kg m s^{-2}
Energy	joule	- J kg m^2 s^{-2} = N m
Power	watt	- W kg m^2 s^{-1} = N m s^{-1}
Frequency	hertz	- Hz s^{-1}
Pressure	pascal	- Pa kg s^{-2} = N m^{-2}
Electric current	ampere	- A
Electric Potential	volt	- V
Electric charge	-coulomb	- C A s
Magnetic induction (field strength)	Tesla	- T

Fundamental Constants

velocity of light	c	- 2.998	10^{8} m/s
electron charge	e	- 1.66	10^{-19} coulomb
electron mass	m	- 9.1	10^{-31} kilogram
proton mass	M	- 1.67	10^{-27} kilogram
Planck's constant	h	- 6.63	10^{-34} joule s
	$h/2\pi$	- 1.06	10^{-34} joule s
Bohr Radius	a_o	- 5.29	10^{-11} m
Bohr (Atomic) magneton	μ_b	- 9.27	10^{-24} A m^2

Nuclear magneton	μ_n	- 5.05	10^{-27}	$A\ m^2$
Boltzman's constant	k_B	- 1.38	10^{-23}	joule / deg

Definitions of Some Common Terms from Physics

Velocity

The rate of change of position or the distance travelled, in a specific direction, divided by the time taken. Speed is the rate of change of position, without reference to a direction, it is thus the magnitude of the velocity. Both speed and velocity have SI units of metres per second, ms^{-1}.

Acceleration

The rate of change of velocity: the change of velocity divided by the time taken by the change. It is measured in SI units of metres per second per second, ms^{-2}.

Momentum

A quantity characterising the motion of any object. Momentum is the product of the mass and the linear velocity of a moving particle. Momentum is a vector quantity, which means that it has both magnitude and direction The total momentum of a system made up of a collection of rigidly connected objects is the vector sum of all the individual objects' momenta. For an isolated composite system, total momentum remains unchanged over time; this is called the conservation of momentum. For example, when a cricketer hits a ball, the momentum of the bat just before it strikes the ball plus the momentum of the ball at that moment is equal to the momentum of the bat after it strikes the ball plus the momentum of the struck ball (neglecting the small effects of air resistance).

Force

Force is any action or influence that accelerates, or changes the velocity of, an object.

Newton's Three Laws of Motion

The First Law

The first law of motion states that if the vector sum of the forces acting on an object is zero, then the object will remain at rest or remain moving at constant velocity. If the force exerted on an object is zero, the object does not necessarily have zero velocity. Without any forces acting on it, an object in motion will continue to travel at a constant velocity. In our everyday world, such perpetual motion is not observed because friction, either from air resistance or contact with a surface eventually dissipates the kinetic energy of the moving object. Thus in its flight through the air, the cricket ball slows down because there is a small resistive force, arising from the impact of air molecules on the ball.

The Second Law

Newton's second law relates force and acceleration. Possibly the most important relationship in all of physics. A force on an object will produce an acceleration or change in velocity. The law makes the acceleration proportional to the magnitude of the force and is in the same direction as the force. The constant of proportionality is the mass, m, of the object thus $F = ma$. The unit of force, 1 Newton, is defined as the force necessary to give a mass of 1 kg an acceleration of 1 ms^{-2}.

The Third Law

Newton's third law of motion states that when one object exerts a force on another object, it experiences a force in return. The force that object A exerts on object B must be of the same magnitude as the force that object B exerts on object A, but in the opposite direction.

Friction

Friction acts like a force applied in the direction opposite to an object's velocity. Dry sliding friction, where no lubrication is present, is almost independent of velocity. Friction results in a conversion of organised mechanical energy into random heat energy. Although the concept of a net frictional force is quantitatively defined in classical physics this is not so in quantum physics, where the frictional (dissipation) process involves an exchange of quanta between one system and a large reservoir of particles, sometimes referred to as a heat bath. Friction actually presents a much more difficult computational problem in quantum physics than in classical physics.

Angular Momentum

The angular momentum of a rotating object depends on its speed of rotation, its mass, and the distance of the mass from the axis. Formally the angular momentum of a particle is given by the product of its momentum mass . velocity and its distance from a centre (often a centre of rotation)

When a skater on (almost) frictionless ice spins faster and faster, angular momentum is conserved despite the increasing speed. At the start of the spin, the skater's arms are outstretched. Part of the skater's mass is therefore at a large radius. As the skater's arms are lowered, thus decreasing their distance from the axis of rotation, the rotational speed must increase in order to maintain constant angular momentum. The conservation of angular momentum, just like the conservation of linear momentum follows directly from Newton's laws. An isolated body experiencing no net torque (torque is the product of force and the distance from an axis) will maintain its angular momentum.

Energy

The quantity called energy ties together all branches of physics. In mechanics, energy must be provided to do work and work is defined as the product of force and the distance an object moves in the direction of the force. When a force, exerted on an object, does not result in any motion, no work is done. Energy and work are both measured in the same units, joules. If work is done lifting an object of mass, m, a distance h, energy has been stored in the form of gravitational potential energy given E= mgh. If the object is released it will acquire kinetic energy by virtue of its acquired speed V, $\frac{1}{2}$ mV^2.

Many other forms of energy exist: electrical and magnetic energy; potential energy stored in stretched springs, compressed gases, or molecular bonds, thermal energy, heat, and mass itself. In all changes from one kind of energy to another, the total energy is conserved. For instance, the work done in raising a ball a distance h, increases its gravitational potential energy by mgh. If the ball is then dropped, the gravitational potential energy is transformed into an equal amount of kinetic energy, thus, $\frac{1}{2}$ mV^2=mgh. When the ball hits the ground and has come to rest all of the organised motional energy of the ball is dissipated into random motions that we call heat. The ball and its immediate surroundings become a tiny bit hotter as a result of the impact.

Power

Power is the rate of change of energy with time. It is measured in watts, 1 watt = 1 joule/s. Domestic appliances, such as light bulbs, are rated by power that they consume 40 watt , 60 watt , 100 watt etc A 100 watt bulb dissipates 100 joules of electrical energy (Volts. Current) each second.

Temperature

Temperature is the property of systems in contact that determines whether they are in thermal equilibrium. The concept of temperature stems from the idea of measuring relative hotness and coldness and from the observation that the addition of heat to a body leads to an increase in temperature, as long as no melting or boiling occurs. The terms temperature and heat, although interrelated, refer to different concepts, temperature being a property of a body and heat being an energy flow to or from a body by virtue of a temperature difference. When a quantity of matter is in equilibrium at a temperature T, the constituent particles are in random motion and their the average energy is approximately $k_B T$ where k_B = 1.38

10^{-23} joule /deg is called Boltzmann's constant.

Field

A field of force is a region in which a gravitational, magnetic, electrostatic, or other kind of force can be exerted on an object. These regions are imagined to be threaded by lines of force, which are close together where the field is strong and are proportionately wider-spaced where it is weaker. The concept of action at a distance or a field of force, was first introduced by Newton in his theory of universal gravitation. It has since become a central theme in modern physics. Each of the fundamental forces, gravitation, electromagnetic and nuclear has a characteristic scale of length. Nuclear forces are appreciable only on the scale of the nuclear radius 10^{-15} m. Gravitational forces at the other hand extreme produce tangible effects right across the galaxy, over distances of 10^{15} m.

Electricity and Magnetism

Physical phenomena resulting from the existence and interaction of electrical charges. When a charge is stationary, it produces electrostatic forces on charged objects in regions (the electrostatic field) where it is present. When a charge is in motion (an electrical current) it produces additional magnetic forces on other moving charges or currents.

Magnetic moment

There are no magnetic charges or monopoles to provide a source of a magnetic force field in the way that the electric charge of the electron or the proton are the source of a surrounding field of electrostatic force. Rather, it is moving charges or currents that produce all magnetic fields. It is however convenient in atomic and nuclear physics to describe the magnetic force field, surrounding an atom or nucleus, in terms of an effective fictitious magnetic charge called the magnetic moment. The units of magnetic moment are electric current times area. Internal motions of charge within tiny objects such as atoms and nuclei create a surrounding magnetic field of force with a characteristic topography called a dipole field. This field is completely specified by quoting the strength and direction of the effective magnetic moment. The dipole field is the same as that arising from a very small loop of wire carrying an electric current. If the loop encloses and area S and the current is I then the magnetic moment is IS Am^2.

Planck's constant

Planck, in this theory of black body radiation, came to the conclusion that light or electromagnetic waves could only be exchanged in discrete amounts, quanta. In particular according to Planck, the energy of a quantum of light is equal to the frequency of the light multiplied by a constant, h = $6.626 \cdot 10^{-34}$ joule-second. Planck's idea was the first step on the road to quantum mechanics. The constant h sets a scale for the graininess of the energy exchanges between atoms.

Uncertainty Principle

The uncertainty principle is one of the foundations of modern physics. It arises from the observation that to "know" something about a material object, one has to interact with that object. At the level of atoms and elementary particles the least interaction is with a suitable photon, such an interaction must necessarily involve an exchange of energy between the light photon and the particle under observation. In turn this will produce a modification of the particle motion. Thus if one wants to find out a particle position then one must bounce at least one photon from that particle and the bounce will impart motion to the particle. Thus one has gained knowledge of position but lost information about particle momentum. This is summarised in the relationship, $\Delta x \ \Delta P > h$ where h is Planck's constant. Decreasing the uncertainty in x, Δx, must necessarily increase the uncertainty in P, ΔP. A similar relationship between time and energy $\Delta E \ \Delta t > h$ limits the simultaneous knowledge of energy and an interval of time.

Intrinsic Particle Spin

All the fundamental particles in nature possess a physical attribute that is called spin. Spin is always quantised in units of $h/2\pi$, all particles of the same type have exactly the same number of quantum units, but this number changes with particle type. Electrons, protons and neutrons all have a spin value of ½ ($h/2\pi$). Photons have a value of zero. If the spin attribute is not zero then the particle will also have an intrinsic magnetic moment, as if there were an internal rotation of electrical charge. Thus, electrons, protons and neutrons all have associated magnetic moments. The magnetic moment of the proton is ultimately the root of MRI.

The statistical behaviour of collections of identical particles depends on whether the spin is a half or a whole integer. Particles with half integer spin are called Fermions, those with zero or whole integer values are called Bosons. The dependence of the statistical behaviour of particles on spin has a strong impact even on everyday phenomena. The peculiar physical properties of all metals depend on the fact that they contain free electrons that are Fermions. Superconducting materials can only arise when electrons can be strongly bound together, with opposing spins to create a single unit that has zero net spin, a Boson.

Pauli exclusion principle

Pauli discovered a law of nature governing the properties of particles with half integer spin; two identical particles with half integer spin cannot occupy the same quantum state. This law is mainly responsible for the different sizes of atom, since only two electrons, with opposing spins can occupy a single Bohr energy level. In multi-electron atoms the electrons have to occupy a range of levels, with increasing average orbit sizes.

Some Common General Terms Used in Medical imaging

Angiogram

An image in which blood vessels are selectively enhanced either by the use of a contrast agent (x-rays) or a flow sensitive pulse sequence (MRI).

Artifact

A feature in any medical image not related to any anatomical structure or physiological process but produced by the measuring process itself.

Biopsy

A small sample of tissue tissue taken from an abnormal tissue mass generally revealed by imaging . The sample is subjected to biochemical analysis in order to determine the types of cell present. For example whether or not the revealed mass is maliganant. It is important to note that medical imaging alone cannot at present make this final crucial assessment.

Contrast

The difference in local image intensity which differentiates one structure or tissue type from another.

Contrast enhancement

A deliberate increase in the image intensity depicting one particular type of structure. Generally, contrast enhancement is brought about by a change in the measurement procedure, for example by introducing a strong x-ray absorbing substance into the blood stream, prior to the exposure. Some contrast enhancement can however be achieved by computer image processing, for example by edge detection.

Dose

An estimate of the amount of ionising radiation actually absorbed by a patient or attendant medical worker.

Exposure

An estimate of the total amount of ionising radiation incident on a patient or attendant medical worker.

Field of View - FOV

The region of the body selected for imaging. In x-ray radiography this is selected by collimation of the incident x-ray beam. In MRI the FOV is selected by the combination of the extent of RF excitation and the range of frequencies accepted by the receiver.

Fluoroscopy

A widely used medical term describing the use of an x-ray image intensifier sometimes called a fluoroscope to produce quasi-continuous imaging of surgical and diagnostic procedures at greatly reduced rates of x-ray exposure.

Grey Scale

A digital visual code relating the range of image intensities or colours to a definite numerical scale of values. In general the scale is chosen to produce the best contrast for anatomical features of interest. In principle the grey scale used in x-ray CT can be absolute so that a particular tissue type has a definite value in Hounsfield Units.

Mammography

A specific combination of imaging procedures designed specifically to allow diagnosis of disases of the female breast. Most commonly the term refers to a specialised x-ray technique made particularly sensitive to small calcium deposits.

MRA

A Magnetic Resonance Angiogram. A 3D image of the blood vessels obtained using just the flow sensitivity of the MR signal.

Pixel

The discrete two dimensional element of a digital image. Typically a medical image will consist of 128 by 128 or 256 by 256 pixels. The size of the pixel sets the spatial resolution of the image, ie 2 by 2 mm. The number of pixels on a side sets the field of view.

Pulse sequence

The pattern in time of RF and gradient field pulses that together result in a particular type of MR imaging.

Scanner

A term often used to describe a particular tomographic instrument. Thus MR scanner, CT scanner, PET scanner. Although frowned upon in some circles the term does actually do justice to the tomographic principle of carving objects into planes and strips.

Screening

The use of an x-ray image intensifier (fluoroscope) as a visual aid in some surgical procedures.

Signal to Noise Ratio

All measurement processes and instruments introduce some extra random contributions to the signal that they produce. In many circumstances an instrument might record unintended random and interference signals in addition to the intended signal. The signal to noise ration is simply the ratio of wanted to total unwanted signal amplitude, at the output of the instrument.

Slice

In most applications of tomography the imaging data is acquired in thin two dimensional blocks or slices. The thickness of the slice is generally chosen to be between 1 and 8mm as a compromise between the needs of spatial resolution and signal intensity in each voxel.

Sonography

A widely used medical term referring to all medical ultrasound techniques.

Tomography

Image reconstruction using a large number of different views of the same object. Invariably these days the term refers to image reconstruction using a digital computer. The first x-ray tomographic imagers were called CAT scanners – Computer Aided Tomography in order to distinquish the technology from earlier "classical " x-ray methods that highlighted a particular plane in the body. Today, the CAT scanner is referred to simply as a CT scanner. Other modalities such as magnetic resonance imaging MRI, positron emission tomography , PET use standard computer tomographic methods.

Voxel

The same as a pixel but with the additional description of the thickness of slice used thus voxels might measure 2 by 2 by 5mm . 2 mm sized pixels 5mm thick.

CONTENTS

INTRODUCTION

An Historical Background to Medical Imaging

Until the beginning of this century the investigative tools available to clinical medicine amounted to the microscope, the thermometer, the knife and the stethoscope. Our understanding of the gross functions of the visually discernible organs of the human body evolved from centuries of careful anatomical studies of the dead and observations of the functional deficits, imposed by injury and disease on the living. Until 1896 no means existed to examine or measure the hidden internal world of the living human body. Roentgen's discovery of the penetrating x-ray, in 1896, started a revolution in medical imaging and began a slow process of reunification of medical science with physics, chemistry and engineering, the so called exact sciences.

From about the time of Newton onwards, the exact sciences were able to make very rapid progress in both experimental technique and the establishment of approximate mathematical laws governing the behaviour of inanimate nature. The mathematical models, although beautiful and deserving of study in their own right, owe their importance to their power of prediction. Whether it is the flight of a bullet, the strength of a bridge or the design of a television set, precise physics and engineering recipes can be looked up or deduced which allow a paper calculation to predict in great detail what the real object will do. The success arises from the well-established underlying unity of the theory and the relative simplicity and reproducibility of the inanimate natural world. Medical science could not match this progress because the available scientific instruments were incapable of isolating simple general principles from the very much more complex behaviour of biological matter. Whereas the exact sciences now rest entirely on unified underlying mathematical theories, medical science is still largely empirical in its approach. Today science and medicine are popularly seen as two separate disciplines, attracting two somewhat different types of scientist. Two hundred years ago the natural philosopher tended to think and write about both areas. The next century may well see a reunification of the two areas, as quantitative measurement methods become as important and revealing in medicine as they became in the exact sciences.

Only time will tell whether or not mathematical models, of the elegance, simplicity and generality of Newton's laws or Maxwell's equations of electromagnetism, eventually appear in the medical sciences, predicting the boundaries between health and disease.

The history of medical imaging closely parallels the history of experimental physics and computing. It would not be terribly illuminating to the average reader for us to provide a detailed chronology of the history and development of each and every component of the imaging technologies discussed in this book. Rather we will try to describe the important historical scientific threads that form the backdrop to the modern era. All the technologies that are now in clinical use, originated in the physics laboratory as a means of investigating the properties first of atoms and then atomic nuclei in inanimate matter. The x-ray, the positron, the cyclotron and the photomultiplier tube were all inextricably bound up with the astonishingly rapid development of atomic and nuclear physics during the first half of this century. The second world war sometimes dubbed "the physics war" was responsible for galvanising many of these new ideas and experimental methods of physics into new technologies which have since dominated and changed the developed world. Four general technologies could be said to have originated during the war; the most dramatic is nuclear engineering, the most pervasive are digital electronics and computing, last but not least is high frequency radio engineering stemming from the crucial wartime developments in RADAR and aerial counter measures. Possibly as important as the developments in "hardware" was an innovation in the sociology and organisation of research and development. Both the development of the atom bomb and RADAR were carried out by specially recruited teams of scientists and engineers managed, mostly harmoniously, on military lines. These two famous stories have attracted many historical accounts. A very large number of graduate scientists, and nearly all academic scientists, whatever their field, were involved in one or another dedicated multi-disciplinary wartime team in both America and the United Kingdom. After the war, these scientists carried their experience back into academia and industry and developed the larger scale scientific and technological enterprises along the same managed lines. All modern commercially produced medical imaging systems owe their existence to this type of science and engineering team organisation. Nuclear medicine, ultrasonic imaging and nuclear magnetic resonance were all developed from wartime technologies between 1947 and the present day.

X-ray Radiography: The First Revolution

Diagnostic medical imaging started just over 100 years ago with the discovery of x-rays by Roentgen in 1895. Within a year of his announcement, $16 home x-ray sets could be bought in the US and a Birmingham doctor in England performed the first x-ray guided operation to extract a fragment of a sewing needle embedded in the hand of a seamstress. Specialised x-ray diagnostic and therapeutic methods developed throughout the century, and up until about 1960, x-ray methods completely dominated non-invasive medical diagnosis. Even today the various specialised modifications of simple projection x-ray radiography are by far the most frequently used diagnostic techniques in medicine. In some areas such as mammography x-rays provide the gold standard for the early detection of breast cancer. Dentistry relies almost entirely on x-rays and the subject as a whole has hardly been altered by the developments whose descriptions occupy the main part of this book.

X-ray CT: The Second Revolution

The second revolution in imaging began in 1972 with Hounsfield's announcement of a practical computer assisted x-ray tomographic scanner, the CAT scanner, which is now called x-ray CT or simply CT. This was actually the first radical change in the medical use of x-rays since Roentgen's discovery. It had to wait for the widespread use of cheap computing facilities to become practical. Although CT uses the mathematical device of Fourier Filtered Backprojection that was actually first described in 1917 by an astronomer called Radon, its practical implementation in clinical medicine could not take place without the digital computer.

Two key improvements in diagnostic x-ray imaging followed from the introduction of CT. First the CT image is a 2D reconstruction of a cross section of the patient anatomy rather than a projection of the shadows cast by overlapping internal organs and bones. Second, CT provides good contrast between different types of soft tissue in the body as well as contrast between bone and all soft tissue. Both make the interpretation of the resulting images very much easier. One important motivation for the development of CT was the location of tumours and stroke damage within the brain. The high level of x-ray absorption by the surrounding skull generally made this task very difficult for the projection radiograph. A measure of the previous difficulty in detecting brain tumours can be judged by a pre- CT method called air encephalography. Patients had a bubble of air introduced into the cerebral spinal fluid, via a puncture in the spine. The bubble was then encouraged to rise into the ventricles of the brain where it displaced the cerebral spinal fluid. The little bit of extra contrast thus obtained between soft tissue and the adjacent air bubble (rather than between tissue and fluid) often enabled radiologists to just spot any enlargements of one brain hemisphere with respect to the other, produced by a space occupying tumour residing in one hemisphere.

X-ray CT on its own has been a remarkably successful innovation in medicine. Even medium sized hospitals routinely use CT for the rapid assessment of structural abnormality resulting from injury or disease throughout the body. CT has also become a standard tool in the planning of cancer radiation treatment. The CT image often provides a clear definition of the extent of the tumour and its disposition with respect to surrounding healthy tissue. The intrinsically digital format of CT allows the radiologist to calculate optimum paths for the therapy beams to deliver a lethal dose to the cancer but spare healthy tissue. CT has some important drawbacks: it entails a relatively large dose of ionising radiation to the patient and there are practical limits on spatial resolution brought about by the very small differences in x-ray contrast between different types of soft tissue. The medical success of CT together with these limitations spurred the development and hastened the introduction of MRI into regular hospital use in the early 1980's.

MRI

MRI is a wholly tomographic technique, just like x-ray CT, but it has no associated ionising radiation hazard. It provides a wider range of contrast mechanisms than x-rays and very much better spatial resolution in many applications. The extremely rapid development of MRI has been possible because it uses many of the techniques of its parent, nuclear magnetic resonance, NMR. In fact, in its infancy, MRI was called Spatially Localised Nuclear Magnetic Resonance but this was changed to magnetic resonance imaging MRI or simply MR both to avoid the long-winded title and to remove any misplaced association with ionising radiation through the adjective nuclear.

Research in nuclear magnetic resonance, NMR started shortly after the war. The phenomenon was discovered independently by two groups in America led by Bloch and Purcell in 1946. Bloch and Purcell later shared a Nobel Prize in physics for this work. The original applications of the technique were to the study of the magnetic properties of atomic nuclei themselves. It was however soon realised that the time taken for a collection of nuclear magnetic moments in a liquid or a solid, to attain thermal equilibrium, depended in some detail on the chemistry and physical properties of the surrounding atoms. Thus NMR very rapidly became and remains a vital tool used in chemistry and biology to study the atomic rather than nuclear properties of atoms and molecules. By 1950, the advantages of pulsed NMR, as opposed to its continuous wave predecessor, were appreciated. Hahn discovered the valuable spin echo technique in 1952. Between 1950 and 1973 the techniques of NMR, especially NMR spectroscopy developed into fully engineered standard laboratory methods used throughout physics, chemistry and biology and also in many industrial test applications. The idea of spatially localised NMR and thus imaging, appears to have sprung up in several groups in both America and Europe in the early 1970's. The first NMR image using magnetic field gradients to image a human finger was published in Nature by Lauterbur in 1973. The history of this paper contains one of science's most delightful ironies. The paper was originally rejected by the referees on the grounds that the idea lacked applications! The first whole-body image was published by Damadian in 1977, who had earlier taken out a patent on the application of NMR imaging to medicine without specifying a precise localising scheme. The earliest images were obtained either in very small laboratory superconducting magnets or whole body resistive (copper) magnets. These gave what today would be considered a very low main magnetic field strengths of 0.1 T. The extremely rapid development of MRI and its widespread introduction into clinical use has been motivated by the complete absence of ionising radiation and the large number of different physical parameters that can be employed to create contrast in soft tissue.

MRI is still a relative newcomer to medicine with many important new developments still to come both in applications and technique. Although standard MRI images require many minutes of scanning, the recent widespread introduction of echo planar imaging EPI allows a single data slice to be acquired in 50 ms. Since 1990 there has been a rapid growth of interest in, and development of, what is called functional MRI or fMRI. It is thought that this very new technique is capable of providing images of brain activity on a time-scale of a few seconds. In the future spatially localised NMR spectroscopy and imaging of nuclear species other than the water proton will undoubtedly become clinical realities. Although MRI is unlikely to entirely displace other imaging techniques it is rapidly becoming the method of choice in an increasing number of clinical investigations both because it is hazard free and because of the quality of its images.

Diagnostic Nuclear Medicine

All of nuclear medicine including diagnostic gamma imaging became a practical possibility in 1942 with Fermi's first successful operation of a uranium fission chain reaction. The intensive wartime effort to produce the first atom bomb, the Manhatten project, laid the foundations of nuclear engineering both in the USA and UK. Shortly after the war had ended, research nuclear fission reactors started to produce small quantities of artificial radioisotopes suitable for medical investigations. Hospitals in the UK were receiving their first shipments of useful tracers in 1947 and the first point by point image of a thyroid gland was obtained by Ansell and Rotblat in 1948. The development of large area photon detectors was crucial to the practical use of gamma imaging. Anger announced the use of a 2 inch NaI/ scintillator detector, the first Anger camera in 1952. Electronic photon detectors both for medical gamma imaging and x-rays are adaptations of detectors developed, in the decades after the war, at high energy and nuclear physics establishments such as Harwell, CERN and Fermilab.

Tomographic, SPECT gamma images obtained by translations or rotations of a gamma camera were first reported in 1963 by Kuhl and

Edwards. The first positron annihilation measurements were reported as early as 1951 by Wren and crude scanning arrangements providing imaging were published in 1953 by Brownell and Sweet. PET is still in its infancy as a clinical tool, largely because it is very expensive to install, but it is firmly established in clinical research. It requires not only an imaging camera but also a nearby cyclotron to produce the required short lived positron emitting radionuclides.

Diagnostic Ultrasound

Although ultrasonic waves could be produced in the 1930's and were investigated as a means of medical imaging , ultrasound imaging did not start properly until after World War 2 . It benefited from the experience gained with SONAR in particular and fast electronic pulse generation in general. The first two dimensional image, obtained using sector scanning was published in 1952 by Wild and Reid. Application to the foetal imaging, began in 1961 shortly after the introduction of the first commercial two dimensional imaging system. Ultrasound imaging is second only to the use of x-rays in the frequency of clinical use.

The Future

Although it is foolish to even try to predict the long term course of medical imaging there are clear trends in the utilisation of existing technologies that provide clear pointers to the next twenty years. These can be summed up by two general factors, a drastic reduction in the use of ionising radiation and the introduction of small specialised, possibly desktop instruments particularly in MRI .

For two thirds of this century, diagnostic x-ray techniques have dominated medical imaging. The small detriment incurred by radiation dose to both patient and staff was, until about 1980, perceived to be a necessary price to pay for the improved diagnosis made available by x-ray methods. Curiously new methods, such as the use of image intensifiers, designed to reduce radiation dose, actually made new interventional procedures possible and these have consistently led to some of the largest doses delivered to the patient. In the past ten years very much more attention has been given to this problem and manufacturers now highlight drastic reductions in patient dose as well as image quality in their advertising literature. The same general perception of the need for further patient dose reductions has also encouraged medical engineering to examine ways in which MR and ultrasound , which do not produce any ionising radiation, could perform diagnostic roles that were traditionally the preserve of x-ray radiology.

Both x-ray CT and MRI, because of their cost and technical complexity, have traditionally been set up in quite large specialised units within the radiology departments of the larger general hospitals. The units employ teams comprising medical, engineering and mathematical staff both to maintain routine clinical work and undertake research into new methods using these machines. Generally patients are referred to such units from other hospitals both for the examination and the interpretation of its results.

MRI is a very flexible and, in principle cheap, technology that lends itself very well to miniaturisation and specialisation. Thus whereas MR investigations of limbs have at present to be performed in the special units, in the future it seems likely that small low field dedicated MR systems will become available, for use in small clinics. The key to these developments is the permanent magnet. Over two thirds of the capital cost, complexity and running costs of the standard clinical MR scanner is taken up by the superconducting magnet This has to be enclosed in a special vacuum enclosure and kept permanently very cold at - 290° C using expensive liquid helium. Apart from the superconducting magnet , the MR scanner is little more than a radio transmitter/receiver linked to a computer. Recent developments in solid state magnetism have made it possible to produce highly uniform fields in the range 0.1-0.3T using shaped permanent magnetic materials. MR systems designed around these magnets can already produce imaging of sufficient quality for clinical diagnosis. Thus MR might well follow the development course of the digital computer, progressing from the very large "main frame scanner " central facility to the desktop "personal scanner".

1 TOMOGRAPHY

Your Majesties, Your Royal Highnesses, Ladies and Gentlemen,
Neither of this year's laureates in physiology or medicine is a medical doctor. Nevertheless, they have achieved a revolution in the field of medicine. It is sometimes said that this new X-ray method that they have developed –computerised tomography – has ushered medicine into the space age. Few medical achievements have received such immediate acceptance and met with such unreserved enthusiasm as computerised tomography. It literally swept the world".
Speech by Professor Torgny Greitz at the Nobel prize ceremony in 1979 at which A. M Cormack and G.N. Hounsfield received the prize for Physiology or Medicine with the citation ,
" for the development of computer assisted tomography"

1.1 Introduction

Very often great ideas have the irritating habit of seeming obvious once they have been explained. Computerised Tomography is no exception. Behind the expensive hardware and the several thousands of lines of computer code that comprise the modern medical scanner there sits a very simple idea, tomography. The word Tomography was constructed using the Greek root Tomos meaning to slice. Computerised Tomography, be it using x-rays, gamma photons or MRI, always proceeds by a series of slicing steps that reduce a 3D volume, first to a collection of 2D sections or slices, then each slice is crosshatched with lines to produce a collection of 1D strips. Each stage reduces the amount of information that ends up in the detectors to a level that a reconstruction algorithm can handle without ambiguity. Although crude methods of tomography existed in the early 1950's, it was Hounsfield's demonstration of a practical computerised tomographic x-ray method in 1972, which marked the start of what we call the "Second Revolution in Medical Imaging". In principle the method could have been carried out using the old fashion hand mechanical calculator, in much the same way that early x-ray crystal diffraction data were manipulated to obtain a crystal structure. In practice however the method demands the use of a digital electronic computer to collect and process the relatively large amounts of data, which contribute to a high-resolution image. The second revolution could not really get going before the introduction of small, dedicated digital computers into industry, science and medicine.

In this chapter we describe the underlying principles of tomography, largely divorced from any one of its several applications within medicine. Each technology has its own combination of special bits of physics and engineering that together create a particular image contrast mechanism. However once the data has been collected, essentially the same mathematical method is used to create the final image in each case. We start the chapter by describing the simplest approximate tomographical reconstruction method that we call simple backprojection. This can be understood, and indeed performed, using just the mathematical operations of division and addition. It turns out that, although this scheme is actually remarkably successful given its simplicity, it is not good enough for high-resolution reconstruction and is not ever used just on its own. In order to go further we have to make use of that branch of applied mathematics called Fourier Analysis. In principle this provides an exact, as opposed to approximate route to image reconstruction. A general introduction to Fourier Analysis is given in appendix A and Appendix B provides a derivation of the most important result for tomography, the Central Slice Theorem. This, in one form or another, is at the heart of all tomographic reconstruction techniques. It is hoped that this chapter, taken in conjunction with the two appendices, will provide the reader with a basic understanding of much of the common jargon of tomographic methods and a single framework for understanding the details of the individual tomographic methods that we describe in later chapters.

1.2 Simple Backprojection Reconstruction

Projections

Nearly all tomographic image reconstruction involves the three basic operations illustrated in figure 1.1. The first step consists of carving the object into a set of parallel slices. In the second step, each two dimensional, 2D, slice is analysed in turn by acquiring a set of projections of that slice. In the final step the set of projections are recombined, according to a particular mathematical algorithm, to reconstruct an image of the 2D slice. Thus our first task is to define what we mean by a projection of a slice.

A projection can be thought of as a shadow. Consider the child's game of creating animal head shadows on the wall, by clasping the hands in a particular way in front of a light bulb. The hands are three-dimensional objects that entirely stop any light that falls directly onto them. The image on the wall however is the shadow cast by the outline of the hands. It is two dimensional and thus can be made to resemble an outline of some other three dimensional object for example a rabbit head with long floppy ears! The contrast in the projection is all or nothing, formed by light absorption in the hands. If we imagine a series of straight lines or rays from the light bulb to the projecting wall, some lines stop at the hands, creating darkness on the wall and some miss altogether creating light on the wall. The boundary between the two regions is the particular outline, formed from the lines that just miss or graze the hands and then continue to the wall. If the hands are rotated, the image on the wall changes, similarly if the hands are held fixed but the position of the light bulb is changed, so the shape of the shadow or projection on the wall changes. In a slightly more serious context this shadow method can be used to reconstruct the 3D surface shapes of objects. Again, we set up a light bulb with an object in front of it, but this time set the object on a rotating table and use a computer to record the shape of the shadow. If we collect lots of shadows, each with the object at a slightly different table angle, then we could use the computer to reconstruct a 3D rendering of the surface of the object.

In the medical applications of tomography we want to see beneath the surface of patients and thus visible light is not terribly useful. If instead, we use x-rays then we will get much more complicated shadows that are formed by objects inside the body, as well as the body surface, since a proportion of x-rays pass straight through. Again if we collect lots of x-ray shadows, taken at different angles, then we can reconstruct the shape of the body, but now we will include details of objects on the inside. In general x-ray shadows for tomography are obtained from one particular thin slice at a time because x-rays are scattered as well as stopped by tissue. X-rays scattered through a wide angle from one region would confound the interpretation of the shadow from another. By concentrating on one thin slice at time, this

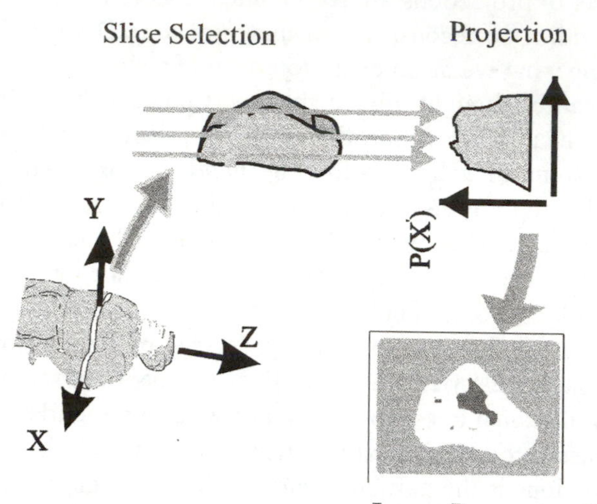

Figure 1.1: The tomographic process generally consists of three steps. First a single slice is selected. Next a complete set of projections of that slice is obtained. Finally the set of projections are combined using a mathematical recipe or algorithm to form a reconstructed image of the slice.

confusion can be avoided. Thus the starting point for most tomographic image reconstruction is a set of shadows or projections of a particular thin 2D slice through the patient.

The Numbers in Box Puzzle

The process of tomographic image reconstruction can be thought of as a way of solving the following puzzle. Consider a square array or matrix of positive numbers whose values we would like to know. Imagine that the array is sealed inside an opaque box so that we can't actually see it but that we have a way of finding the sums along the rows, columns and principal diagonals. In other words we have a means of obtaining a set of projections of the box of numbers.

a	*b*	*c*	*d*	*e*	**0**
f	*g*	*h*	*i*	*m*	**25**
k	*l*	*m*	*n*	*o*	**5**
p	*q*	*r*	*s*	*t*	**5**
u	*v*	*w*	*x*	*y*	**5**
5	**5**	**20**	**5**	**5**	

Figure 1.2: The numbers in a box problem. Each set of numbers outside the box represents one projection. Each number within a projection is a sum through the box along a horizontal or vertical line.

What we want to do is reconstruct the array of numbers inside the box, using just the projections. Figure 1.2 illustrates this idea. All the medical technologies produce essentially the same hidden number puzzle. In each technology the mechanism of image contrast is of course different so that the numbers in the box (a,....,y etc) arise from very different physical mechanisms and represent very different aspects of the underlying tissue. In x-ray CT, each number corresponds to the linear x-ray attenuation coefficient averaged over the particular little volume of tissue. In SPECT and PET each number represents the concentration of radioactive isotope in the little volume. In MRI it is the local free-water proton density, weighted by local spin relaxation times. In practice each slice must have a small but finite width and thus each of our numbers actually represents a small element of volume. With small arrays it is clearly possible to solve the puzzle using trial and error beginning at the corners. For the large arrays,

128*128 or larger, used in medical tomography the guessing could go on for a very long time before convergence sets in. Fortunately there is a very simple procedure that will rapidly produce an approximate answer, and a more exact method that uses the properties of Fourier transforms.

Simple Backprojection

In figure 1.2, each of the numbers outside the box is the sum of box numbers, along a particular direction, through the array. Since we are going to consider a number of projections, each one comprising a number of elements, we need a systematic labelling convention for a particular element in a particular projection. A projection is labelled P_i where i =1,2,3,4 etc denotes the angle at which the projection was taken. Thus the first element, counting from the left, in projection 1 is written $P_1(1)$,the second element in projection 1 is $P_1(2)$ and so on. In the bottom outside row of figure 1.2 we see that

$$P_1(1) = 5 = a + f + k + p + u$$
$$P_1(2) = 5 = b + g + l + q + v$$
$$\ldots\ldots\ldots\ldots\ldots\ldots$$
$$P_1(5) = 5 = e + j + o + t + y$$

and in the right hand outside column we have that

$$P_3(1) = 0 = a + b + c + d + e$$
$$P_3(2) = 25 = f + g + h + i + j$$
$$\ldots\ldots\ldots\ldots\ldots\ldots$$
$$P_3(5) = 5 = u + v + w + x + y \qquad \ldots 1.1$$

Each set of external numbers represents a single projection through the box and of course it contains information about the values and the arrangement of the hidden numbers. In the x-ray case the analogy with our puzzle is quite clear. The first number in projection 1 is obtained by shining a fine pencil beam of x-rays through the box from the top to the bottom along the leftmost column. The relative number of x-ray photons that survive the trip through the box is here represented by $P_1(1) = 5$. Each little volume will attenuate the beam by an amount determined by the local attenuation, so that each number in the projection represents the total attenuation through the box, along a particular line; the sum of box numbers along that line. $P_1(2)$, the second number in the first projection, is obtained by shifting the beam by one column to the right. The different projections arise from different choices of beam direction.

If we pick any box number, m, for example, it is clear that it contributes to all projections, since at every angle there will be one line of sight that passes through the m cell. Since this is true of all the box numbers, we can think of the reconstruction algorithm simply as a way of solving a set of simultaneous equations. In this case we have 25 numbers and this requires 25 equations to get a solution. If we were only dealing with very small arrays then the reconstruction could proceed in this manner. In a real application, providing a spatial resolution of a few mm in a cross section of the human torso, the size of the array will be 128 * 128 or larger. This would require the simultaneous solution of sixteen thousand equations. Although this is by no means impossible it would be extremely time consuming and fortunately is unnecessary.

Simple backprojection begins with an empty matrix that represents the object slice and the idea is to fill in, or reconstruct, the unknown number in each cell. Consider the first number $P_1(1)$, in the first projection. During the data collection this was built up from contributions from all cells, along the line of sight that ends at this point in this projection. As a first approximation we can think of each cell along this line as contributing the same average amount to the sum. Thus we put back into each of the corresponding empty cells the number

$$\frac{P_1(1)}{5} = \frac{a+f+k+p+u}{5} \qquad \ldots 1.2$$

and repeat this for all the other numbers in the first projection. The backprojection process is repeated for the second and subsequent projections, but at each stage we add the new average numbers from the current projection to those already in the matrix. After 2 projections have been backprojected, the centre cell of the reconstruction array will contain

$$\frac{P_1(3)}{5} + \frac{P_3(3)}{5} = \frac{c+h+m+r+w}{5} + \frac{k+l+m+n+o}{5}$$

We see that the real contribution to the centre cell, m, enters twice into the temporary answer whereas all the other contributions only enter once. After

N projections we will have N.m / 5 coming from the target cell together with a sum of other contributions. Figure 1.3 shows the reconstruction process interrupted after 1 and 2 and 4 stages of backprojecting the sums. As we proceed, the image gradually condenses into a reasonably good approximation to the original array.

1 projection

2	1	1	2	0
2	2	1	1	2
1	2	2	1	1
0	1	2	2	1
0	0	1	2	2

2 projections

3	2	5	3	1
3	3	5	2	3
2	3	6	2	2
1	2	6	3	2
1	1	5	3	3

4 projections

3	4	6	4	3
11	9	11	9	11
4	5	9	5	4
3	5	9	5	3
4	4	7	5	4

Figure 1.3: Reconstructing the numbers in a box. At each stage the sums or projections have been averaged over the number of cells along the line of sight and the corresponding nearest integer added to each cell, backprojected along that line of sight. The left-hand box is the backprojection of the sum taken with the lines of sight along a diagonal. The other backprojections correspond to projections taken at successive angular increments of +45 ° to the horizontal axis.

Simple backprojection has a severe limitation as far as real imaging is concerned. Regions of the phantom that contribute no signal to the projection actually end up with a definite positive intensity after backprojection. We noted above that the simple process puts back into each box a multiple of the real original number together with other contributions. These unwanted contributions are all positive and thus keep on adding to our answer as we step through the projections. This leads to an incorrect, gradual fogging of the reconstructed image that is unacceptable. A partial remedy would be to calculate the average value of all the entries currently in the reconstruction matrix at each stage and subtract the average from each entry. In general simple backprojection on its own is seldom if ever used, since the better Fourier method or its derivative Filtered Backprojection produces a very much more reliable result. We have dwelt on the simple process in order to show that tomographic reconstruction is basically very easy to understand and requires no mathematics beyond averaging and addition.

1.3 The Fourier Method

Fourier methods are frequently used in science and engineering, particularly in the solution of differential equations. One reason for their use is that difficult problems in real space can often be solved, using simple algebra, in Fourier space or K space. The idea is very similar to that of using log tables to do multiplication and division. It will be recalled that the logarithm of a product of two numbers is just the sum of the individual logarithms. Thus to multiply two numbers we take their logs, add the logs together then take the antilogarithm of the sum to obtain the product. Fourier methods have similar steps: the original problem is transformed into a new space, K space, where some algebraic manipulation takes place. Finally the answer is obtained by an inverse Fourier transformation back to real space. A more detailed description of waves, Fourier analysis and K space is given in Appendix A. The Fourier solution to the tomography problem relies on the Central Slice Theorem. Appendix B gives a simple mathematical derivation of the theorem. Here we will show qualitatively how this is used to provide in principle an exact method of image reconstruction.

The Central Slice Theorem

Once a real image slice has been selected, we have our numbers in a box problem but on a very much larger scale. We superimpose an imaginary grid on the slice and the local contrast value in each grid square is denoted by μ. This varies continuously with position X, Y, and so we write $\mu = \mu(X, Y)$. One element of a projection, that is to say one of the numbers outside the box in figure 1.2 consists of a sum of all the values of $\mu(X,Y)$ along a particular

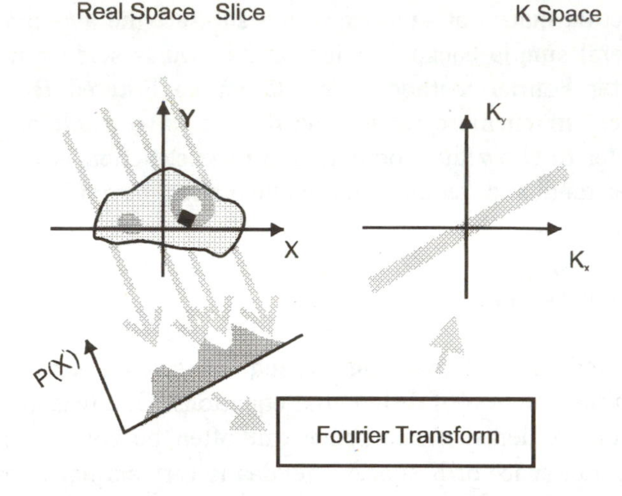

Figure 1.4: The Central Slice Theorem. The projection, P(x'), of the slice through the object could be obtained by moving a fine x-ray beam across the slice at the angle shown in the left panel. A 1D Fourier transform of P(x') yields the 1D Fourier transform, F (K_x,K_y) and this is one profile through the 2D Fourier transform of μ (X, Y) as shown in the right panel. The line in Fourier space makes the same angle to the K axes as the original projection direction does to a set of axes established in the real space slice.

line of sight chosen for that projection, across the slice . If we denote the position of this number within the projection as x' Then we can write P (x') = ∫μds, where the element ds is an element of length along the particular line of sight across the slice. As we move along the projection, we vary x', and build up the profile, P(x') of this particular projection. The next step requires Fourier analysis. Using the Central Slice theorem, we can relate the Fourier transform of this particular projection to one line in the 2D K space formed by the 2D Fourier transform of μ (X,Y). Figure 1.4 illustrates this process. The lengths in K space correspond to $2\pi/\lambda$, where λ is the wavelength of a spatial wave. As we describe in appendix A, the Fourier transform process amounts to decomposing a function in real space into a set of component spatial waves, each one having its own wavelength. Thus small lengths in K space correspond to large wavelengths in real space. The smallest K is

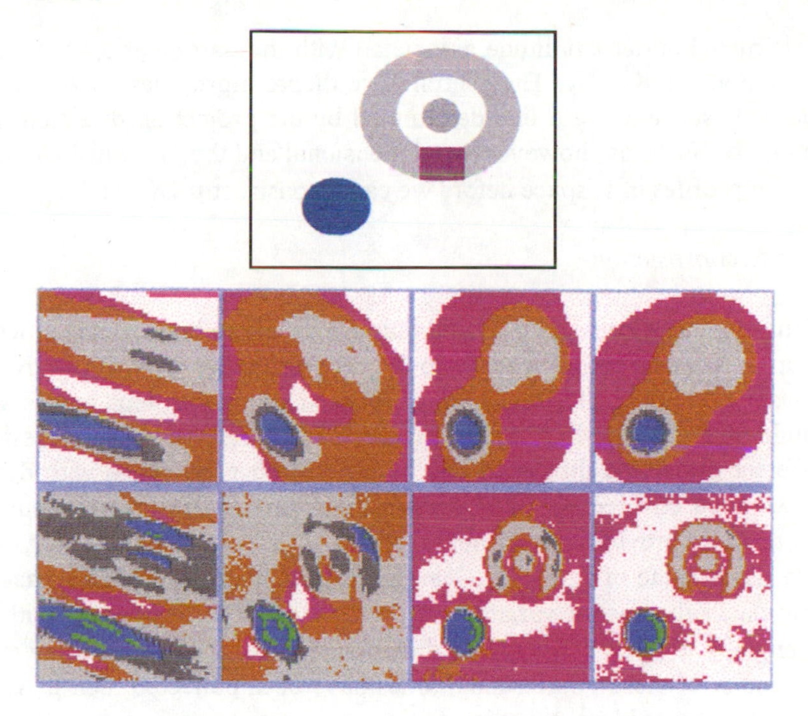

Figure 1.5: A simple phantom model used throughout this book to illustrate tomographic reconstruction . The model consists of an homogeneous disc, and a toroid-disc combination. Different values of "attenuation coefficient" correspond to different colours.

Figure 1.6: Tomographic reconstruction of the model slice shown in figure 1.5, using the simple and the Fourier method. The top four panels show the simple method and the bottom four show Fourier method, (Filtered Backprojection). In each case the reconstruction of the model phantom has been interrupted after 10, 30, 70 and finally 90 projections obtained at intervals of 3° have been added. An early crude reconstruction of the form of the phantom is clearly visible in even the left hand panels. Using the simple scheme, the final image reproduces the gross features of the model but the details, especially within the toroid are lost.

K is determined by the overall outline of the slice. On the other hand, large values of K correspond to short wavelengths and these relate to the fine detail within the slice. At each value of K_X, K_Y there is a value, $F(K_X, K_Y)$, the particular Fourier amplitude associated with the particular spatial wave with components K_X, K_Y. The central slice theorem provides one Fourier profile in K space along a line determined by the projection direction, see Appendix B. Slices are however two dimensional and thus we need an array of Fourier profiles in K space before we can reconstruct μ (X, Y).

Fourier Reconstruction

Our original slice was in 2D and thus we expect to have to construct a 2D map in K space before we can do the reconstruction. This is relatively straightforward. By collecting a set of projections at different angles we can build up a 2D map in K space. Figure 1.7 illustrates how a succession of different projection directions can be used to cover K space with data. Once we have covered K space with profiles, then a 2D inverse Fourier transformation provides an estimate of μ (X, Y), the quantity that we require. Each value of F (K_X, K_Y) in the Fourier map provides a measure of how much the spatial variation in μ (X,Y) matches a sinusoid with a wavelength $2\pi/\sqrt{(K_X^2 + K_Y^2)}$ and a direction (K_X, K_Y). When a sufficiently large number of these Fourier amplitudes have been collected then μ(X,Y) is recreated by adding together the component waves each with its appropriate amplitude and phase. This in words is the inverse Fourier transformation process.

Digital Reconstruction

In practice all medical tomography is carried out with discrete data using a digital computer. This leads to a number of slight changes from the description that we have just given and some constraints. Each projection is measured using detectors that have a finite resolution. In x-ray CT the x-ray photons are recorded using an array of detector elements, each of which is about 5-10 mm across. Thus each profile is actually sampled at a finite number, 256 say points. Similarly in MRI the projection profile is obtained from a time signal that is sampled at definite, discrete intervals in time. In addition there will be a finite number eg. 256, different projections that together span K space. In x-ray CT the projections are lines through the origin of K space and so the complete sampling is achieved by assembling a set of "spokes" with a set of "beads" on each spoke. Each bead corresponds to one value of K_X, K_Y and at each of these points there is one value of $F(K_X,K_Y)$. The complete set constitutes a sampled version of the 2D Fourier transform of $\mu(X,Y)$. In MRI there is considerably more flexibility in the way in which K space is actually sampled. Conventionally in MRI, the values of $F(K_X,K_Y)$ profiles are collected along lines parallel to K_X at constant values of K_Y, each projection then corresponds to a different value of K_Y. Rapid imaging MRI schemes are now used that have a variety of more exotic paths through K space, including spirals. In x-ray CT, the order in which discrete profiles are gathered is immaterial but in MRI, particularly in rapid imaging schemes, the order and trajectory in K space can have important consequences on the resulting image. This is discussed further in chapter 6.

The final step of inverse 2D Fourier transform of the sampled version of $F(K_X,K_Y)$ is in principle straightforward. In practice there some important points, arising from the way the data is collected, that need to be considered if we want to understand the final image and be able to distinguish real from spurious features or artefacts. Each technology using tomography has its own particular set of limitations brought about by engineering compromises. These will be discussed in the appropriate subsequent chapters. Here we will give brief description of some general limitations created by sequential data gathering, digitisation and the intervening excursion to and from K space.

Reconstruction Variations

Although all the medical imaging technologies use a reconstruction method that is based on the Fourier method, there are differences in its implementation in PET, SPECT and CT on the one hand and MRI on the other. We will show in chapter 6 that in MRI the measured signal is in fact already a 1D Fourier transform corresponding to one line in K space. Thus in this case there is no need for an explicit forward transformation of profile

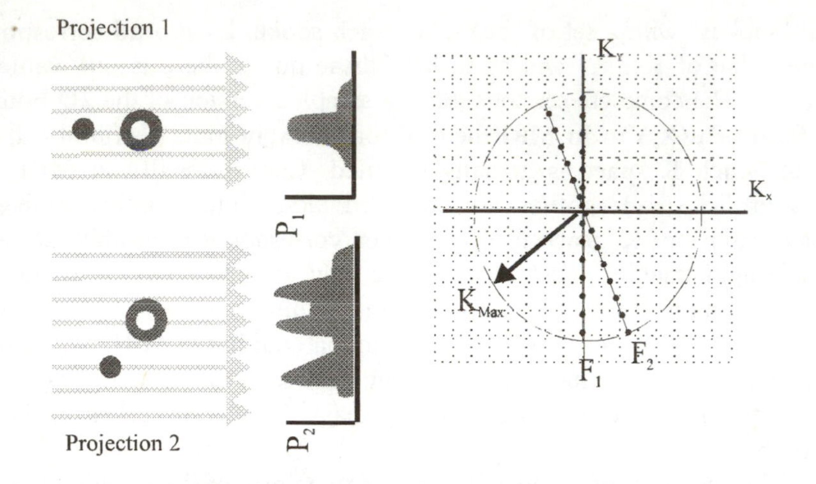

Figure 1.7: Two projections P1, P2 of the slice in the left panel provide two lines in K space as shown in the right panel. Each of the filled "beads" in Fourier space represents one Fourier amplitude obtained from the 1D FT of the corresponding projection data sampled at discrete intervals. Notice that at large K, Fourier space is less well sampled than at low K. This leads to a loss in spatial resolution. The circle in Fourier space represents the maximum possible value of K that can be obtained when real, detectors of finite width or finite sampling intervals are used.

data. Rather the slice data is collected line by line and then a single final inverse 2D transform provides an estimate of the reconstructed image. Thus reconstruction in MRI follows most closely to the Fourier method .

CT, SPECT and PET use a slightly different method that has particular advantages for these technologies. Here the method is Filtered backprojection. This is described in appendix B and its results are illustrated in figure 1.6. Filtered backprojection is rather similar to our simple scheme since it doesn't involve the explicit use of a Fourier transform–inverse manipulation. Instead it makes use of a mathematical operation called convolution In filtered backprojection each projection is first multiplied by (convoluted with) a mathematical function before backprojection. The function, in its simplest form, is the Radon filter shown in figure B.3. More complex filter functions are used in gamma imaging in order to remove the effects of noise on the final image. Some common composite filters are illustrated in figure B.4.

Filtered backprojection is computationally more efficient, especially for CT, since reconstruction can begin as soon as the first projection has been obtained, allowing the data acquisition and reconstruction steps in the imaging process to be interleaved in time. The method also lends itself rather easily to modification, along the lines described in Appendix B, in order to deal with the effects of statistical noise that are common in gamma imaging.

Sequential Data Collection and Patient Movement

The projection reconstruction scheme that we have just described necessarily requires that the data be acquired over a finite period of time, dictated by mechanical, physical or even physiological constraints that enter a particular method. The original Hounsfield EMI CT scanner took about an hour to collect enough data for one slice. Each element in each projection required the mechanical positioning of an x-ray source and an x-ray detector and then a counting time at each position. Modern x-ray CT machines are very much faster and collect slice data in about 0.2 s. Standard MRI on the other hand can still take about five minutes per slice for some examinations. During the time that data is being acquired in a sequential fashion, great care has to be taken to limit movement of the patient. Any movement will alter the relative positions of a subset of the projections. At best this will ultimately limit the spatial resolution of the image, leading to some smearing. At worst completely fictitious features can appear in the image because some of the projection data is misplaced with respect to the rest. The Fourier reconstruction process can be unforgiving of this type of data error.

Digitisation of the Projections

As we have already said the projection profiles, P(x') are not obtained as continuous functions of position but rather as a finite number of samples. In the case of x-ray CT the distance between successive samples in any one projection is determined by the size of the x-ray detector element used. Each detector sums the collected x-ray intensity over its width and averages over any spatial variations whose scale is smaller than the detector size. It thus acts like a low pass spatial filter. Finite detectors or indeed any finite sampling of the projections sets a definite lower limit on the final spatial resolution in the image. The finite sampling interval in real space corresponds to the maximum values of K_X, K_Y at which data are collected and thus the fine spatial detail that can be resolved.

In MRI the data is acquired as a time signal that is sampled at a discrete number time points. Here care has to be taken to avoid the aliasing problem, (see appendix A) by ensuring that the signal, before sampling, does not contain any temporal frequency components that are more than twice the sampling rate. Failure to observe this will result in gross image artefacts as a direct consequence of the Fourier scheme.

The Number of Projections

In every technology a particular expanse of a slice, called the field of view, FOV is selected for image reconstruction. Whatever the scheme used to sample Fourier space, each projection will span the field of view in one direction with say N discrete points. If the FOV has dimensions $L \times L$, then the interval between points, along one direction will be L/N. The smallest K value will be $K_{min} = 2\pi/L$ and the maximum value of K will be $K_{max} = 2\pi N/L$. Generally K space will be sampled in the other orthogonal direction in increments of K_{min}. The aim is to sample K space as uniformly as possible so that there are approximately equal intervals between points in both the K_x, and the K_y directions. Once the digitisation step along a projection is established, by for example the choice of detector size, then the interval between successive projections in K space is arranged to achieve uniform sampling. Any under sampling in a particular direction will introduce artefacts again as a direct result of the Fourier process.

Signal to Noise Ratio

All medical imaging schemes suffer to a greater or less extent from the effects of random noise. Both x-ray and gamma ray detection is a random process which has an inevitable associated statistical uncertainty. If N counts are collected in a detection process then there is an intrinsic uncertainty of $\sqrt{N}$ in the number of counts. Thus the signal to noise ratio is $N/\sqrt{N} = \sqrt{N}$. Thus the relative accuracy of any atomic counting measurement increases with the total number of counts collected. X-ray CT is the least effected because extremely large numbers of x-ray photons are collected to give each element in a projection. Both SPECT and PET, on the other hand, are limited by random fluctuations in the photon counts making up each projection element. In MRI the radio frequency signal is intrinsically weak in comparison with thermal noise as shown in chapters 2 and 6 and thus it too is eventually limited by random fluctuations in the signal. In addition to degrading the imaging signal, random noise can also upset the intervening Fourier steps in tomography, introducing extra problems. For example large amounts of noise can mimic high frequency signals, leading to aliasing and eventually to image artefacts as well as a general reduction in the sharpness of the final image.

Tomographic Image Artefacts

Tomography, can introduce false detail and distortion, that is to say image artefacts, into the final image as a result of faulty equipment, errors in the stored data and patient movement. Because of its digital nature, tomography can also produce generic artefacts. Patient movement is always a problem when data acquisition times exceed a few seconds. The movement can simply blur the image or introduce more troublesome discrete artefacts, when the patient motion modulates the projection data with a particular temporal frequency. The interpretation of any tomographic image can be confused by partial volume effects. These arise directly from the use of finite image voxel dimensions. Even a very high resolution image, made up from cubical cells about 1mm on a side may well comprise some cells which cover a range of tissue types, bone, blood vessel, muscle and fat. The signal from these cells will reflect the average rather than just one tissue type. At the edges of very high contrast, such as bony structures in CT, there will be very abrupt changes in the the numerical values making up the projections. Image reconstruction using a finite number of projections cannot faithfully reproduce the sharp edge, rather it can creates a halo close to the edge in the image.

Questions and Problems

1. a) In the context of X-ray CT explain the term projection
   b) Describe the method of simple backprojection
   c) State the central slice theorem and explain how it may be used to reconstruct tomographic images. (see appendix B)
2. A phantom object used to check x-ray CT consists of a solid cylinder of bone of radius R cm, embedded in a very weakly attenuating block of resin . Show that the path length ,L, of an x-ray beam passing across the cross section of the cylinder, a distance x from the cylinder axis is given by

   $$L = 2\sqrt{R^2 - x^2}$$

   Sketch the profile of a projection obtained by scanning a pencil beam across the diameter of the cylinder and use a superposition principle to deduce, without further calculation, the profile of a hollow cylinder with inner and outer radii 5 and 8 cm.
3. A 6*6 square matrix contains one positive integer in each cell (the numbers are not all different but they are all less than 9)
   The sum over rows gives:- 6 8 10 16 8 6 (T to B)
   The sum over columns gives :- 6 8 13 13 8 6 (L to R)
   The sums perpendicular to the diagonals give:-
   1 2 3 6 8 11 11 6 3 2 1 (T R to B L)
   1 2 3 6 8 11 11 6 3 2 1 (T L to B R)
   What are the numbers in the matrix? (Hint) start at the corners and work inwards
4. The empty reconstruction matrix for a cross section of the human thorax consists of a square array of 512 *512 cells. If the field of view just contains the entire thorax estimate the dimensions of each cell. Can blood vessels be imaged with this field of view ?
5. The detectors used in a certain CT machine have a width , w.
   a) How many samples are required to obtain the highest possible resolution in a projection of an object with a maximum dimension D?

2 ATOMIC AND NUCLEAR PHYSICS

"I have settled down to the task of writing these lectures and have drawn up my chairs to my two tables. Two tables! Yes; there are duplicates of every object about me — two tables, two chairs two pens. ....
One of them has been familiar to me from earliest years. It is a commonplace object of that environment which I call the world. How shall I describe it? It has extension; it is comparatively permanent; it is coloured; above all it is substantial. By substantial I do not merely mean that it does not collapse when I lean on it; I mean that it is constituted of "substance" ....
My scientific table is mostly emptiness. Sparsely scattered in that emptiness are numerous electric charges rushing about with great speed; but their combined bulk amounts to less than a billionth of the bulk of the table itself. Notwithstanding its strange construction it turns out to be an entirely efficient table. It supports my writing paper as satisfactorily as table No. 1; for when I lay the paper on it the little electric particles with their headlong speed keep hitting the underside, so that the paper is maintained in shuttlecock fashion at a nearly steady level. If I lean upon this table I shall not go through; or to be strictly accurate, the chance of my scientific elbow going through my scientific table is so excessively small that it can be neglected in practical life..........."
Sir A. S. Eddington " The Nature of the Physical World " 1928

2.1 Introduction

Diagnostic imaging in medicine began with Roentgen's discovery of the penetrating radiation, x-rays. The discovery came near the beginning of a period of intense experimental and theoretical scientific activity which ended with the overturn of the established nostrums of classical physics and the rise of the "new physics", quantum mechanics. All through the nineteenth century, physics had advanced in a state of supreme confidence. Armed with the seemingly infallible trinity of classical physics, Newtonian mechanics, electromagnetism and thermodynamics, it seemed possible to make accurate predictions of the behaviour of any aspect of inanimate nature. If a particular phenomenon did not quite fit into the grand order of things, then it would be only a matter of time before the wrinkles were smoothed out and the majestic progress of the theory continue. Towards the end of the century, technology was creating new, more sensitive scientific instruments, which opened new possibilities for experiments. The efficient vacuum pump provided the means to study the properties of gases at very reduced pressures. Batteries and an electrical supply allowed electrical and magnetic experiments to be performed in any laboratory. The development of ship borne refrigeration plants for meat, transported from Australia, provided a spur to the liquefaction of air and thence the study of matter over a wide range of temperatures. The generation and detection of, Hertzian (radio frequency) waves excited the imagination with the possibility of instantaneous long distance communication and provided the seed from which electronics grew. These technological developments themselves fitted snugly into Newton's grand design. After all, the technologies were themselves the practical confirmation of the underlying classical principles. In addition some of the phenomena studied, Hertz's waves for example, were more confirmation of the theory, in this case Maxwell's electromagnetism, developed forty years before.

This cosy state of affairs did not see the celebrations that marked the new century. By the time the new century was only five years old it had become abundantly clear that there were areas of nature, particularly in the realm of atoms, where classical physics was consistently giving completely the wrong predictions. By 1926 the old certainty had completely vanished, and was replaced by a revolutionary mathematical theory of matter, Quantum Mechanics, whose profound philosophical and theoretical implications are still today, the subject of academic debate and new scientific papers. Our opening quotation expresses Eddington 's concern, in 1928, the time of most hectic change, about the state of affairs in physics as he saw it. Conceptually the quantum physics of atoms and particles just did not fit with the physical intuition accumulated over a lifetime dealing with everyday objects. Physicists had been forced into using two different sets of equations, one for small things and one for large things. This was an anathema to the classical physicist who, not long before, had had a single unified set of concepts and a single calculating engine. In some respects the slightly messy situation in physics has not really changed since 1928. Eddington, Einstein and others imagined that the confusion was only temporary, we now know it is rather long lived. We are now convinced that quantum mechanics is the underlying truth and it has long been known how quantum mechanics can be merged into the classical laws, when dealing with the properties of large numbers of particles, at everyday levels of total energy. We still however use the mathematical methods of calculation found on Eddington's everyday table. If we are interested in designing a rocket, building a bridge or even assembling the circuit for a TV set or computer we will almost certainly make little or no direct reference to the quantum physics. Quantum physics will however dominate the thinking on the strength and corrosion resistance of the steel in the bridge and determine both the design and manufacture of the transistors inside the electronic chips. Our present day choice of conceptual "table" is then largely decided by the size of the items that we place on it. The physical principles underlying both the technology and methods of medical imaging span almost the complete spectrum of modern physics, covering both conceptual tables. A large number of the techniques invented by physicists to probe the properties, first of atoms and then their subatomic constituent particles have been adapted by medicine to create standard clinical diagnostic and therapeutic methods. It follows that an understanding of the principles of medical imaging has be predicated on an understanding of physics itself, if we are to avoid treating the subject as collection of black boxes. It would be impossible and inappropriate to cover the background "physics" in few introductory pages with sufficient depth to allow the non-physicist, starting from scratch, to fully comprehend all the physics that is harnessed into the various imaging technologies. We can however, through words and pictures provide a summary of the most important principles, show how these form a coherent and consistent theoretical account of nature and through simple examples introduce the non-physicist to the culture of physics. Our experience has taught us that physics, to the non-specialist, is shrouded in mystique because of its dependence on mathematics. Often (but not always) the mathematics is really expressing a quite simple concept or a convenient way around an esoteric difficulty. Bragg once claimed that all good science should be accurately explicable to a schoolboy. Unfortunately this is no longer always possible and so the reader is asked to forgive our occasional inconsistencies that arise from our use of analogy and picture rather than mathematics. In the following sections we provide very brief accounts of those branches of physics that impinge most heavily on medical imaging. We start with the trinity of classical physics because these are still fundamental for a practical understanding of the everyday sized objects on Eddington's first table. We then try to bridge the gap between the two tables and describe those parts of atomic and nuclear physics which underpin x-ray radiography, gamma imaging and MRI.

2.2 The Scope of Physics

What is Exact About Physics?

Physics, like any other area of science, depends on the rigorous use of the scientific method, a logical thought process, which we have inherited from early clinical medicine. The process is circular and always involves the same three steps of experimentation – hypothesis building leading to prediction –confirmation or rejection of the hypothesis through further experimentation. This universal scheme seems to guarantee that only enduring truths survive the test of time and allow these to be refined. Physics and the other so called exact sciences, chemistry and engineering have developed a slight but very important mutation to the middle step of this generic process. The physics hypothesis is expressed not only in words but in symbolic mathematical relationships, equations, which can then take on a life of their own to make quantitative predictions of further experiments, sometimes predicting the existence of entirely new phenomena. It is this mutation which gives rise to the adjective "exact" in the phrase "exact science" but it must not be taken too literally. The primary job of physics is to find the underlying general principle(s) governing a particular natural phenomenon and cast that principle into a mathematical form. When the most important factors have been established then the secondary, tertiary, ... effects can be isolated and added to ever more refined models which should provide better and better predictions for a widening class of real related phenomena.

Physics as described, even in undergraduate textbooks, is rarely if ever exact as far as everyday phenomena are concerned. Rather it is the known and very valuable collection of general principles governing idealised and simplified cases. The job of refining the models for real world cases is often considered to be the province of engineering. Physics, in the popular imagination, is sometimes castigated for its "light inextensible strings", for its "ideal gas" or its neglect of air resistance. Idealisation or simplification in the right context, is not at all ludicrous but rather allows the key influences to be established. If it turns out that secondary effects are important then there is often ample mathematical machinery to incorporate these into the model.

Quite apart from the intellectual kudos associated with the successful general description of the inanimate universe, both physics and engineering are valued because detailed quantitative predictions can have immense economic importance in an industrialised society. Occasionally (very occasionally) the mathematical model derived from one area will predict an entirely new behaviour or phenomenon that can create new technologies. More commonly a physics principle is used as the basis for a particular instrument or household gadget and then detailed mathematical modelling of that principle together with the real life effects, such as friction, allows refinement. The paper and pencil or more likely the computer calculation allows existing methods to be refined to build better bridges, make computers run faster and improve and cheapen the manufacture of gadgets, without having to go through extensive, time consuming and expensive trial and error experimentation.

Classical Physics

Classical physics is concerned with the properties of everyday objects, Eddington's first, apparently common place table. Here "everyday" means a large collection of interacting particles. Classical physics concerns itself with the description and prediction of the bulk properties of the object as a whole. The trinity of classical physics comprises mechanics, thermodynamics and electromagnetism, all of which were essentially developed by 1865.

Mechanics

The whole of physics rests on a bedrock of mechanicslaid down by Galileo and Newton. The fundamental concepts and definitions of mass, force, length, time , velocity and acceleration were established by these two men and have remained almost unchanged since their time. After these definitions, in order of importance, come Newton's three laws of motion and his theory of universal gravity. Not only did these provide our most basic calculating tools but, they also established the philosophy of physics. In describing universal gravity Newton invented the idea of a field of force or an "action a distance". We commonly think of force in the context of one object pressing against another, in direct physical contact. To explain gravitation Newton had to imagine an influence, transmitted across empty space, to cause two distant bodies to "feel" each other's presence. All of modern physics, whether it is dealing with macroscopic electromagnetic fields or microscopic quantum interactions relies completely on this idea of a field of force and action at a distance. Modern physics also assumes that the same physical situation, anywhere in the universe, will be governed by exactly the same principles and equations as have been deduced from experiments or observations made on the earth.

The Conservation Laws

Newton was perhaps the first scientist to rely heavily on the mathematical model and he invented calculus to assist him. Most of Newton's work concerned the motions and interactions of simple solid objects such as planets and billiard balls. His equations are however far more general than this, so that for two hundred years after Newton's death, scientists and mathematicians were able to extend the application of Newton's laws to ever more complex objects (all of course of everyday size) and very much more complex motions. Although the derived equations express precisely the content of Newton's laws, they are often extremely complex and very difficult to solve. In the days before computers or indeed mechanical calculators, scientists were ever on the look out for mathematical tricks and dodges that allowed them to obtain answers or predictions from such hard equations with the minimum of tedious calculation. The best tricks often found a wide application and became standard methods of analysis. There is one broad category of such devices that is taught to all young physicists; it is the use of the conservation laws. We will start our description with a simple idealised example. Suppose we wish to calculate the maximum height, h, reached by a cricket ball thrown vertically upwards with an initial speed of v m/s. To a first approximation we can neglect air resistance and just consider the effect of the constant force of gravity, g, acting on the mass of cricket ball, m. The calculation can proceed by finding the details of the ball's motion, that is to say to solve the equation of motion and obtain a detailed quantitative description of the ball's position at any time, h(t). It is however simpler to use the conservation of energy. The cricket ball has an initial an amount of kinetic energy which, as it rises and decelerates, is turned into gravitational potential energy. When it reaches its maximum height the ball is, by definition, at rest and therefore all of its kinetic

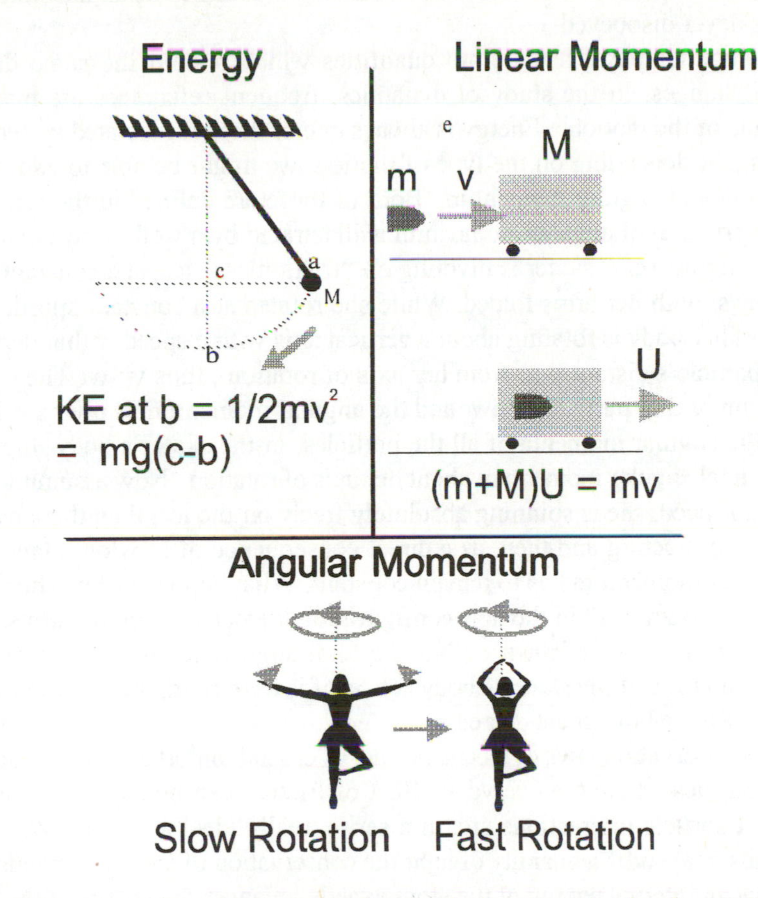

Figure 2.1: The conservation laws of physics. A simple pendulum illustrates the conservation of energy. A bullet fired into a large mass on wheels illustrates the conservation of linear momentum. The pivoting skater illustrates the conservation of angular momentum. In each case the conservation law allows us to relate "before and after" states of motion often without the need for an incremental analysis.

energy has been turned into static potential energy. We do not need to consider precisely how the ball moves, as it decelerates, but only consider its initial and final states and equate their energies in order to relate the final height to the initial upward speed. The conservation of energy is taken to be an absolute law, which is never disobeyed.

Throughout physics there are quantities which remain the same during physical changes. In the study of dynamics, frequent references are made to "constants of the motion". Energy is always one of these for isolated systems of particles, but depending on the type of motion, we might be able to add linear momentum and angular momentum. Both of these are defined in the glossary. The conservation of angular momentum is illustrated by a well-known example shown in figure 1c. A skater is pivoting on "frictionless" ice, at a constant rate, w radians/s, with her arms folded. While she rotates at a constant speed, each particle of her body is rotating about a vertical axis with a speed v that depends on that particle's distance, r, from her axis of rotation , thus v= wr The linear momentum of that particle is mwr and the angular momentum is mwr x r. If we add up the angular momenta of all the particles in the skater's body then we have her total angular momentum about her axis of rotation . Now assuming that, once up to speed, she is spinning absolutely freely on the ice, then there can be no net torques acting and then, as a direct consequence of Newton's laws, her total angular momentum has to remain constant. What happens when she flings out her arms sideways? In the new configuration the particles of her arms have increased their values of mwr x r but the total angular momentum must stay constant and thus, if the skater's body acts as if it were rigid, then w, the rate of rotation of the whole must decrease.

The conservation laws of mechanics are of crucial importance in atomic and nuclear physics where the everyday effect of friction can be made sufficiently small that particle interactions are, in a sense, really ideal and pure. When an atom emits or absorbs a quantity of light the conservation of energy demands that the change in internal energy of the atom exactly balances the energy of the light received or emitted. The conservation of linear momentum is responsible for the two gamma rays, in positron annihilation, travelling away from the nucleus in opposite directions, see chapter 5. Angular momentum and torque are central parts of our description of nuclear magnetic resonance in chapter 6.

Thermodynamics

The principles of thermodynamics are of fundamental importance to all branches of science and engineering. These principles describe the physical properties of large collections of particles, that is to say gases, liquids or solids in terms of measurable bulk properties such as pressure, volume and temperature and their relationships, during physical changes brought about by the addition or subtraction of heat or work energy. Several other useful but more complicated quantities can be defined such density (mass/ volume), specific heat and compressibility that change in a predictable manner and are often used to characterise a particular substance. Thermodynamics plays an important role in science for a number of related reasons. First it gives precise definitions of temperature, heat, work. Then being a completely general theory, it often provides a particularly useful test for new microscopic theories of matter. If a new theory makes predictions that are at variance with the laws of thermodynamics then the theory is deemed to be either wrong or at best incomplete. Finally it is particularly good at finding definite energy limits on physical processes. At the trivial level it shows categorically the impossibility of a perpetual motion machine. More usefully it allows an engineer to place an upper bound on the amount of useful work that can be obtained from any sort of heat engine be it coal or nuclear powered.

The central concepts of thermodynamics are summarised in four laws, called zeroth, first, second and third. The zeroth law defines temperature as that common property of two bodies that are placed in contact and allowed to settle down or come into thermal equilibrium. The property, temperature, can of course be measured using a mercury column or electrical thermocouple. Two identical blocks of copper one placed in the fridge and the other in the oven eventually acquire properties that we describe as "hot" and "cold". If we now put the two blocks together then the hot block looses a certain amount of heat energy which, assuming no losses, is precisely the same as that gained by the cold block. After some time the two blocks will reach an equilibrium temperature that is the average of the initial "hot" and "cold" temperatures. The first law of thermodynamics is no more than a restatement of the conservation of energy but with precise definitions of the terms work and heat as different forms of energy. Heat is the energy which "flowed" from the hot block to the cold block in our previous example. Work most simply is the energy involved when the point of application of a constant force, F, moves through a distance l , so that Work = F x l . Thus we could calculate the work done on a quantity of gas enclosed in a cylinder, when a close fitting piston is moved by an external force. Thermodynamics is often applied to electrical systems where the definitions of work changes to Charge x Voltage.

The second law provides a definition of the important concept of entropy or disorder and deals with the question of the efficiency of turning heat energy into useful (mechanical) work. Although work can be turned into disordered energy, heat, with 100% efficiency the converse is not true. Any heat engine, in one cycle, takes in a quantity of heat, Q from a hot reservoir at a temperature T, does a quantity of organised work, W and must reject a quantity of heat q to a colder reservoir of heat at a temperature, t . The conservation of energy requires that Q =W + q , but this tells us nothing about the relationship between the two quantities of heat, Q and q. This is crucial if we want to calculate the efficiency of the engine, (Efficiency =Work out / Heat in = W/ Q). The relationship was discovered by Carnot; in the process he invented the quantity entropy. In a hypothetical completely reversible cycle the quantity, Q/T + q/t =0. The quantity, Q/T is the entropy change associated with the exchange of heat between the hot reservoir and the working substance (gas) of the engine and q/t the entropy change at the lower temperature. If the engine cycle could be made to be completely reversible then the total entropy change of the working gas would be zero since in each cycle since it would return to the same thermal state.

In its most cryptic form the second law simply states that in any physical change, the change in entropy is either greater than or equal to zero. In general a zero entropy change is an idealisation in the context of heat engines but can be realised in more constrained situations. Entropy is of great conceptual utility in the theoretical account of the alignment of nuclear magnetic moments in large magnetic fields, the starting point for nuclear magnetic resonance. The degree of nuclear polarisation, achieved in a given magnetic field represents a balance between magnetic forces that encourage directional order amongst the nuclear moments (just like the alignment of magnetic compass needles) and random thermal energy changes whose net result is to produce complete disorder. The third law of thermodynamics is concerned with the unattainability of an absolute zero of temperature and does not need to enter into our discussions.

The Microscopic Picture of Thermodynamics

Once we accept that all matter is made up of discrete atoms and molecules then our microscopic picture has to be the one described by Eddington is our opening quotation, "*substance consists ... mostly of emptiness. Sparsely scattered in that emptiness are numerous electric charges rushing about with great speed...* " At first sight this change to a microscopic picture of substances seems to encumber us with a great deal of extra complexity; we might anticipate having to account individually for the motions of about 10^{23} atoms in a substance. In fact this is not the case and the switch in picture, provides us with an enormous advantage. The laws of motion should apply as well to molecules as billiard balls, allowing us to calculate the motions of a representative or average particle. Any real particle will have a speed and direction that is constantly fluctuating from this average but the vast numbers of particles comprising even just 1cc of substance ensure that the random fluctuations cancel out, leaving the average result as representative of the whole. This type of reasoning leads to a complete microscopic description of the thermal properties of substances that is to say solids, liquids and gases.

The temperature of a substance is related directly to the average molecular speed of the constituent molecules. Heat is simply the transfer of high speed molecular motion from a high speed "hot" region into a "cooler" low speed region. In mathematical terms the average kinetic energy of an atom with mass, m and average speed , $<v>$ in a gas, at temperature T is obtained by writing $\frac{1}{2} m<v^2> = k\,T$, where k is a physical constant called Boltzmann's constant,(see the Glossary). A correct microscopic theory will allow us to calculate the average square velocity $<v^2>$ and thus we have a direct relationship between a calculable microscopic quantity and an observable macroscopic quantity T. Adding heat to the substance increases the kinetic energy of the atoms and hence the temperature is increased.

The average force, or pressure, that an enclosed gas exerts on a piston, results from the sum of molecular impacts on the underside of that piston. The internal energy of a gas is the sum of all the molecular kinetic energies. Finally entropy is proportional to the degree of disorder of the gas molecules, that is to say, the number of equivalent microscopic configurations that molecules could have and still have exactly the same total internal energy. This microscopic picture is crucial to our understanding of bulk matter. The overall picture survives the transition to quantum physics but the detailed results are altered because we have to use quantum mechanics to deal with the motions and interactions of tiny molecules and atoms.

Electricity and Magnetism

Electricity and the electric force arise from a general property of matter, quite distinct from its quantity or mass, which we call electric charge. There is a force between two stationary separated electric charges that acts along the line between charges, falls off as the inverse square of the distance between them and is attractive for opposite and repulsive for like charges. Magnetism, although initially considered to be a completely separate type of force, was shown by Faraday and then Maxwell to be intimately related to the electric force. Magnetic forces arise solely from charges in motion, that is to say electric currents since there are no magnetic charges. For nearly 150 years, since Maxwell, science has considered electricity and magnetism as different aspects of the same property of nature. Stationary charges attract or repel each other, electric currents also attract and repel each other but the direction and strength of the force between currents depends on the relative directions and speeds of the charges as well as distance and strength of charge. The classical picture of both electric and magnetic effects embraces Newton's concept of action at a distance. Static electric charges are the source of the electric force field which is felt by other charges in the vicinity. Electric currents give rise to a field with a different geometry that we call a magnetic field of force that is felt by other moving charges. Faraday discovered that a magnetic field, changing in time, produces an electric field (induction) and so the two types of effect are intimately linked one to another. In 1865 Maxwell brought together the known laws of electricity of magnetism. These comprised Ampere's law relating magnetic field strength to electric current, Coulomb's law relating electric force to electric charge and Faraday's law of induction. Maxwell put these into a systematic and consistent form and added a vital extra term that was necessary for consistency. The collection of four equations are called Maxwell's equations. All the engineering concerned with the production of magnetic fields in MRI stems from exclusively from this set of equations. This however is not the most important result of Maxwell's equations. Rather it is the direct prediction of electromagnetic waves that carry energy and travel with the speed of light. In other words Maxwell's equations show directly that light is an electromagnetic disturbance that carries an energy that can be converted into other forms when it is absorbed by matter. Maxwell's equations were perhaps the single most important discovery in physics in the nineteenth century. They provided the first proper theory of light and of course led to the invention of radio and all that follows from it.

The Need for Quantum Mechanics

Towards the end of the C19, theoretical physics rapidly became the victim of its own success. The trinity allowed enormous technological progress to be made which in turn led to the manufacture of new and better scientific instruments. Furthermore the three components seemed to be completely consistent. However, during the last two decades of the century the results of new experiments, using the new instruments, showed that at the microscopic scale, nature simply did not obey the laws of classical physics. The discrepancies were not just minor questions of degree but huge matters of principle that pointed to a near complete failure of prediction in all three areas of classical physics when applied to the internal structure of atoms.

Light coming from natural objects is generally composed of a continuous range or spectrum of frequencies. Natural sunlight that we call white light comprises the visible spectrum " Richard Of York Gave Battle In Vain ", red, orange, yellow and so on. In spite of this well known natural phenomenon, spectroscopic experiments performed on the light emitted, by pure gases produced overwhelming evidence that atoms only emitted and absorbed very discrete light wavelengths (colours), rather than a continuum. Classical physics had no explanation for this fundamental discreteness in nature. On closer examination even the continuous spectrum from hot bodies, the so-called black body radiation, could not be described quantitatively by classical physics. When monochromatic light was used to eject internal electrons from isolated simple atoms, the photoelectric effect, it was found that the energies of the electrons did not appear to obey the classical laws. The photoelectric effect provided a key paradox and eventually a vital clue to what was wrong. According to Maxwell's theory the energy of a beam of light is proportional to the intensity of the beam. Thus it was predicted that electrons, ejected from an atom by an intense light source would have a great deal more kinetic energy than those ejected by a dim source. In fact the energy of the electrons remained constant but the numbers of ejected electrons varied with the light intensity. Only when the frequency or colour of the light was changed did the energies of the ejected electrons alter. A glaring inconsistency was discovered even in the physics of bulk substances. Classically the specific heat of all substances was predicted to be constant, independent of temperature with a magnitude given by the empirical Dulong and Petit law. Experimentally it was found that the heat capacity of all substances decreased dramatically as the temperature was lowered to the boiling point of air (-220°K). These discoveries of course prompted even more experimentation showing a large collection of other more subtle errors, all pointing to the same conclusion that, at the microscopic level of atoms, classical physics just doesn't work. There was an urgent need for a new and radically different theoretical framework which, eventually in 1926, came to be known as quantum mechanics. The name conveniently summarises the two important ingredients of the new theory. Quantum, because all the fundamental processes of energy exchange between atoms and light can only take place in discrete lumps. Mechanics, because the theory requires a new way of describing and calculating the motions of atoms and more particularly describing their internal structures.

Photons – Quanta of Light

The notion of the discreteness of light first arose from the work of Planck (1889) and then Einstein(1905). Planck was able to account very accurately for the shape of the emitted black body radiation spectrum using a formula that he derived quite simply, but using a radically new hypothesis. Planck found that the vibrations of the atoms, and thus the light frequencies emitted and absorbed by a hot body, had to be quantised, that is to say they could only occur at intervals, rather than forming a continuous spectrum. The energy of a "particle" of light was, he guessed, proportional to its frequency. Einstein took up this idea and applied it to the photoelectric effect. If, he argued, the energy of a light beam is proportional to its frequency (1/wavelength) and the processes of electron ejection from an atom of a metal corresponds to the impact a single particle of light on a single atom, then the results could be understood. Each ejected electron corresponds to just one light-atom encounter and the energy transferred from the light particle to the atom only depends on the frequency of the light. More intense light beams contain more particles of light and thus more electrons are emitted, from more atoms, all at the same energy. Quite apart from dealing brilliantly with a major paradox, Einstein, in this analysis, had an enormous influence on the general progress of the theory. Light could behave as both particle and wave. Fundamental complex interactions could be understood if they were broken down into individual encounters between particles.

In this book we are mainly concerned with electromagnetic quanta or photons. The energy of the photon or light wave is given by

$$E_{\text{Photon}} = h\,f \qquad \ldots 2.1$$

where $h = 6.626 \times 10^{-34}$ joule s. is Planck's constant and f is the frequency of the wave. The frequency of a visible light wave (photon) is about 5×10^{14} sec^{-1} and thus the energy associated with such a quantum is about 3×10^{-19} joule. This is clearly an exceedingly small packet of energy, described by a rather unwieldy number. Throughout physics these small numbers are expressed in a rather more convenient format using the unit of electron- volt, eV,

$$1\text{ eV} = 1.602 \times 10^{-19}\text{ joule} \qquad \ldots 2.2$$

where, as the name implies, 1 eV is the energy acquired by a single electron when it is accelerated by an electric field, through a voltage difference of 1 Volt. In these units the visible light quantum has an energy of about 2eV. A x-ray quantum has an energy of about 50×10^{3} eV = 50 keV, and the energy of a radio wave quantum, arising in MRI, has an energy of 10^{-5} eV= 0.00001 eV. Although each of these photons has a completely different energy and frequency they all travel at the speed of light and they all belong to the same species - electromagnetic photons or radiation. For all of them the frequency, f, wavelength, λ, and speed of light, c, are related by $c = f\lambda$, and $c = 3.10^{8}$ m/sec. Planck's constant ,h, sets the graininess of natural phenomena. At the atomic and nuclear level the graininess of nature is clearly apparent and the equations of quantum physics have to be used to predict the future. This is the realm of Eddington's atomic table. Every day we deal in billions of atoms and billions of billions of photons. This is the realm of Eddington's everyday table. Here, h, is minute on the overall scale of the collective action of the particles and it is here that Newton's laws and all of classical physics operates as successfully now as it did a hundred years ago.

The sensitivity of our skin to temperature changes and our vision gives to us a direct perception or experience of light quanta only in a very narrow range of energies (0.001 to 5 eV). We have no direct primary sense of electromagnetic quanta outside this range or of the quanta of other fields; rather we have to use the indications of man-made instruments to detect their presence. Figure 2.2 provides an overview of the range of electromagnetic quantum energies encountered in both medical imaging and everyday applications of science and technology. At first sight this diagram might seem a little confusing, even counter intuitive. For example a domestic oven, which is manifestly a very hot place, is listed as having a rather small quantum energy, not much larger than that associated with liquid air, manifestly a cold substance. This apparent paradox illustrates two important points. First on the cosmic scale of quantum energies, hot ovens and liquid air are actually very close together, even though our direct senses put them poles apart. Second the total energy of an object like an oven describes how many quanta are being produced as well as how big those quanta are. Quantum physics is thought to be true across this entire range but as we go down the list we can get away with classical models and predictions with an increasing degree of accuracy.

The energy range of photons is extraordinarily large, stretching from very long wavelength radio waves with energies of 10^{-9} eV, to high energy gamma rays found in cosmic radiation, with energies of 10^{10} eV. Oscillations in electrical circuits produce the very low energy radio waves. Chemical energy exchanges produce photon energies ranging from about 0.001eV in the infra red to 100 eV. in the ultraviolet. X-rays photons, produced by bombarding a solid with electrons, range in energy from 100eV to 200keV, depending on the applied tube voltage . Nuclear processes produce photons ranging from about 100keV to 1000MeV.

Other Quanta

Physics recognises four separate fundamental forces, gravitation, electromagnetic, weak nuclear and strong nuclear. Each of these has its own quantum, each with a characteristic scales of energy and length. The quantum energy associated with the fundamental forces is largest in the atomic nucleus and weakest in gravitation. It might seem a little pedantic to include gravitation in this overview since the human frame hardly has the galactic dimensions where gravitation is thought to be a dominant factor. It should however be remembered that gravitation is indeed important in shaping and sizing our bodies and determining very basic mechanical functional phenomena such as the direction in which food, drink and waste products travel through our digestive systems. Gravitation has almost no role to play in medical imaging simply because we have no means, apart from space travel or high speed aircraft manoeuvres, with which to manipulate the force.

The two nuclear forces, the Weak and the Strong operate only over the very small distances of the atomic nucleus (10^{-15} m). The Weak interaction is responsible for beta radioactive decay (electron or positron emission from the atomic nucleus). The Strong interaction is responsible for holding the nucleus together in spite of the huge electrostatic repulsion between the protons present. The quanta of the Nuclear forces are particles with typical energies of tens to hundreds of MeV and are generally extremely short lived. Such rare and energetic objects are only seen on earth in high-energy experiments around the particle accelerators at CERN and Fermilab and in cosmic rays coming from outer space.

As a result of the enormous differences in characteristic energy and length scales atomic and nuclear phenomena generally only slightly affect each other. The relatively low energies (eV) of atomic chemistry can only have a slight impact on nuclear properties,(MeV). Similarly, the short range of the nuclear forces means that they cannot reach out into the atom and make significant changes to the motions of the orbiting electrons. Within the atom, the positive electrically charged nucleus and the surrounding negatively charged electrons interact through the long ranged electrostatic forces of attraction between the nuclear protons and the atomic electrons, not the nuclear forces. Although the energy and length scales of the atomic and nuclear phenomena differ enormously and the techniques used to examine them seem superficially rather different, in fact the whole physics enterprise (both tables) relies on the relatively small number of general principles and laws, many of which are inherited from classical physics. The conservation of energy is thought to reign supreme in all interactions. Both linear and angular momenta are still conserved for complete isolated atomic and nuclear systems.

It is thought that quantum mechanics is capable in principle of explaining all physical phenomena, at all energies and over all distance scales. The rules and techniques of quantum mechanics should in principle be capable of making accurate predictions of the future behaviour of any material object, no matter how large or small. To be absolutely rigorous and consistent, scientists should always use just this single set of ideas and techniques since they embody the truth, as we know it. Practically this is impossible since at the level of macroscopic objects the mathematics of even simple calculations can become totally unwieldy. Furthermore there are aspects of macroscopic behavior, which have still not been sufficiently well described in quantum terms to allow quantitative predictions to be made. Friction is perhaps the best example of a very common phenomenon whose description in quantum terms does not provide a very useful or practical means of calculation. Fortunately, the older classical rules and techniques that

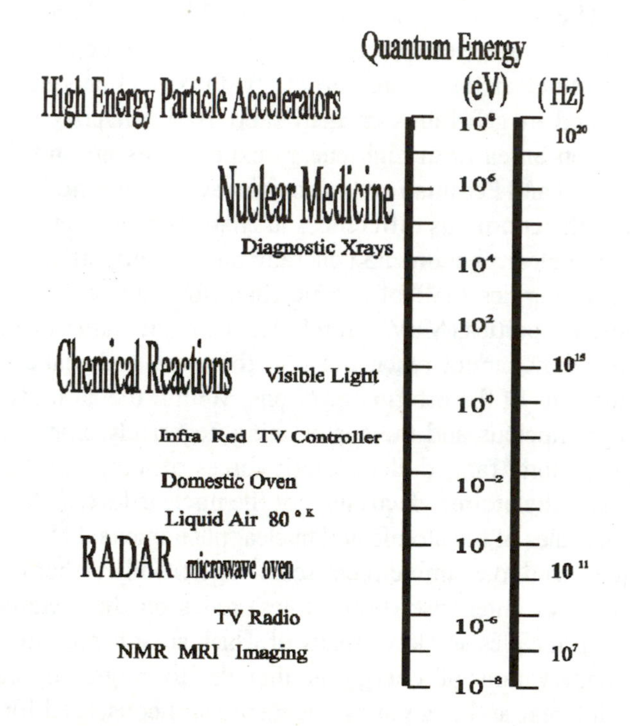

Figure 2.2: The Quantum energy spectrum and the ranges of energy covered by processes important to medical imaging and everyday applications of physics. The quantum energy is in electron volts. The corresponding frequencies are measured in hertz Hz (cycles per second). The vertical size of the typeface describing the application or phenomenon gives a rough guide to the range of energies covered. Some phenomena such as NMR use a very definite frequency whereas for others, such as boiling liquid air, the indicated position corresponds to a typical, mean quantum energy.

provide a good approximate description of the behaviour of very large numbers of atoms acting together, still apply. Nearly all physicists will use the conceptual table that provides the quickest route to a desired prediction without too much regard for mathematical rigour or philosophical consistency.

Atomic Physics

States of Atomic Matter

All of terrestrial matter comprises assemblies of discrete atoms. A typical atom has a diameter of about 10^{-10} m and mass of 10^{-26} Kg. Clearly a single atom is a truly tiny object which we cannot perceive directly, rather in our everyday lives, we handle billions of billions of atoms linked together to form solids and liquids or only tenuously associated in the air that we breathe. The three states of matter, solid, liquid and gaseous, form a continuum in the number density of these atoms and strengths of the forces that bind one atom to another. A given chemical substance comprises a particular set of atomic types or elements and that substance will have particular bulk mechanical and chemical properties such as density, strength and ductility that are determined by the internal structure of the constituent atoms and the manner in which those constituents are bound together. Nearly all substances can exist in each of the three states of matter providing that the temperature can be made sufficiently high, to form a vapour or low, to form a solid.

A typical atom or molecule in a gas maintained at a temperature T will , on the average, have an energy $E=3/2\ k_BT$, where k_B =Boltzmann's constant = 1.38×10^{-23} joule/deg. Room temperature, on the absolute temperature scale, is about $300°$ K and this corresponds to an energy of 2.5×10^{-2} eV. This, as we shall see, is small in comparison with the internal energies of all atoms but can be large in comparison with some types of binding force between atoms. When the random energy, k_BT, is large in comparison with the inter-atomic binding energies, then the collection of atoms behaves as a gas of free atoms. At the opposite extreme, the binding energies outweigh thermal energies and keep atoms in almost the same relative positions, so that the whole behaves like a rigid lump or solid. Water is the most familiar example of this range of states in matter. Above $373°$ K, the boiling point of water, individual water molecules rush around as a gas, which we call steam. If we lower the temperature to below $373°$K the collection of water molecules forms a liquid, water. With a further reduction in temperature to below 273°K a solid, water ice is formed. The liquid is essentially incompressible and has a definite density, but offers no resistance to changes of shape. The solid also has a definite density and is incompressible but in addition it offers considerable resistance to changes in shape. The characteristic temperatures, boiling point and freezing point of any substance provide rough measures of the energies involved in the attraction between the individual molecules of that substance. Thus, since the boiling point of water is about 373°K we can estimate that the attraction between water molecules involves an energy of about $373\ k_B$ J and so when steam condenses into liquid water each molecule has to lose an amount of energy of about this size. In warm water molecules can tumble on their own axes and diffuse easily from one part of the volume to another but have only a small chance of escaping from the bulk. This is a low viscosity liquid. As we approach the freezing temperature of water, the individual molecular kinetic energy becomes less than the weak binding forces between molecules causing them to clump together. Now the diffusion of one molecule through the liquid becomes increasingly difficult. This is a high viscosity liquid. Below 273°K the binding forces hold each molecule in a definite position and orientation with respect to its neighbours against the random thermal energies and we have a solid. The molecules of water ice are not completely stationary at or below 273°K, rather they are in a state of perpetual vibrational motion, about the average positions determined by the bonds between molecules.

Except at the very high temperatures of the stars, 10,000° K, most matter, be it solid, liquid or gaseous, goes about its business with its constituent atoms retaining their essential identity but exchanging energy in packets or quanta that are electromagnetic quanta, photons or electrons. Thus any substance has two levels of organisation corresponding to the internal structure of the constituent atoms and the manner in which those atoms are joined together. Contrast in medical imaging is determined these two levels of atomic organisation. Thus x-rays and gamma rays interact with matter by a sequence of encounters with individual atoms. Whether the x-ray is absorbed, scattered or simply passes by, depends primarily on the internal atomic structure of the particular atoms present rather than the larger scale arrangements of atoms. MR and Ultrasound imaging on the other hand derive contrast from subtle variations in the way that atoms are joined together. In other words the latter are sensitive to the chemistry and bulk properties of biological matter whereas the former are more sensitive to the internal structure of individual atoms. Thus rather more than half of all medical imaging depends on an understanding of the internal structure of atoms.

Atomic Structure

All atoms have the same basic structure. A very small, heavy and positively charged central region, the nucleus, is surrounded by a much more diffuse cloud of the very much lighter and negatively charged electrons. This conclusion arose from the work of Rutherford and his students in 1911. He was investigating the scattering of alpha particles from thin gold foils. Alpha particles are helium nuclei emitted from radioactive isotopes like radium, which was discovered by Mme. Curie in 1898. Alpha particles have enormous kinetic energies and thus most of them passed, almost undeviated, through Rutherford's very thin foils. However occasionally (about 1 in 8000) he found particles that bounced back from the foil. In picturesque language, Rutherford described the results as

"about as credible as if you fired a 15-in shell at a piece of tissue paper and it came back and hit you "

The colour of the language reflects just how unexpected was this result at the time. Negative electrons and positive protons were known to be constituents of atoms. Neutral atoms were assumed to be composed of equal amounts of these positive and negative charges. Their arrangement within the atom was however unknown. In fact classical physics required that the entire volume of a single atom should contain a fine mixture of positive and negative charges. Rutherford concluded correctly that the overwhelming majority of an atom's mass must be concentrated in a very small region that we now call the nucleus, the rest of the atom being essentially empty space.

Whole isolated atoms of any type are electrically neutral. The positive charge of the nucleus is exactly balanced by the total negative charge of the electrons. Electric charge is quantised, that is to say it comes in discrete, well defined amounts exactly equal to multiples of the charge of 1 electron , $1e = 1.66 \times 10^{-19}$ coulomb. Different atom types, that is to say elements, are characterised by different whole numbers, Z, of electrons in the outer cloud and precisely the same number of whole protons in the nucleus. Most of the atom remains as a stable object at terrestrial temperatures because of the strength of the electrical attraction between the negatively charged electrons and the positive nucleus. In some respects an atom is a microscopic solar system with the nucleus as the sun and the electrons the planets. The key difference however is that the planetary electrons carry a significant electrical charge in comparison with their mass, ($e / m = 1.759 \times 10^{11}$ coulomb / kg). This was the insurmountable problem for a classical account of atoms after Rutherford. On the classical table we would say the atomic electron, orbiting a central positive nucleus, executes a roughly circular orbit. It is therefore constantly being accelerated and, because it is charged, it will emit electromagnetic radiation. Thus classically all atoms will continue to lose energy until all the electron orbits have collapsed into the very small nucleus. Since a typical atomic radius is ten million times larger than its nucleus, the classical prediction of atomic collapse to nuclear dimensions is simply wrong.

The Bohr Atom

Bohr presented the first quantum theory of the atom. The theory retained the classical ideas of force and definite particle position but included a new assumption, that the angular momentum of the orbiting atomic electron could only have whole multiples of h, Planck's constant. The model is sufficiently simple to allow us to derive its main result without recourse to difficult mathematics. The argument used by Bohr is typical of the train of thought used throughout physics and thus it provides an insight into the ways of thinking in physics.

We start with the idea of an atomic solar system with a negatively charged electron with mass, m, and charge, -e, executing a circular orbit with an orbital speed V, about a small heavy positively charged nucleus having a charge +Ze. Here we ignore any motion of the atom as whole. If the electron is in a stable orbit then there cannot be any net radial force acting on the particle. There are however two forces acting along a radius of the circular orbit which must be balanced if equilibrium is to be maintained. These are the attractive electrostatic between electron and nucleus

$$ElectricForce = \frac{+Ze \times -e}{r^2} = \frac{-Ze^2}{r^2} \quad \ldots 2.3$$

The negative sign arises from the attraction between unlike charges and acts inwards on the electron. There is also the centripetal force acting outwards, arising from the electron's circular motion

$$Centripetal\ Force = \frac{mV^2}{r} \qquad \ldots 2.4$$

For the orbit to be stable, the net radial force must be zero, otherwise the radius of the orbit would, according to Newton's second law, be changing with time. Thus we have the orbit equation

$$\frac{Ze^2}{r^2} = \frac{mV^2}{r}$$

$$hence \quad r = \frac{Ze^2}{mV^2} \qquad \ldots 2.5$$

Up until now the argument has been strictly classical and is precisely the same as the one used to calculate planetary orbits, the only difference being that electrical attraction between charges has replaced gravitational attraction between masses. The result so far says that any orbital radius, r, is possible, provided that the speed of the electron, V, and hence its kinetic energy, $\frac{1}{2}mV^2$ can take on any value. Bohr introduced a restriction on this energy by assuming that, in its motion, the electron would move with an angular momentum, that could only take on multiples of Planck's constant. Thus, the angular momentum $L = mVr = n\,h/2\pi$ with $n = 1,2,3$ etc). With this restriction on the orbit equation we have, after a little algebra

$$r_{Bohr} = r_o = \frac{n^2 \hbar^2}{Z m e^2}$$

$$and\ Energy = E_n = \frac{1}{2}mV^2 + \frac{-Ze^2}{r} \qquad \ldots 2.6$$

$$E_n = \frac{-mZ^2e^4}{2n^2\hbar^2}$$

The total electron energy is made up from a positive kinetic energy and a negative potential energy. The resulting net Bohr energy of the electron is negative. This means that an electron in an orbit around a nucleus has less energy when it is far away from the nucleus. Energy must be supplied to the atom to free the electron from its orbit. Putting in the numerical values for the atomic constants for hydrogen with Z=1, the lowest energy state, n=1, has an energy -13.6 eV. This is the resting internal state of a hydrogen atom. As n increases the atomic energies get closer together and less negative. By the time the electron energy is close to zero, the spacing between levels is so small that energy transitions could be taken to be continuous. When the charge on the nucleus is increased, as we go from hydrogen, Z=1 through helium, Z=2, lithium Z=3, beryllium , Z=4, the Bohr model predicts that the internal energies become more negative (because of the Z^2 factor). Electrons become more tightly bound to heavier nuclei. Atoms with many orbiting electrons such as He, Li, C, O etc. were assumed to be assembled by allowing the right number ,Z, of electrons,to occupy the definite Bohr energy levels, one electron at a time.

The main purpose of the Bohr model was to account for the internal properties of atoms when they gained or lost internal energy by the emission or absorption of quanta. It was known that particular elements could be identified by the pattern of light wavelengths that they emitted (and absorbed) when put into a gas discharge tube. In the tube, an electric field accelerates charged electrons and ions and these collide with other atoms, exchanging energy in the process. With a sufficiently strong electric field, the gas inside the tube becomes hot and the tube glows with a characteristic colour, reflecting the particular spectrum of photon frequencies emitted by a particular element. Sodium atoms emit the strong yellow colour familiar to us in street lighting. Hydrogen emit a dull mauve colour, neon the bright red of advertising signs. The key point is that atomic spectra comprise a relatively small number of discrete "lines" at very precise frequencies, and a particular set of lines is characteristic of a particular element.

Bohr assumed that any alterations in the internal atomic structure must involve electron jumps from one quantised state to another. The atom can only absorb whole light quanta with energies such as 13.6-3.4 = 10.2 eV (n=1 to n=2) or 3.4-1.5 =1.9 eV (n=2 to n=3) and so on. Conversely when excited atoms relaxed to their resting state, exactly the same energy quanta would be emitted. This is illustrated in figure 2.3. The model provided the first way of calculating the frequenciess of light quanta emitted by hydrogen but also gave us an enduring cartoon picture of atomic structure that is still used today in all elementary accounts of atomic structure and atomic chemistry.

Quantum Mechanics

Unfortunately the Bohr model turns out to have insuperable problems as a general atomic model. Its assumptions cannot be justified within classical physics and it makes the wrong predictions for the spectra of helium and higher elements. Finally the theory provides no means to calculate the relative intensities of the various component lines of atomic spectra. For over 10 years after 1913, physicists tried to justify the original assumptions and improve the model by including relativity and the motion of the nucleus, but their efforts were, in the end, unsuccessful. Heisenberg and Schrödinger presented a completely new theory in two slightly different forms in 1926. This together with subsequent work by Born, Bohr, Dirac and Pauli between 1926 and 1932 laid down the main features of a revolutionary new and wholly successful theory of atomic structure which rapidly came be seen as a universally applicable theory of all matter, both atomic and nuclear. This new theory is Quantum Mechanics.

Quantum mechanics arises from the discrete nature of atomic matter, the uncertainty principle and the dual wave and particle appearance of both light photons and material particles such as electrons. What does this duality mean? The paradox was resolved, at least from a theoretical, point of view, by the new mathematical theory, given its final physical interpretation by Born. He suggested interpreting the waves, not as real physical vibrations in a sub-aether, but rather as measures of uncertainty or probability. When energy is absorbed or emitted, a whole quantum worth is always involved and the quantum energy exactly fits a difference between atomic energy levels. Whether or not this happens for a particular encounter is a matter of chance; the odds being decided by the amplitude of associated wave. The success of quantum mechanics lies in its unerring ability to provide the means to calculate both the quantum energies and the probabilities. Creating a picture of what the mathematics is saying, often taxes the intuitive imagination. For example the probability distributions calculated for an electron in hydrogen reveal that although the electron energy is exactly specified, its position is certainly not. Furthermore there is a finite probability for finding the bound electron both at the nucleus and well outside its Bohr radius. In other areas of quantum physics the new theory makes predictions that are totally impossible on classical grounds. For example in the nucleus there is a very large energy barrier keeping all particles inside its walls. Alpha particles get out of the nucleus not by hopping over the wall, but by tunnelling through. This is illustrated in figure 2.9 and is discussed in section 2.5 below. In general the probability wave of quantum mechanics is not an observable quantity at the level of a single electron. Born encouraged the statistical idea that when very large numbers of similarly prepared atoms are subject to the same experiment then the probability wave makes itself manifest as a distribution of atoms in particular states or configurations. The Bohr atom illustrated in figure 2.3 has a

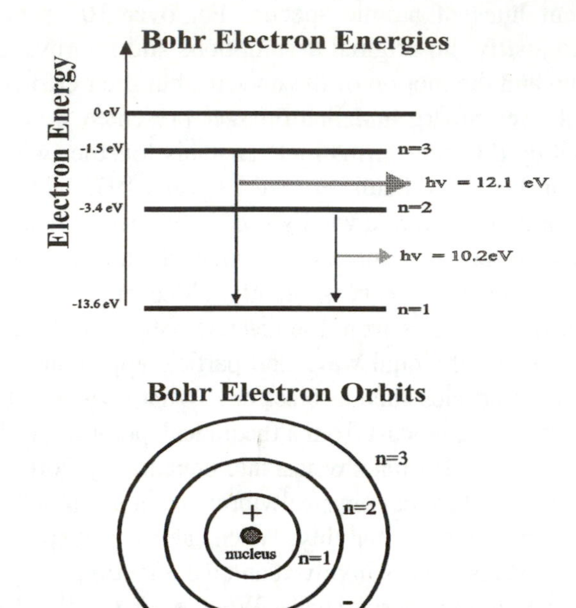

Figure 2.3 Electronic Energy levels and orbits of the Hydrogen atom according to the Bohr model. Each horizontal line represents one stable electron configuration. A positive amount of energy (+13.6 eV) must be given to the electron to allow it to leave the atom. Photons are emitted and absorbed by atoms when an electron jumps between stable energy levels. The frequency of the radiation is exactly, $\nu = \Delta E / h$. All atoms eventually relax to their ground state energy, n=1.

definite radius for the electron orbit. The now accepted atomic picture replaces this with a smeared out distribution of probabilities. The distribution is not taken to mean that the electron itself is smeared, rather it reflects the fact that we do not know, at any given time, exactly where it is; we can only assign a set of probabilities for finding it here or there in its orbit. If we could look, over a long period of time, at a single atom or, alternatively, at millions of identical atoms, then the shapes of the probability distributions would manifest themselves as the number of times an electron was found at a particular location. Both physicists and chemists use pictures such as this as an aid to atomic intuition, often speaking of the shapes as though they had a reality more permanent than the statistical interpretation given to us by Born. To be absolutely rigorous we should have long ago turned our backs on pictures and tuned our minds to the realities embodied in the equations rather like musicians are able to conjure the music from the printed score. It is perhaps not too surprising that the vast majority of physicists, being humans, are unwillingly to do this and, at least in private, retain slightly false and inconsistent geometrical pictures of individual atoms.

The Periodic Table of Elements

During the course of nineteenth century chemists became aware of the existence of a vast number of different chemical combinations of about eighty different atomic types or elements. Inevitably patterns of properties were observed and simplifying generalisations were sought to explain these patterns. Mendeleef (1869) made a systematic study of the simpler problem of the properties of the elements and proposed a periodic table. He listed the elements in order of increasing atomic weight but grouped together in columns those elements that were similar in their physical and chemical properties. Thus, lithium, sodium, potassium, rubidium the so-called alkali metals, (soft metals that react violently with water to form a slimy caustic alkaline substance) appeared together in one column, while the equally reactive halogens, fluorine, chlorine, iodine and bromine appeared in another. Mendeleef, of course did not know the full complement of stable elements nor anything of atomic structure and thus his table was incomplete. His periodic ordering left several gaps and had to list a few of the elements out of atomic weight sequence. Quantum mechanics provides a complete account of the periodic table of elements. Most simply the explanation can be seen as a process of filling up Bohr levels with electrons.

We can imagine the process of assembling a multi-electron atom from the nucleus upwards by filling the hydrogen-like states (shown in figure 2.3) modified by the increased nuclear charge. The hydrogen-like energy level diagram provides a sort of scaffolding into which additional electrons can be fitted. There is however more to atom building than this. We have to consider the orbital angular momentum of the electrons, their mutual electrostatic repulsions and finally the restriction on the number of electrons per state demanded by a new quantum concept, the Pauli exclusion principle. Each hydrogen-like level in a

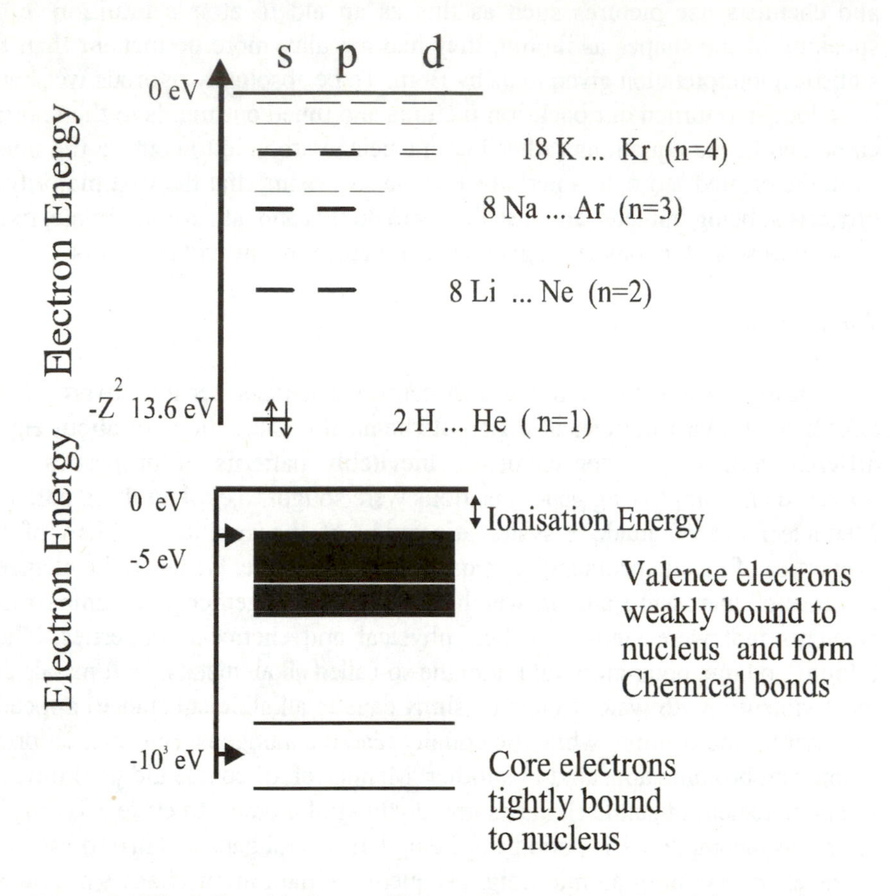

Figure 2.4 The shell model of atoms. When angular momentum, L, and electron spin, s, are included, each Bohr level, n, is found to comprise n^2 possible levels, each can accommodate 2 electrons (Pauli Principle). According to a spectroscopic tradition, levels with angular momentum l=0 are called s (sharp), l=1, p (principal), l=2, d (diffuse). In hydrogen all of these have the same energy, but in multi-electron atoms the effects of the electron-electron repulsions split these levels apart by significant amounts. In a typical atom the core states, deep inside the atom (E~ - keV) take no part in chemical bonding. The outer valence or chemical electrons are only weakly bound (E~ -5eV).

heavy atom becomes a group of relatively closely spaced levels (n^2 of them). Each of these grouped levels corresponds to a different state of angular momentum, L and two electrons can occupy each of these with their intrinsic spin directions opposed. The ground state of a multi-electron atom is characterised by paired electrons occupying the scaffolding states. The electrons occupying the lowest energy states are called core electrons and those in the outermost shells are called valence electrons. The distinction arises because the core electrons are too tightly bound to their parent nucleus to take part in chemistry, whereas the valence states are very much closer in energy to E=0 and thus do not need to gain or loose very much energy in order to enter into associations with other atoms, figure 2.4.

In general the electronic structure of atoms retains an image of the hydrogen energy level scaffolding and it this which accounts for the periodic table of chemical properties. The groups of closely spaced levels around each Bohr-like level are called shells. As each successive shell is just filled the chemical properties repeat themselves. Successive shells are generally separated by relatively large energy gaps. An element with atomic number Z has 2 electrons in the lowest, n=1 shell, 8 in the n=2, 18 in n=3 state and so on until all the, Z, electrons are accommodated. Elements with Z=2, 10, 18, 54, that is to say helium, neon, argon and xenon are said to be closed shell atoms. The next empty shell is a long way off and these atoms find it exceedingly difficult to bind either with each other or with other atoms. The halogens on the other hand, with Z= 9, 17, 35,..., have one electron less than a full shell and there are plenty of nearby energy states that allow these atoms to bind very strongly with other atoms. There are large tracts of chemistry that are not explained by such a simple picture and require a very much more detailed account of the splittings within shells. Perhaps the most important of these is the central role that the atom, carbon Z=6, plays in terrestrial life forms. Its ability to form an enormous range of organic molecules in association with hydrogen , oxygen and phosphorous is almost unique within the periodic table. This versatility arises from the ease with which the four electrons in the n=2 shell can collectively interact with other atoms to form a very wide range of chemical bonds with other atoms. Quantum calculations show that here there is a delicate balance between the outer electron's attraction to the nucleus and repulsion from the other outer electrons. This is a very good example of an important natural phenomenon that is truly incomprehensible without the precise quantitative results of detailed quantum mechanical calculations.

2.3 The Interaction of Photons with Atoms

X-ray Scattering and Absorption by Atoms

Imaging with x-rays depends on the relative numbers of photons, emitted from an x-ray tube, that traverse a patient body undeviated by interactions with atoms of the body. This fraction depends on efficiency with which x-ray photons are absorbed or scattered within the body. The scattering and absorption of x-rays and γ–rays by matter are quantum phenomena. We can describe each of several processes in terms of collisions between atoms, which have both a finite size and an internal structure and photons. The collision can be one in which the photon loses no energy but merely alters its direction. This is the elastic scattering that is familiar to the snooker player. Alternatively the photon can loose some or all of its energy to an atom, whose internal structure is disrupted in the process. This is inelastic scattering and absorption. All photon-atom interactions must conserve both energy and momentum. The processes, which we describe in this section, are of great importance to both x-ray and γ-ray methods of imaging. In the case of x-rays it is these interactions which are the essential mechanisms in x-ray image formation. This is not the case for γ-ray methods in which the imaged quantity is the concentration of radioactivity throughout the body. The γ-rays emitted by the radioactive tracer have to pass through tissue on their journey to the detectors and the photons are absorbed and scattered en route in exactly the same way as x-rays. Now however these processes degrade the spatial resolution and accuracy of the γ-ray image. In other words in this case photon interactions with matter are an unavoidable nuisance.

In general for a given atom and a given incident photon energy the efficiencies of both inelastic and elastic processes vary with the photon energy and the atomic number of the atom. In medical diagnostic imaging we are concerned with the interaction between x-rays, γ-rays and biological tissues which contain a wide variety of atoms. In these circumstances it is almost impossible to provide accurate first principle calculations of the precise amounts of scattering and absorption that take place. Instead empirical values of attenuation coefficients, obtained by direct experiment, are used to calculate the amounts of radiation that are absorbed or pass through the body with or without scattering. Taking all processes together it is found that a well-defined beam of photons is attenuated by an amount characterised by two numbers, the Mass

Attenuation coefficient for the tissue concerned and the path length of the beam within the patient. In general the amount of radiation transmitted through the patient decreases exponentially with the path length of the beam through the patient. There are two important general points to understand. First, the various attenuation mechanisms are independent and each one adds independently to the overall attenuation coefficient. Second, the magnitude of each contribution depends in a predictable manner on the atomic number of the atoms and the energy of the incident radiation. This allows the radiographer to manipulate the degree of contrast between different tissue types both by altering the x-ray energy and by introducing radiopaque dyes that will "stain" particular areas or organs for the duration of the radiological examination.

Variations in photon scattering and absorption with photon energy are actually everyday experiences. They are responsible for both the blue colour of a cloudless sky and the red sunset. In both these cases more blue light (short wavelength, higher photon energy) is scattered by the atmosphere than red light (longer wavelength smaller photon energy). If we look at a cloudless sky, away from the direction of the sun, it is the blue light scattered by the atoms and molecules of the atmosphere that enters our eyes. Conversely, the disc of the setting sun appears red because the blue light has been scattered away from the direction of the sun's beam.

Elastic Scattering

Classically we can think of x-rays as electromagnetic waves consisting of electric and magnetic fields changing in time at the frequency of the wave. When such a wave encounters an atom the wave's electric field accelerates the negatively charged atomic electrons and, to a much lesser extent, the positively charged nucleus. These accelerations will have the same frequency as the incident wave and, in turn, the charges become sources of electromagnetic waves that propagate away from the atom, in expanding spherical waves. The electrons absorb some energy from the incident wave and then reradiate it in all directions. After the interaction with the atom , the scattered wave has the same energy as the incident wave but its direction of propagation or travel has been altered. The conservation of the wave energy gives rise to the generic term elastic scattering. When the wavelength of the x-ray is smaller than the dimensions of the atom each electron re-radiates independently and thus the total amount of re-radiation or scattering from any atom increases linearly with the number of electrons, the Z value of the atom. The efficiency of elastic scattering is nearly constant over the range of photon energies used in x-ray radiography.

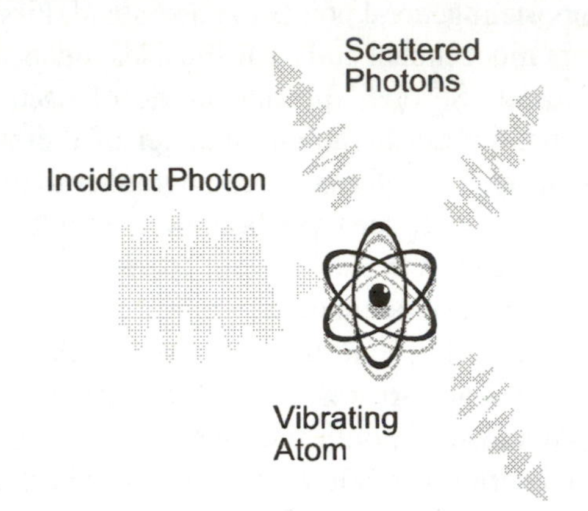

Figure 2.5 . Elastic scattering by an atom. The electron cloud is accelerated by the electric field of the photon. The accelerated charges re-emit photons in all directions, which classically is interpreted as a spherical wave.

Inelastic Scattering

If an atom absorbs a photon of sufficient energy then one or more electrons can acquire sufficient energy to escape the atom altogether, creating free electrons and leaving behind positively charged atoms called ions. The minimum energy required to ionise an atom depends on the energy of the highest filled electron state in the Bohr model scaffolding. This varies between 5 and 25 eV across the periodic table. The smallest ionisation energies are found in the alkali metals Li, Na, K etc and the largest in the inert gases He, Ne, Ar etc. Visible light photons, hf = 2eV, are thus unable to ionise any isolated atom, but ultraviolet photons, hf = 100 eV can certainly ionise any atom. When we consider x-rays, hf =50 keV or γ– rays, hf = 100 -1000 keV then these photons are capable of ejecting not only the outermost but all atomic electrons from all but the heaviest elements in the periodic table. When an x-ray liberates any electron from its parent atom the escaping electron generally carries away a large kinetic energy since, after reaching the escape energy, there is still about 50keV to be taken up from the photon. This energy is shared between the ion and the photoelectron. Energetic photoelectrons, because they are electrically charged, interact very strongly with other atoms and can go on to knock more electrons out of surrounding atoms thus creating a cascade of electron -ion pairs. Ions and free electrons play a crucial role in all chemical reactions as a result of their electric charges. In radiology, it is the ionised atoms and energetic free electrons, created by the x-ray or γ–ray radiation that damage tissue through the resulting uncontrolled chemical reactions. Ionising radiation gets it generic title from this process and the ionisation cascade produced by high energy photons is what makes x-rays and gamma rays rather dangerous. Any process in which photons exchange energy with atoms falls under the generic title of inelastic scattering.

Compton Scattering

In 1912 Compton discovered that x-ray beams, after transmission through thin metal foils, contained a new component with less energy than the incident beam. These x-rays had been inelastically scattered by the atoms of the foil. The quantitative explanation of this effect provides convincing evidence for the particulate aspect of electromagnetic radiation and hence the existence of photons. Compton's simple but successful treatment for x-ray scattering from free electrons shows that the x-ray loses an amount of energy that depends only on the scattering angle.Compton scattering takes place preferentially from the outermost electrons of an atom that are most weakly bound to the parent nucleus. For a typical atom in biological tissue (Z= 7) at diagnostic x-ray energies, the relative amounts of elastic (Thompson) and inelastic (Compton) scattering are about equal. As the photon energy increases into the gamma range, Compton scattering can occur from more of the atomic electrons and thus the Compton process gradually dominates the scattering process. Generally the Compton process is accompanied by a photoelectron which carries away from the atom a kinetic energy approximately equal to the difference in energy of the incident and scattered photons. All of this energy ends up in nearby tissue hence contributing to patient dose.

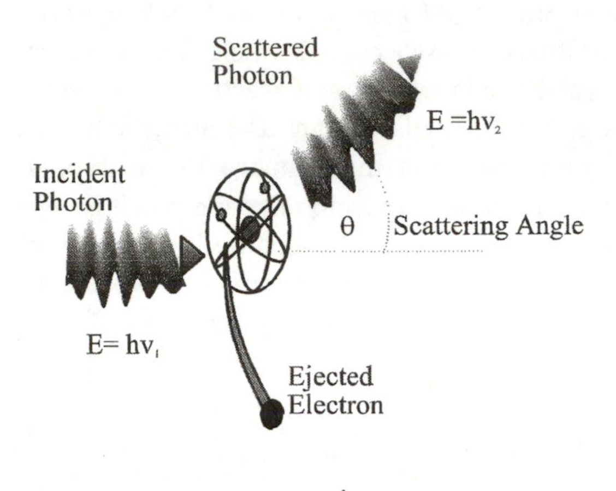

$$1/ hv_2 - 1/hv_1 = 1/ mc^2 (1- \cos(\theta))$$

Figure 2. 6 Compton Scattering of photons involves electrons that are relatively weakly bound the nucleus. Only part of the incident photon energy is imparted to the scattering electron but generally this is more than sufficient to eject that electron from the atom. The scattered photon has suffered both a change in direction and a reduction in energy. Compton scattering contributes to energy absorption and hence patient dose in imaging.

The Photoelectric Effect

We have already met the photoelectric effect in our general discussions of quantum physics. In the diagnostic x-ray energy range, photon absorption is dominated by the photoelectric effect. Typical biological atoms like carbon, hydrogen and oxygen, all relatively light, low atomic number atoms ($Z< 8$) atoms have ionisation energies for their outermost electrons of less than 25eV. Thus if an atom is able to absorb just this much energy from a 50KeV photon then the atomic structure will be disrupted and a charged electron ejected into the surrounding space. In fact these atoms are able to absorb nearly all the photon's energy and thus it is possible for any of the atomic electrons to be ejected with considerable amounts of kinetic energy. The ejected electron loses it energy to the nearby tissue by creating more ions in a cascade. A calculation of the photoelectric cross section requires a detailed treatment using quantum electrodynamics. Two important points arise from such a calculation. First the atom as a whole has to absorb momentum from the photon and this becomes less and less possible as the photon energy increases. The absorption efficiency therefore decreases with increasing photon energy. Second, the chance of an electron being ejected from an atom increases dramatically with the number of atomic electrons, Z. Thus, the photoelectric absorption efficiency for an atom is given approximately by

$$\sigma \approx \frac{Z^5}{E^2} \qquad \ldots 2.6$$

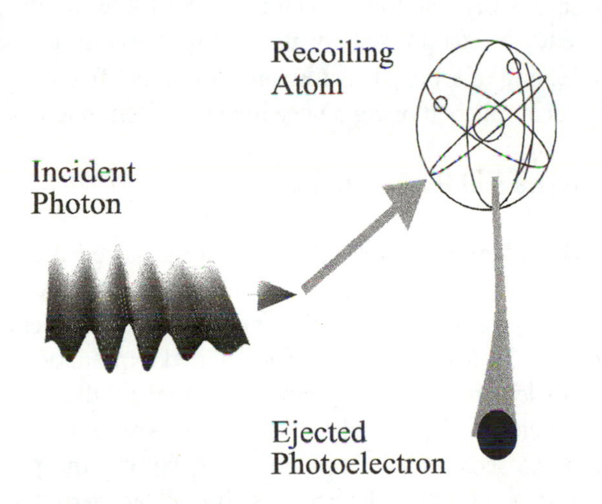

Figure 2.7: The Photoelectric effect. The atom absorbs all the energy of the photon and an electron ejected. Since an entire photon is absorbed, energy and momentum can only be conserved if the ionised atom recoils. At x-ray and gamma ray energies all atomic electrons can be given sufficient energy to leave an atom.

This approximate relationship is vital for radiology. The increasing absorption with decreasing photon energy ($1/E^2$) has important consequences for patient dose in imaging and the choice of radiation in therapy. The rapid dependence on the element (Z^5) forms the basis of contrast enhancement used in x-ray radiography to image soft tissue and very small structures such blood vessels. Adding a small quantity of a heavy element such as iodine to the blood stream increases the x-ray absorption by a very large amount, allowing fine detail to be imaged.

X-ray tubes generate a continuous spectrum of x-ray energies and the lower the photon energy the more likely is its complete absorption within the patient by the photoelectric effect. This low energy fraction of the diagnostic x-ray tube's output must be removed, before it hits the patient. Rather neatly, this is accomplished using the photoelectric effect itself. A light metal such as aluminium is used as an exit window filter on the x-ray tube. The low energy photons up to about 15 keV are almost entirely removed from the beam by the metal foil but the higher energy fraction is transmitted with only minimal attenuation.

All x-ray and γ-ray photon detectors exploit the photoelectric effect. Detection efficiency is maximised by using the heaviest atoms (largest Z). Similarly the shielding of tubes, detectors and personnel from stray x-ray and γ-ray radiation is accomplished using a very heavy element such as lead, Z= 82.

2.4 Nuclear Physics and Radioactivity

The detailed explanation of atomic structure required the introduction of quantum ideas and the establishment of a completely new way of doing mechanics. Given the differences in dimensions, masses and energies, nobody would have been in the least surprised if a detailed explanation of the internal structure of the nucleus required a second revolution and the introduction of yet another set of principles. Fortunately, as far as we know, this is not the case. Quantum mechanics seems to be capable of explaining the properties of the nucleus and indeed beyond that, the zoo of sub-nuclear particles. Although the underlying quantum mechanics remain the same, its application to the nucleus is computationally very much more difficult. The difficulty arises from the competing and immensely strong forces that operate within the nucleus. All the protons repel each other, and being confined to the very small nuclear dimensions, the strength of this repulsion is enormous by atomic standards; MeV rather than eV. There has to be an equally powerful attractive glue to keep the nucleus together. This, roughly speaking is the neutron. Attraction between neutrons and protons, the nuclear force, has a completely different mechanism from the electrostatic repulsion between protons. Any given nucleus is then the result of a balance between two different powerful physical forces which, within the nucleus, are still not completely understood. Medical imaging makes use of two sorts of nuclear property, radioactivity in gamma imaging and permanent nuclear magnetic moments in MRI. Radioactivity arises in those nuclei for which the forces of repulsion and attraction are so finely balanced that small extra energy inputs trigger a complete disruption of the nucleus and the consequent emission of very energetic photons and nuclear fragments. Nuclear magnetism arises from the details of the internal structure of particular stable nuclei. Both of these require some understanding of nuclear structure.

Nuclear Structure

Each element is characterised by its nuclear charge, Z, the number of protons in the nucleus. Any element can have a variable number, N, of neutrons in the nucleus. Species with the same Z but different N are called isotopes. Some of these isotopes are unstable, giving rise to the phenomenon of radioactivity. In medical applications, these nuclear species are called radionuclides. Each nuclide is specified by Z and its mass number, A = N+Z. The shorthand notation is ^{A}X, so that we will make reference to fissile uranium, Z=92 as ^{235}U and an important isotope used in gamma imaging called technetium , Z=43, as ^{99}Tc. All isotopes of a particular element, say carbon, are chemically almost indistinguishable, since they all have the same number of protons in the nucleus surrounded by the same number of electrons. The nuclear properties and of course the atomic masses of isotopes are slightly different but their atomic physics is almost unaltered.

When an element has several abundant isotopes its chemical atomic weight will differ from the mass spectrometer value (~ AM_P, where M_p is the mass of a proton) since chemical methods of analysis necessarily involve large numbers of atoms. In general a sample of even a pure element will contain a mixture of isotopes. A few elements have only one stable isotope eg fluorine ^{19}F, its atomic weight is $19M_P$. Others have many, for example the metal zinc has five stable isotopes ^{64}Zn,^{66}Zn,^{67}Zn,^{68}Zn,^{70}Zn with natural abundances of 49%,28%,4%,19% and 0.6% respectively. The chemical atomic weight of zinc is 65.37 M_P, the average of all the natural isotopes. The stable and thus most abundant isotopes of a given element tend to have Z = N.

Nuclear Binding Energy

When individual atoms are "weighed" in a mass spectrometer it is found that even the mass of pure isotopes is smaller than $A.M_P$ by an amount that varies in a systematic fashion with mass number A. For light elements such as fluorine, A=19, the difference is about 0.1% . As A increases so the difference increases to a maximum of about 10%, near Cd (A = 100) and then decreases slowly with further increases in A. This is called the nuclear mass deficit and it is a direct reflection of the stability of the particular nucleus. We saw in the Bohr atom that the total energy of an intact hydrogen atom was less than that of an isolated proton and electron by ΔE= -13.6 eV, the binding energy of the atom. The analogous binding energy of the neutrons and protons in a nucleus is typically 10 - 100 MeV. The nuclear binding energy, because of its large size, reflects itself as a measurable difference in mass as a result of Einstein's famous relationship $E=mc^2$.

A very important aspect of nuclear binding emerges if we consider not the total binding energy but the binding energy per nucleon, B/A. This is plotted in figure 2.8. Two important points arise from this curve. The majority of the

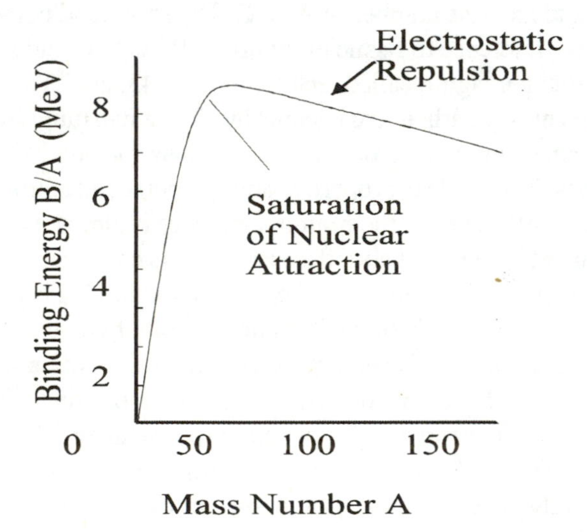

Figure 2.8: The variation of nuclear binding energy per nucleon, B/A with atomic mass for all the stable elements in the periodic table.

elements have roughly the same value, B/A = 8MeV. This implies that the nuclear attractive force is short ranged, it is said to saturate. The slow decline at large A is the result of the long ranged electrostatic repulsion between all the protons, which is gradually diminishing the effect of the short ranged attractive nuclear force. Many of the properties of nuclei that are of importance to us follow from this curve. On the left-hand side, at small A, there is no energy advantage in a nucleus loosing some of its parts. At the heavy end of the periodic table, nuclei can lose internal energy and therefore become more stable by losing some bits of mass and releasing the energy difference as photons or energetic particles. Thus all the elements beyond lead, Z= 82, A=202 are actually unstable either to alpha emission or fission. In both decays the result is a nucleus or nuclei with lower mass and thus a slightly higher binding energy per particle. One might well wonder why there are any stable nuclei at all beyond iron, ^{56}Fe, since, all heavier nuclei could increase their nuclear stability by loosing an alpha particle and reduce Z by 2 and A by 4. Here we are approaching the end of the usefulness of these simple deductions from the gross nuclear property of mass. We have to supplement the picture with the effects of individual nucleon motions within the nucleus.

The Nuclear Shell Model

Figure 2.9 illustrates the likely form of the nuclear potential experienced by a proton. A neutron will experience only the inner negative part of this potential. The nuclear shell model uses this potential to calculate the stable energy states of just one nucleon moving in this potential. This is just like the Bohr picture of hydrogen, with definite possible energy levels for the electron. In the nucleus there are two separate sets of states, one for protons and one for neutrons. These are imagined to be filled with neutrons and protons until the full complement for a given nucleus is competed. Both protons and neutrons have half integer intrinsic spin. Two like nucleons are thus forbidden by the Pauli exclusion principle from cohabiting the same state with parallel spins. Thus the energy scaffolding is filled with pairs of neutrons and protons in each state. It turns out, just as in the periodic table, that there are especially stable nuclei. These are at the so-called magic numbers with Z or N equal to 2,8,20,28,50,82 and 126. Super stable elements are doubly magic with both Z and N a magic number. These values are the nuclear analogues of closed shell atoms He, Ne, Ar etc. In these nuclei the next available unoccupied state is relatively far off in energy and thus it takes more energy to upset the structure.

The intrinsic spin angular momenta , ½ (h/2π) of both the proton and the neutron means that both particles give rise to tiny magnetic fields, as if they possessed an internal electric current. This field is characterised by quoting the strength of the particle magnetic moment,see equation 2.11. When protons and neutrons are associated in a nucleus, the resulting total nuclear magnetic moment is a complicated vector sum of the constituent particle magnetic moments. Some

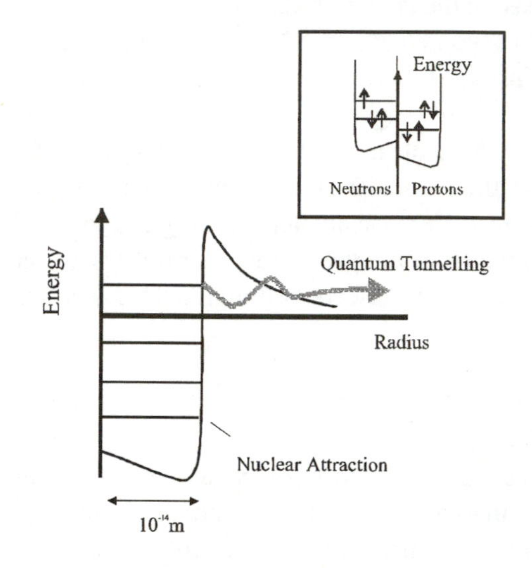

Figure 2.9 The schematic form of potential and total energy for a proton in the nucleus. Inside the nuclear sphere the strong nuclear force attracts the proton to the rest of the nuclear matter. Beyond this range the potential energy becomes positive as a result of the electrostatic repulsion between the proton and the rest of the nucleus. The neutrons experience only the inner, attractive portion of this curve, the protons experience both forces. The potential barrier prevents particles from escaping from the lower nuclear energy levels, that is to say low Z elements. As the number of protons and neutrons increases across the periodic table, so the levels fill up. Eventually there are nucleons occupying levels whose total energy is slightly positive. Alpha particles can leak out of the nucleus from these states by quantum tunnelling through the nuclear potential barrier. The inset illustrates the filling of separate neutron and proton nuclear energy levels by pairs of particles with opposed spins. In this illustration there is one unpaired neutron which would give rise to a nuclear magnetic moment.

nuclei have a finite magnetic moment others precisely zero. The pairing of spins within each nuclear state provides an explanation of the variation of nuclear magnetic moments with Z and A. If a nucleus is doubly magic then all the spin angular momenta are paired off, one up the other down, giving zero net spin and no nuclear magnetic moment. This means that that there is no net rotation of charge within the nucleus and thus no external magnetic field, see section 2.6. If either Z or N is odd however then there will be unpaired nuclear spins that can give rise to a definite external magnetic field characterised by the strength of the nuclear magnetic moment. The nuclear moment of the proton is + 2.79 but that of the neutron is -1.9 in units of the nuclear magnetons defined in section 2.6. The range of both magnitudes and signs of nuclear moments is thus potentially rather large since these spins, together with the nuclear orbital contributions to the moment can be combined in many different ways. In table 2.2 there are examples of stable abundant nuclei, ^{12}C ^{16}O which have no nuclear moment .

Radioactive Decay

When an unstable nucleus decays, very large amounts of energy can be released as a result of enormous binding energies, 8 MeV per particle in the nucleus. The quanta emitted in radioactive decay can be in the form of very energetic particles; neutrons, electrons, positive electrons called positrons, helium nuclei consisting of two protons and two neutrons (alpha particles) or electromagnetic photons (x-rays and γ–rays). Nuclear reactions are the result of particle bombardment or of nuclear fission. The earliest experiments with nuclear properties made use of naturally occurring spontaneous fission, arising from one of the isotopes of uranium, Z=92, ^{238}U. It is thought that all possible isotopes were formed in the big bang at the beginning of the universe, about 3 billion years ago. Most of the unstable nuclei have long since vanished in transitions that have left the stable isotopes that we recognise in the periodic table. Natural radioactivity can arise only from three causes. Unstable nuclei, like uranium, decay very slowly on the timescale of the universe and thus have survived from the big bang. Very small quantities of other unstable nuclei are produced continuously by bombardment from the cosmic radiation that continually hits the earth from outer space. Finally, there are since 1945, short lived isotopes left over from atomic explosions and releases from nuclear power plants.

Since 1932 it has been possible to produce unstable isotopes of all elements by artificial means. By bombarding nuclei with high-energy charged particles, produced in an accelerator such as the cyclotron, it is possible to induce nuclear reactions and to produce unstable isotopes of many elements. With the advent of the nuclear reactor in 1942, very high densities of slow, thermal, neutrons can be

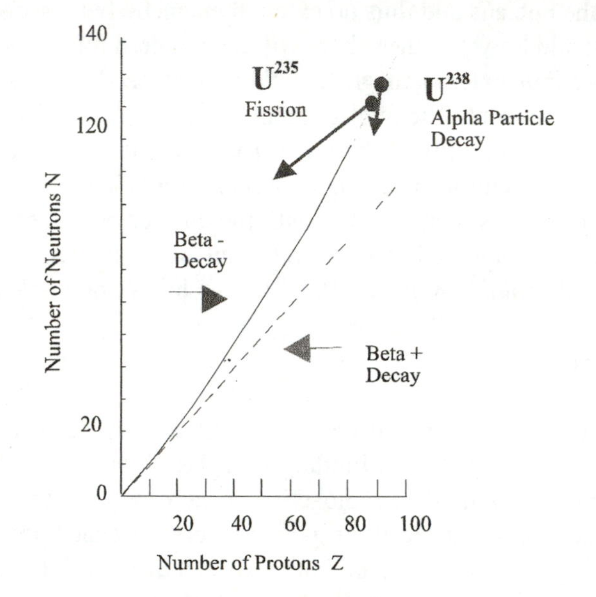

Figure 2.10 : The line of Nuclear Stability. The solid line represents the stable elements. This line is nearly linear in Z up to the element phosphorous, Z= 32. Many elements have several, naturally occurring stable isotopes. The very heavy naturally occurring nuclei such as ^{235}U lie well above the line of stability and decay via alpha emission in a series of intermediate steps ending at a stable isotope of lead Z= 82. ^{235}U undergoes nuclear fission after capturing a slow neutron. The fission products comprise two large fragments with Z~ 45 together with a few, high-energy neutrons. Radioactive isotopes produced artificially by bombardment or neutron capture lie either above or below the line of stability, Z =N. Proton rich (neutron deficient) nuclei, ie ,^{15}O , lie below the stability line and decay to a stable nucleus by positron emission. Proton deficient (neutron rich) nuclei such as ^{99}Mo lie above the line and decay by electron emission.

produced and these are rather readily captured by many nuclei, again producing unstable isotopes. Radioactivity is used in medicine in two distinct ways; therapy and diagnostic imaging. In the former the idea is to deliver a lethal dose of ionising radiation to a tumour. Ideally this would be achieved best using a locally administered isotope that emitted an energetic charged particle, whose mean free path in the surrounding tissue, would only be a few millimetres. Diagnostic imaging has a completely different requirement of the isotope. Here the image is formed by penetrating gamma rays that leave the body and enter external detectors. Any charged particle emission from the isotope would increase the patient dose without contributing anything to the image. In general nuclear reactions involve complex nuclear decay processes involving both gamma and particle emission but there are a minority of isotopes which are pure gamma emitters. These are the ones that are chosen, and carefully purified, for diagnostic imaging.

Unstable Nuclei

The stable nuclei lie on a curve of N versus Z that is nearly straight for the lighter elements but which bends upwards as the number of protons increases, Figure 2.10. Unstable nuclei fall into four main categories; they can be neutron rich, proton rich, mass rich or have the stable number of protons and neutrons existing temporarily in an excited state of the nucleus. Each of these decays to a stable state by one of the following changes

Proton Rich, via positron emission $Z \dashrightarrow Z - 1$
$N \dashrightarrow N + 1$

A proton is transformed to a neutron. A remains constant

Neutron Rich via electron emission $Z \dashrightarrow Z + 1$
$N \dashrightarrow N - 1$

A neutron is transformed to a proton but A remains constant

Mass Rich via alpha emission $Z \dashrightarrow Z - 2$
$N \dashrightarrow N - 2$
$A \dashrightarrow A - 4$

Excited state via gamma emission Z, N, A all constant

In each case considerable amounts of energy are released from the nucleus as a result of small changes in mass and the enormous nuclear potential energy. The ideal radioactive isotope for medical imaging is the last of these, which releases only the most penetrating electromagnetic photon, a γ– ray. The radionuclide, ^{99m}Tc is widely used because its decay is ideally suited to imaging. It decays with a half-life of just 6 hours, this is an ideal time scale for investigations. The decay is by the emission of a single gamma ray with E= 140keV. There is no associated charged particle emission.

Radioactive Decay Law

Medicine, industry and archaeology make extensive use of radioactivity. In all diagnostic applications the emitted radiation is detected and counted. The nature and energy of the emissions together with the rate at which they are counted generally provides the useful quantitative information. The number of nuclear disintegrations, N, occurring in unit time from a single pure isotope follows the. exponential law ,$e^{-\lambda\tau}$. More formally we write :-

$$\frac{dN}{dT} = Q_o e^{\frac{-0.693\,t}{T_{1/2}}} \qquad \text{......2.7}$$

Where $T_{1/2}$ is the half-life of the process, and Q_o is the initial rate of disintegrations measured at whatever point in time an experiment or medical procedure started. Exponential functions arise naturally in a wide range of natural phenomena, such as x-ray attenuation with penetration distance, population growth and curiously, where it was discovered, in finance- the law of compound interest ! In all cases the law arises from a change whose rate is proportional to the amount currently present. Thus in radioactive decay, the rate of change of the numbers of nuclear disintegrations is proportional to the number of unstable nuclei that remain.

Frequently, in radioactivity, more complex total activity time decay functions are observed. These arise because the number of nuclei of a particular unstable species (the symbol Q_o) is itself varying with time. In medical imaging a radioactive substance is administered to the patient or subject by injection or inhalation. Immediately after the administration of the radioactive tracer, a remote part of the body will not yet have received any radioactivity and initially the activity arising from this place will rise with time as the blood circulation transports tracer to that area. After an initial period the amount of radioactive substance present will saturate, then a pure exponential decay will be observed,

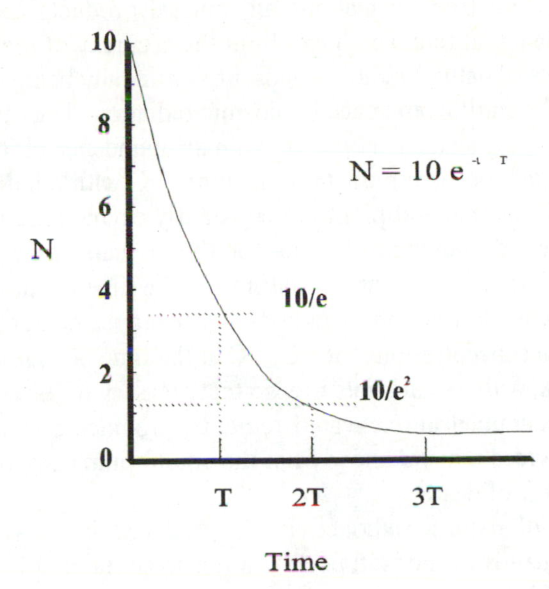

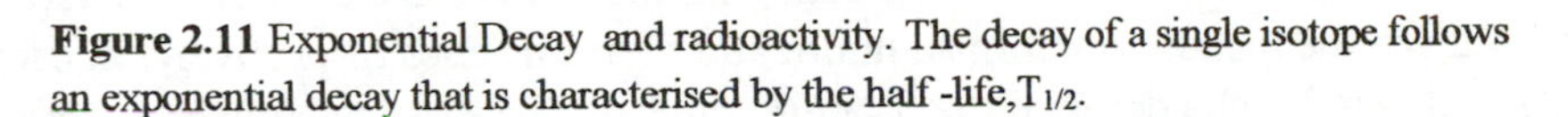
Figure 2.11 Exponential Decay and radioactivity. The decay of a single isotope follows an exponential decay that is characterised by the half -life, $T_{1/2}$.

as long as natural clearance processes have characteristic times much longer than the half-life of the tracer

Natural Radioactivity

We have already mentioned that all possible isotopes were once formed at the beginning of the universe, in the big bang. The vast majority of the unstable isotopes have half-lives very much less than the age of the universe. These isotopes have long since vanished. There are however a few isotopes, around the element uranium, which decay exceedingly slowly and thus there is a small amount of natural radioactivity in the environment. The radioactive gas radon is a natural radioactive product, which is formed in the earth and then diffuses up through the rock. There is some concern about the health hazard created by very small concentrations of radon gas which become trapped in the relatively hermetically sealed modern house. Natural radioactivity is also a nuisance in nuclear experiments and medical applications of gamma imaging, since this natural activity arising from just about all mineral products creates a small background photon count rate which can limit the accuracy of the experiment.

Tiny quantities of natural radioisotopes are continually being created by the bombardment of the earth from space by cosmic radiation. The element carbon Z=6 exists mainly as the stable isotope, ^{12}C with an abundance of 98.89%. There is however a naturally occurring unstable isotope ^{14} C with a half-life of 5570 years. All life, both animal and plant are constantly exchanging carbon atoms with the environment during their lifetimes but the exchange process stops with death. This is the basis of radiocarbon dating. At the time death the animal or plant corpse contains the fraction of radiocarbon isotopes characteristic of the current equilibrium concentrations of ^{12}C, ^{14}C at the time of death. The heavier isotope then decays, with its long half-life, so that an assay of the composition of carbon isotopes, contained in excavated remains, provides a means of dating those remains, provided that we know what the equilibrium composition should have been at the time of death.

The majority of natural radioactivity is produced by uranium. Natural uranium consists mainly of the isotope ^{238}U but also contains 0.7% of ^{235}U and tiny amounts of ^{232}U. These are the starting points of the three natural radioactive series given the names, uranium actinium and thorium series respectively. In each series an initial alpha decay of the uranium nucleus gives rise to another unstable nucleus that decays, either by alpha particle or beta, electron emission, to produce a third and so on. Each of the three series contains at least ten intermediate steps that finally reach a stable isotope of the heavy element lead Z= 82. This process has been going on for so long that each series is now in equilibrium with each intermediate nucleus being present in the earth in an amount that is not changing perceptibly with time. The isotopes ^{238}U and ^{235}U emit alpha particles with a decay half-life of about 1 billion years, about the same as the age of the universe. The earth's crust on the average contains about 5 parts per million of uranium. One gram of pure, natural uranium produces about 150,000 disintegrations per second, 150 kBq (see equation 3.6). In each event the alpha particle releases about 4MeV in kinetic energy from the nucleus. All this energy eventually gets turned into random heat energy since the particles are stopped within the earth. One gram of natural uranium produces about 10^{-7} W of heat energy . This is the main source of heat that keeps the core of the earth molten. The emitted alpha particles are helium nuclei. The alphas quickly acquire two electrons to become helium atoms. Helium is a small constituent of natural gas,(.1-2%) and all industrial helium is recovered from natural gas wells. Helium atoms are so light and chemically inert that, once released into the atmosphere, they diffuse up through atmosphere and escape into space. If it were not for its continuous production by uranium decay, within the earth's crust there would be no helium on the earth and no means to produce the cold environment of the superconducting magnet in MRI. Commercial quantities of helium gas are recovered as by-products from a small number of gas wells in the US and Poland.

Nuclear Fission

^{235}U has a special place in nuclear physics and technology. If this isotope ever captures another neutron to form ^{236}U, fissile uranium, then it becomes very much more unstable and rapidly undergoes nuclear fission. Fissile uranium is then just on the edge of stability .The temporary capture of another neutron tips the balance. In the fission process the uranium nucleus splits into two approximately equal parts and shortly after the split more neutrons boil out of the two heavy unstable fission fragments. This provides the possibility of a chain reaction. If a suitable mass of ^{235}U can be assembled very quickly, then an explosive chain reaction results. Each fission gives rise to about 3-4 new neutrons and each of these has the potential to cause more nuclear fissions as long as they do not escape from the block of material first. This is the physics of the uranium bomb. The same process is put to more worthwhile use in the nuclear reactor. One of the nuclear reactions producing fission can be written as

$$^{235}_{92}U + n \rightarrow\ ^{236}_{92}U \rightarrow\ ^{99}_{42}Mo +\ ^{133}_{50}Sn + 4n \qquad \ldots\ldots 2.8$$

A nuclear reactor consists of a carefully constructed matrix of rods of uranium fuel and neutron absorbing material such as graphite of boron. The whole matrix is immersed in a bath of water. The neutron absorbers maintain the chain reaction at a controlled level by reducing the neutron flux to a steady level rather than allowing an uncontrolled runaway process. The water acts as a so-called moderator. By a series of collisions with the water molecules the neutrons loose their large initial kinetic energy and rapidly come into equilibrium with the temperature of the water. This process is of course another energy conversion in which the neutrons dissipate kinetic energy and raise the temperature of the water moderator. This heat energy is used in power reactors to drive steam generators, which in turn create electrical power for distribution. The process of thermalisation, in which the neutrons slow down, increases the probability of their capture by the uranium fuel and thus increases the overall efficiency of the process. A schematic drawing of a fission reactor is shown in figure 2.12

Artificial Radionuclide Production

Artificial radionuclides are produced in four ways; a) neutron capture, b) nuclear fission, c) charged particle bombardment and d) local radionuclide separation. Each production method has technical and economic advantages such as efficiency, cost and purity of product. Some radionuclides such as ^{99}Tc, ^{131}I can be produced by more than one method, either in the hospital or at a remote central reprocessing site. Others, such as the short-lived positron emitter ^{15}O, are practically only produced within the hospital using a cyclotron. Methods a) and b) take place within a nuclear reactor and these are used to produce either long-lived isotopes or precursor isotopes (see table 2.1). Method c) involves the use of an accelerator such as a cyclotron to produce high-energy protons or alpha particles. This is essential in the context of PET functional imaging where short half-life positron emitters are required. In method d) a convenient isotope such as ^{99}Tc is produced locally when required, from a simple chemical separation column. The column is replenished periodically with a longer lived precursor, ^{99}Mo in this example, which is produced in large quantities at a central reactor facility using method a .

Neutron Capture

Just like the fission process itself, neutron capture takes place very much more efficiently when thermal neutrons with energies in the range 0.02 to 100 eV are used. The core of a nuclear reactor, in steady state operation, has a very high density or flux of thermal neutrons. If a block of a suitable stable precursor element such ^{98}Mo is inserted into the reactor core, then after a time, depending on the neutron flux, a fraction of the target, Mo in our example, will have become ^{99}Mo. Excess binding energy is immediately carried off in the form of a γ–ray.

Table 2.1 Radionuclides frequently used in medical imagingThe entries marked with β^+ actually emit a positively charged electron or positron. In biological tissue the positron rapidly associates with an electron , the pair annihilate to produce 2 gamma rays, each of energy 511 keV emitted nearly 180 ° apart

Nuclide	Half Life	Photon Energy (keV)	Method of production
Molybdenum ^{99}Mo	66 h	740 γ	Fission
Iodine ^{131}I	8.07 d	364 γ	Fission
Xenon ^{133}Xe	5.27 d	81 γ	Fission
Caesium ^{137}Cs	30 y	662 γ	Fission
Carbon ^{11}C	20.4 min	511β^+	Cyclotron
Nitrogen ^{13}N	9.96 min	511β^+	Cyclotron
Oxygen ^{15}O	2.07 min	511β^+	Cyclotron
Fluorine ^{18}F	109 m	511β^+	Cyclotron
Indium ^{111}In	67.9 h	159,171,245 γ	Cyclotron
Iodine ^{123}I	13 h	159 γ	Cyclotron
Technetium ^{99m}Tc	6 h	140 γ	Chemical generator
Gallium ^{68}Ga	68 min	511 β^+	Chemical generator

$$ {}^{98}_{42}Mo + n \rightarrow {}^{99}_{42}Mo + \gamma \qquad \ldots 2.9 $$

Radionuclides produced in this manner generally have low specific activity since the unstable nucleus was produced from the chemically indistinguishable stable element itself and, in a finite time, only a fraction of the initial charge of nuclei will be converted. The isotope ^{99m}Tc is widely used in medical imaging, it is a

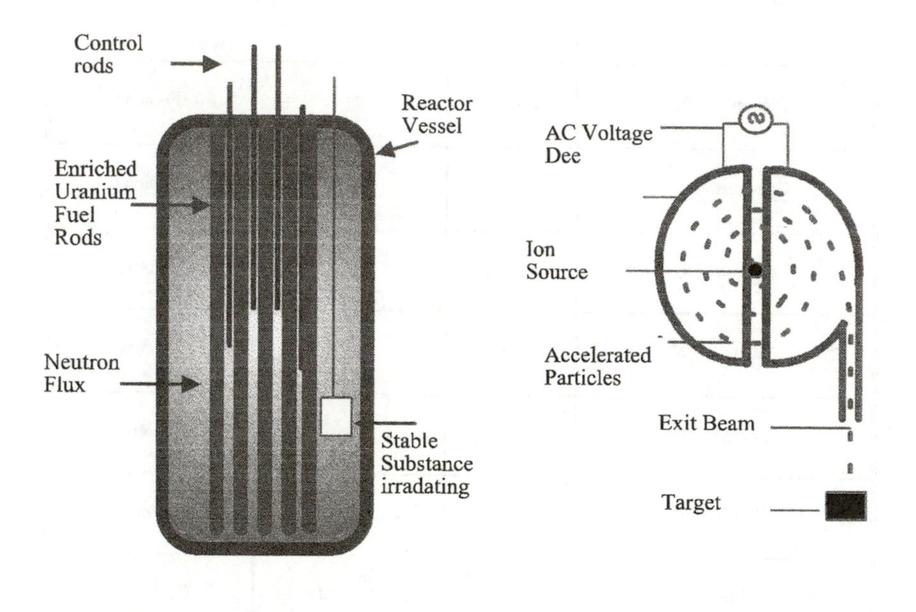

Figure 2.12: Methods of radionuclide production. a) The Fission reactor. Very large numbers of neutrons released from the uranium fission in the fuel rods are slowed down by to thermal energies by collisions with the water moderator. These neutrons have a very high probability of capture by nuclei and thus are efficient producers of isotopes. Samples are introduced into the reactor core where the constituent nuclei are transformed by neutron capture. Useful radionuclides can also be recovered from reprocessed fuel rods at the end of a reactor cycle.b) The Cyclotron: Charged particles

released in the centre of the "Dees" are accelerated by a high frequency electric field and confined by a strong vertical magnetic field to move in an expanding spiral. The exit particle beam is directed at a target to produce radionuclides. This is the method for producing the very short-lived positron emitters used in PET such as ^{15}O.

pure γ ray emitter with a half life of 6 hours. Its precursor, ^{99}Mo decays into Tc by beta emission, with a half-life of 66 hours. Hospitals using Tc obtain, primary radioactive source containing Mo from a central supplier and then chemically separate out the required Tc when it is required, within the hospital. This is only possible because of the relatively long half life of the Mo precursor.

Fission Products

Radionuclide production within a nuclear reactor also takes place as a natural consequence of the fission process itself. Many different fission products are produced in all nuclear reactors, the longer lived and useful isotopes are recovered from spent fuel rods in reprocessing plants such as Sellafield. To extract an isotope such as ^{99}Mo for medical imaging, the processing has to be done very carefully in order to remove contamination arising from remaining uranium fuel and unwanted radioactive material. This of course would only take place at the end of a reactor cycle when fuel rods have expended their useful uranium. The isolated isotopes have very high specific activity (the ratio of radioactive isotope to stable element) since all these nuclei were created from uranium, not stable versions of the element. Fission products tend to be neutron rich and thus they generally decay by electron emission. Such isotopes can be used in therapy or as, in the case of Mo, a precursor to a daughter, purely gamma emitting isotope for use in medical imaging.

Particle Bombardment

Radionuclides such as ^{15}O, with half-lives of about 2 minutes, could not be produced at a central facility because the majority of the activity would have decayed long before the source reached the hospital. In these cases the isotopes are produced within the hospital itself. A small cyclotron (figure 2.12) is used to produce high-energy (1- 100 MeV) protons which are allowed to bombard a target and induce radioactivity there. The proton energy has to be rather high so that it has sufficient kinetic energy to overcome the electrostatic repulsion and penetrate the nuclear potential energy barrier, see figure 2.9.

2.5 Nuclear Magnetism and Magnetic Resonance

In medicine, nuclear magnetic resonance, NMR and magnetic resonance imaging, MRI , are both used as probes of local atomic chemical environments in tissue. The technical details of NMR and MRI are discussed at length in chapter 6. In this section we discuss the origins of nuclear magnetism, the nuclear magnetic moment and the resonance process in terms of nuclear magnetic energy levels and the absorption and emission of photons. Magnetic resonance is actually a generic technique, quite widely used in physics and chemistry. Whole atoms, whole nuclei, free electrons, neutrons and muons all have intrinsic magnetic moments which can be utilised in different types of resonance experiments. NMR is just one of these but it happens to be particularly well suited to biological investigations and of course non-invasive imaging. The origin of atomic and nuclear magnetism is actually best illustrated using the simple Bohr atom and thus we begin with this description of atomic rather than nuclear magnetism.

Atomic Magnetism

In our daily lives we have direct experience of magnetism through compasses, toy magnets and electric motors, magnetic door catches and more indirectly, magnetic recording media in both entertainment and computing. In all of these everyday applications a solid object containing ordered atomic magnetic moments is being used to exert a macroscopic magnetic force on another magnetised or magnetisable object. The type of solid material used in these devices is called a ferromagnet, which is characterised by strong microscopic internal electronic interactions that maintain the individual atomic magnets all pointing in the same direction. Individual atomic moments can only exert a tiny force that would be totally imperceptible to human senses. Only by making 10^{22} such atomic magnets all work in the same direction can any perceptible magnetic effect be observed. Our seafaring ancestors, who used crude lodestone compasses for navigation, regarded magnetism as a somewhat mysterious natural force. Since the work of Faraday and Maxwell in the last century, the magnetic field has been regarded as an inevitable consequence solely of the motion of electric charge. Although the theory of magnetic forces associated with electric currents was completely unravelled in the last century, the extension to magnetised bodies such as lodestone required atomic physics and quantum mechanics for an explanation. First there is the problem of the atomic analogue of the electric current, then there is the question of how, in a lump of iron, these atomic currents spontaneously organise themselves to produce a tangible external field of force. In our concern with nuclear magnetism, the second question does not arise since nuclear magnetism, at least at room temperature, does not involve any spontaneous ordering of the nuclear magnets. At least a rudimentary understanding of the atomic and nuclear origins of magnetism is however crucial to an understanding of both NMR and MRI.

In principle any atom can create an external magnetic field since within itself there are rapidly circulating electrons that is to say electric currents. If we take a simple Bohr picture of an hydrogen atom (see section 2.3) and confine the electron orbit to a single circle of radius a_o then we could expect a magnetic field to be created. The electron charge is -e and, according to Bohr, the angular momentum ,L, of the electron motion around the circle is quantised, thus, $L = r_o mv = nh/2\pi$. This gives rise to a current

$$i_{electron} = -\frac{e}{T_{orbit}} = -\frac{Ve}{2\pi r_o} = \frac{-en\hbar}{2\pi m r_o^2} \qquad \ldots\ldots 2.10$$

This loop of atomic electric current creates an external magnetic field whose lines of force mimic those of an electric field created by a pair of opposed electric charges, Q, placed a small distance, d, apart. This is called a dipole field. The electric dipole field is completely characterised mathematically by just two quantities; the direction of the line joining the charges and the product, Qd , which is called the dipole moment. The strength of the field at any distance, R, from the charge pairs is proportional to Qd/R^2. The magnetic dipole field has the same mathematical form but the magnetic dipole strength, μ, of a small loop with area,πr^2 and current I is given by $\mu = \pi r^2 I$ .The direction of the magnetic dipole moment is perpendicular to the plane of the current loop. The magnetic moment associated with the electron circulating in the Bohr atom is given by

$$\mu_{Bohr} = \pi a^2 \times i_e = -\frac{e}{2m} \times n\hbar$$

$$\mu_{Bohr} = -\frac{e}{2m} \times L \qquad \ldots 2.11$$

These formulae are vital to any discussion atomic or nuclear magnetism. First the strength of the atomic magnetic field is proportional to the ratio e/m and to the angular momentum of the circulating charges. The ratio e/2m is called the magnetogyric ratio (ratio of magnetic moment to angular momentum),γ. We will see below that this parameter sets the resonance frequency in NMR and MRI. Atomic magnetic moments are generally written in units of the Bohr magneton,

$$\mu_b = e\hbar/2m = \gamma\hbar\ . \qquad \ldots 2.12$$

If we put some typical numbers into these formulae then we find that the magnetic field strength, 1cm away from a single atomic magnetic moment is about 10^{-25} T. This is vanishingly small in comparison with the earth's magnetic field of about 5.10^{-5} T and the magnetic field 1cm away from the north pole of a toy bar magnet, about 0.1 T. Any observable consequences of atomic or nuclear magnetism must then arise from very many aligned atoms or nuclei. Isolated magnetic dipoles, sitting in a uniform applied magnetic field, experience a turning moment or torque, rather than a translation force, just as a compass needle turns on its axis to point towards magnetic north. In general the motion of a compass needle is heavily damped by friction at the pivot and from the air. When disturbed and then released the needle will make a few wobbles before settling along magnetic north. Atomic and nuclear moments are relatively weakly damped and so when they are disturbed from equilibrium their motions continue for many periods of revolution about the direction of applied magnetic field before coming to rest. In order for any magnetic alignment, be it of compass needles or atomic moments, there must be a loss of energy from the magnetic objects to the environment. The compass needle can be said to dissipate energy through frictional forces, the atomic moment loses energy by emitting photons that are absorbed by the surrounding atomic chemistry.

Although any atom is capable of forming a magnetic moment, only a few elements in the periodic table actually do exhibit permanent magnetic dipole moments. Atomic hydrogen does not have a permanent atomic moment arising from its orbital electron motion. The reason for this is that the electron orbit is not confined to a single circle.Rather it occupies a spherical region and, in the ground state (1s state with L=0), the motions average orbital angular momentum to zero. Atomic hydrogen does however have a magnetic moment arising from the intrinsic spin of the single electron. Elements in the middle of the periodic table, collectively called the transition metals (Cr, Mn, Fe, Ni ,Co), all have electronic configurations in which there are unpaired electrons in 3d (L=2) states and all of these exhibit permanent atomic magnetic moments. The atomic gyromagnetic ratio reflects the details of the atomic electronic motions.

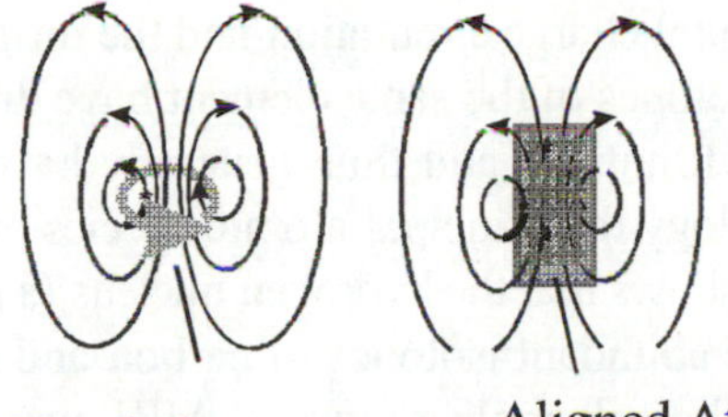

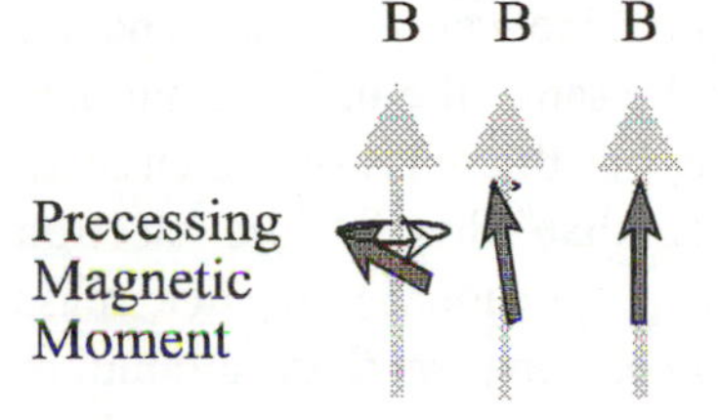

Figure 2.13 a) The lines of magnetic force surrounding a small current loop such as that of the Bohr orbit. The shape of the field is said to be that of a magnetic dipole. b) The macroscopic external field created by aligned magnetic moments in a bar of iron. Here there are strong internal electronic interactions that maintain the alignment of the atomic moments. This is called ferromagnetism. c): The behaviour of a single dipole moment, in a static external magnetic field. The moment precesses about the field direction gradually losing energy until finally its direction becomes aligned with the field. The precession frequency is called the Larmor frequency and is given by $f_L = \gamma B$.

Nuclear Magnetic Moments

As we saw in section 2.4, each nucleus has a unique number and configuration of protons and neutrons each carrying a spin angular momentum of ½ (h/2π) and , because of their motion within the nucleus, each particle can have an orbital angular momentum. The combined internal motions of nuclear electric charge results, amongst other things, in there being a definite external magnetic field associated with some nuclei. Magic nuclei create no external magnetic field because there is no net orbital charge circulation and the intrinsic nucleon spins are paired off. Different isotopes of the same element have different numbers of neutrons, different internal motions and thus generally have different nuclear magnetic moments. In biology the principal atomic species of interest are H, O, C, P, K, Na, Cl. Table 2.2 shows that the hydrogen nucleus (a single proton) has a nuclear moment, but the abundant isotopes of carbon and oxygen do not. At present very nearly all clinical applications of MRI use nuclear magnetic resonance of hydrogen nuclei . In the future MRI will, in all probability, be extended to other nuclei of medical interest. The nuclear charge currents produce an external magnetic field in the same way as atoms. The unit of charge on both the electron and the proton is the same, the units of angular momenta for both atoms and nuclei are the same, but the mass of the circulating particles in the nucleus is 1000 times larger than that of the electron. Thus since the gyromagnetic ratio is inversely proportional to particle mass, nuclear magnetic moments are about 1000 times smaller than atomic moments. At room temperature there is no analogue of atomic magnetic ordering and thus nuclear magnetism is an extremely weak effect.

Nuclear Magnetic Resonance

On their own, nuclear magnetism and nuclear magnetic resonance are of particular interest only to nuclear and high energy physicists. In biology, NMR and MRI are not at all concerned with the details of nuclear magnetism but rather

Table 2.2: Nuclear magnetic properties of biologically important nuclei. The last column lists the imaging sensitivity of each listed nucleus with respect to hydrogen (water proton). These numbers take into account the isotopic abundance, the concentration within the human body and the field for resonance

Nucleus	Abundance (%)	Spin h/2π	γ/ 2π (Mhz/T)	Relative Imaging Sensitivity
^{1}H	100	1/2	42.577	1
^{12}C	98	0	-	-
^{13}C	1.1	½	10.71	$1.8.10^{-4}$
^{17}O	0.04	5/2	-5.77	2.10^{-4}
^{14}N	99.63	1	3.08	1.10^{-3}
^{23}Na	100	½	11.26	$9.3.10^{-2}$
^{31}P	100	½	17.23	$6.6.10^{-2}$

what the motions of nuclear moments can reveal about the chemistry of the surrounding atoms. The magnetic resonance phenomenon is used as a probe of atomic chemistry. Earlier in this chapter we suggested that nuclear and atomic phenomena carry on largely independently of one another. This is certainly true in the case of NMR but now it is precisely the weak atomic-nuclear moment interactions that are of interest. In essence what both NMR and MRI do is to examine how weak atomic magnetic fields, set up by the surrounding atomic chemistry, alter the response of nuclear moments to well defined external applied fields. The origins of the interaction are quite simple in principle but extremely complex in reality. Any nucleus sits inside an atomic environment which in turn, in biological tissue, is embedded in a complex arrangement of molecules. Under the conditions of a resonance experiment the entire body of tissue is subjected to man-made static radio frequency fields. These are designed to set nuclear spins in motion in a well-defined and predictable manner. The nuclei are at the same time subject to magnetic and electric fields, some fluctuating and some relatively static, that are created by the atomic environment of the nuclei. These intrinsic

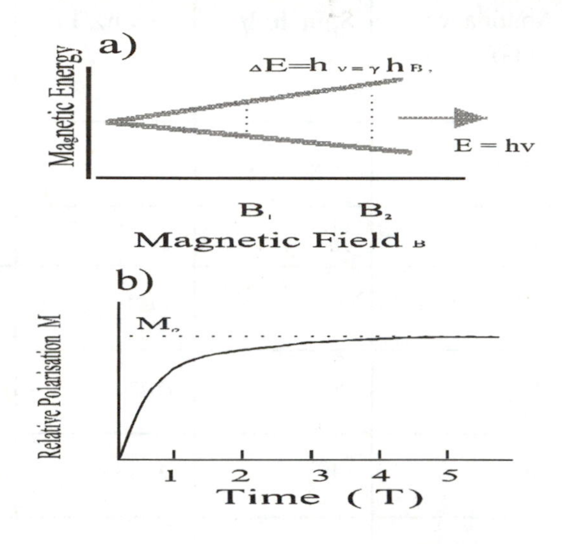

Figure 2.14 a) The magnetic energy splitting (Zeeman effect) for nuclear moments in an applied magnetic field. In real substances each nucleus is subjected to the same value of B_o, the applied external field and a smaller random atomic field which we have represented in exaggerated form by the width of the lines. This represents here the T_2 process which contributes a spread in resonance frequencies (the line width in figure 2.17). **b)** The spin lattice relaxation process, T_1, corresponds to the random thermal flipping of nuclear moments from up to down and vice versa by the exchange of photons with the atomic environment. This is how thermal equilibrium is established in a time of about 5 T_1 The solid line is the function $M = M_o(1 - \exp(-t / T_1))$ The saturation value of the nuclear magnetic polarisation, M_o, varies with magnetic field and temperature approximately as M_o B/T. The arrows represent the growth of the polarisation with time after the external field has been first established.

atomic fields change the nuclear spin motion and these changes provide the valuable information about the atomic chemistry. This is described in more detail in chapter 6. In order to gain this information a large number of nuclear spins has first aligned and then set into a coherent motion in order to create an observable signal.

The first step in NMR and MRI is to create a bulk nuclear magnetic magnetisation or plolarisation, using a strong static magnetic field. Classically we picture the nuclear moments reacting to this field just like damped compass needles, oscillating, gradually losing energy and finally coming to rest along the direction of the field. The quantum picture is described as the nuclear Zeeman effect illustrated in figure 2.14. In the applied field, the energy levels of each nucleus acquire an extra small amount of energy that depends on the strength of the field, the size of the intrinsic nuclear magnetic moment and finally the relative orientation of the magnetic moment. The essential quantum result is that there are only two possible stable orientations for the proton moment: parallel and anti-parallel to the field . In this case each proton energy level is split in to two, the anti-parallel moments occupying the higher level and the parallel spins the lower level, The energy splitting between the two is given by the magnetogyric ratio,γ, multiplied by the strength of the applied magnetic field and Planck's constant, thus ,$\Delta E = \gamma(h/2\pi)B$. At the absolute zero of temperature all the spins would finish up in the lower energy, down, state corresponding to a complete alignment of nuclear moments with the external field. At room temperature however, thermal energy, kT, is very much larger than the energy spacing , ΔE and this results in a dynamic equilibrium, with only a tiny difference between the populations of the two magnetic energy levels, see section C.1. The thermal distribution of a total of N nuclear moments at a temperature, T in a static applied magnetic field, B, results in a net magnetic polarisation, M given by Curie's law.

$$M = \frac{N\gamma^2\hbar^2 B}{4kT} \qquad \text{......2.13}$$

As the magnetic field is increased so the degree of alignment increases as the magnetic energy, $\gamma\,(h/2\pi)B$, gains ground with respect to thermal energy, kT. At room temperature, with a field of 1T, the degree of nuclear polarisation is about 1: 10^6. See Appendix C. Although small, this is enough to provide a measurable signal when the aligned moments within a region of about 1 mm^3 can be made to change their magnetic energies simultaneously.

Changes of spin direction require the gain or loss of energy quanta with frequencies $\nu = \Delta E/h$.. Since, ΔE varies linearly with magnetic field strength so does the photon frequency. This is the key to MRI. In NMR and MRI typical photon frequencies are in the range 1 to 100 MHz , these are all radio waves.

When a magnetic field is first applied to a piece of tissue the polarisation is not created instantaneously, but proceeds towards magnetic equilibrium exponentially with a characteristic time, T_1, the spin lattice relaxation time. The

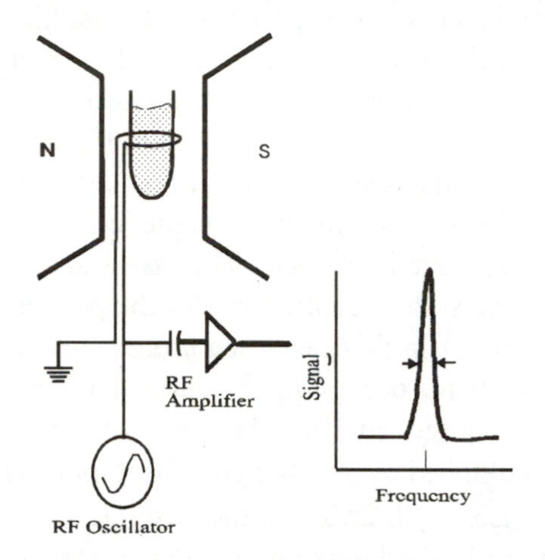

Figure 2.15 A simple NMR experiment (continuous wave NMR). Here the magnetising field is provided by an electromagnet. The RF field is applied to the sample through a transverse coil. The same or another similar coil can used as the detector when connected to a high gain electronic amplifier. As the frequency of RF field is swept through resonance the signal amplitude rises and falls through a resonance curve. The width of this resonance curve or line is determined by a function of T_1 and T_2 which varies with RF power level.

relaxation process, described by T_1, involves the exchange of the radiofrequency photons between the atoms and the nuclear moments. The speed and efficiency of this energy exchange varies with the effective viscosity of the liquid and the numbers and types of atoms in the immediate vicinity of the nuclear moment. This is one of the contrast mechanisms of MRI. Values of T_1 in biological matter range from 4000 ms in pure water down to 5 ms in bone. These are in fact rather long times for quantum processes and are themselves the result of the small photon energies involved in polarisation and the weakness of the nuclear moment-environment interaction.

Once thermal magnetic equilibrium has been established, after a time of about $5T_1$, we have obtained some nuclear magnetisation but we still don't have a measurable signal. In order to get information out of the aligned state we have to disturb it and watch it return to equilibrium. In both NMR and MRI this is achieved by using another applied field, this time a time varying, radio frequency, RF, field whose frequency exactly matches the energy separation of the possible nuclear magnetic energy levels. The nuclear moments can absorb just the right quantum of energy from this field to flip from up to down. Once disturbed from equilibrium the moments remain relatively coherent in their motions and so create their own organised fields varying in time at the resonance frequency $f_L = \gamma B/2\pi$. These fields constitute the NMR signal, which is recorded using a small pick-up coil and amplifier set close to the sample or patient under investigation. If the frequency of the applied radio frequency field is allowed to wander off the resonance value then the flipping effectively stops and the signal disappears in just the same way that radio reception waxes and wains with adjustments of the tuning dial on the radio set . Each nucleus has its own very precise frequency to field ratio and this can be used to analyse the nuclear species content of an unknown sample using the technique of NMR spectroscopy, see table 2.2.

In this section we have given a preliminary description of NMR using the earliest and simplest experimental arrangement of a fixed static magnetic field together with a continuous, but variable frequency, RF field to set the nuclear moments in motion. This arrangement is called continuous wave NMR or CW NMR. Very early in its development, standard NMR methodology changed over to the use of pulses of RF instead of continuous waves. Although the two schemes are equivalent in terms of the underlying physics, the pulse method has a number of engineering advantages. The key difference is that using a pulse of RF, the aligned magnetic moments are given a single sharp kick from an RF pulse to set them in motion. After the pulse has finished the subsequent "free motion " of the nuclear moments is governed by the nature of the atomic fields.

Questions and Problems

1. What do the terms, quantum, photon, atom, nucleus mean ?
2. What does classical physics predict for the sizes of atoms??
3. Describe the atomic processes involved in the emission of light from a domestic light bulb
4. Estimate the number of light photons per second hitting a 1 cm^2 area of floor 3m from a 60 watt domestic light bulb. Assume that 10% of the electrical power is converted into visible light with a photon energy of 1eV. Given that working light bulbs feel warm what other photons are emitted by the bulb?
5. Estimate the frequencies and wavelengths of photons with energies 0.0001 eV, 0.1eV, 1 eV. What names are given to these ranges of frequency?
6. Calculate the energies of the lowest four energy levels of a hydrogen atom. Explain how this accounts for the light emitted and absorbed by hydrogen atoms.
7. Describe the ways in which x-ray photons interact with atoms. Which of these will to lead to energy deposition in a patient.
8. Explain what is meant by the term half life.
9. Calculate the fraction of the original amount of radioactive material left after 1, 10 and 100 half lives.
10. The half lives of the uranium isotopes, U^{235}, U^{238} are $7.\ 10^{8}$ and $4.51.10^{9}$ years, with present natural abundances 0.72% and 99.27% respectively. Assuming that the abundances were equal when the earth was formed, estimate the age of the earth.
11. When an electron and a positron annihilate, the electron and positron rest masses disappear to produce two gamma rays. Using the conservation of energy and Einstein's equation , $E = m c^2$ estimate the energies of the photons. Compare your result with table 2.1. What are the directions of the two gamma photons?

3 RADIATION PROTECTION

"Various physiological effects have been observed with radium rays; they excite phosphorescence in the interior of the eye; when an active product is brought near to the temple, a sensation of light is perceived. They act upon the epidermis and profoundly disorganise the skin, as do x-rays. The effect is produced without any sensation being felt at first and it only develops after several weeks; it then produces more or less deep lesions which can take several months to heal and which leave scars . At present an effort is being made to utilise this action in the treatment of lupus and cancers. Radium rays have an active effect on the nerve centres and can then cause paralysis and death; they seem to act with particular intensity on living tissues in the process of evolution .."

A. H. Becquerel Nobel Prize Lecture 1903

3.1 Introduction

The benefits of the penetrating power of x-rays were obvious from the moment of their discovery in 1895 and medicine was quick to develop x-rays into the most important diagnostic tool of this century. Acute biological effects, similar to sunburn, of direct exposure to high levels of x-ray radiation also became apparent rather quickly to both researchers and medical practitioners. The International Commission on Radiological Protection (ICRP) was set up in 1928 to evaluate the risks to man of all exposures to ionising radiation and to set limits on maximum permissible levels of exposure. The exposure limits decreed by this and other similar bodies have dropped continuously over the past sixty years, as the more insidious effects of exposure to radiation have been revealed both by basic biological research and statistical studies. The most important of the latter arose from the long term effects on the survivors of the Hiroshima and Nagasaki atomic explosions at the end of the second world war.

The biological effects of exposure to ionising radiation can be divided into two main regimes that depend on the level of exposure and the time interval between the radiation insult and the appearance of symptoms. At the very high doses received by early radiation workers and many of the Japanese atomic bomb victims, death and/or acute biological effects appear or reappear in seconds to weeks. At very much lower doses other effects such as genetic damage, life shortening and carcinogenesis only appear after many years. At whole body doses in excess of 2-5 Sv there is a direct, deterministic link between the level of exposure and a spectrum of illnesses (radiation sickness). There is a relatively simple relationship between the severity of an illness, its prognosis without medical intervention and dose level. In addition there appears to be a fairly well defined exposure threshold below which these symptoms never appear. At whole body doses in excess of 20 Sv, death is almost always the outcome in a matter of weeks to months. The acute effects of high radiation dose are not a matter for concern for the general public since, barring nuclear accident or nuclear war, the public does not come into contact with sufficiently powerful sources of radiation. They are potentially a matter of extreme concern for many workers in both the medical and industrial areas and radiation protection measures are universally in place to prevent accidental exposure to high levels of radiation. The beneficial use of ionising radiation, particularly in cancer therapy, necessarily exposes parts of the patient to lethal doses of radiation. Side effects in healthy tissue are an inevitable consequence. Of more concern, both to the general public and to radiologists, are the long term effects of an accumulated exposure to very low levels of radiation that are routinely used in, for example, dentistry. This is an extremely difficult problem. It is now assumed, although it took fifty years to demonstrate and become accepted, that there is no lower exposure threshold for long-term effects such as carcinogenesis. Rather there is a very small but calculable chance of even a single, trivial x-ray examination producing a fatal cancer sometime in the patient's subsequent lifetime. The International Commission on Radiological Protection, ICRP has, in over fifty reports and publications, stressed that all exposure to radiation should be justified and guided by the three basic tenets of radiological protection. First that the benefits of any radiation examination or procedure should outweigh the likely detriment. Second that in any procedure the ALARA principle should be adopted. ALARA stands for "As Low As Reasonably Achievable". Finally only in exceptional circumstances should the dose limits set by the ICRP ever be exceeded.

The entire field of low level radiation protection is fraught with difficulties. The effects are generally delayed by many years, thus creating logistical difficulties for the epidemiologist. There appears to be a strong element of chance and genetic predisposition involved. Cosmic rays and more particularly natural radioactivity, in the form of radon gas, produce a geographically variable, low level of background radiation, (see section 2.4), that may have significant effects on morbidity in local populations. These are often called the stochastic effects of radiation exposure. Our basic ignorance about the mechanisms of carcinogenesis mean that although there is a proven link between radiation exposure and cancer induction, the mechanism is not fully understood. The final problem has a political/historical origin and arose more in the medical than in the industrial applications of radiation. Industrial radiation protection is driven by the need to harness cheap and abundant source of power provided by nuclear power. The general public has come to associate radiation exposure at any level with the horrors arising from the atomic bomb explosions and the several nuclear power accidents, most notably the 1986 Chernobyl reactor explosion. Nuclear Power generation can only survive in a democracy if it is seen to be as safe as humanly possible. If we can discount accidents, then very good safety in the nuclear industry has actually been achieved by the rigorous application of protection measures and a level of general radiological education amongst radiation workers that makes for truly safe working practices. Curiously the medical area is in a rather different situation. The tangible benefits of exposure to radiation from diagnostic and therapeutic procedures is experienced daily by both medical workers and patients, indeed in some circumstances, patients expect a radiological examination to form part of their diagnosis. Possible long-term harm is of course not immediately visible. Patient exposure to ionising radiation cannot be regulated by legislation, rather it depends on voluntary codes of practice, education and peer review. The ALARA principle encapsulates this code. Although it is generally accepted that the benefits of ionising radiation in medicine have consistently outweighed long term harm, there is continuing pressure to further reduce medical exposure and a developing culture of tighter voluntary control. Twenty years ago it became clear that essentially the same diagnostic yield from x-ray examinations was being obtained in different hospitals in the UK using exposure levels that differed by factors of 30 to 100

(Shrimpton 1990). Since that time there has been significant progress in encouraging the adoption of technical measures to reduce patient dose, the more efficient recovery of existing patient imaging records and the education of clinicians in the choice of investigation.

If we are to have the undoubted benefits of diagnostic radiation in dealing with both life threatening and more benign medical conditions, we have to accept that there is a small but finite associated risk of long term harm, no matter how trivial the medical examination, no matter how well the technology can be improved to reduce the radiation dose. Given the widespread use of radiation, even a small probability of a single individual cancer resulting from a medical examination translates into an economically and politically significant population of individuals, whose working lives are shortened and whose families are bereaved. Not too surprisingly all countries, but particularly the industrialised countries have put in place statutory controls on the use of ionising radiation by workers in industry and medicine and set up organisations devoted to all aspects of radiation protection. In the UK this is carried out by the Environmental Agency. It is responsible for registering and keeping a check, through the use of personal radiation dose meters (film badges) on the accumulating dose received by all radiation workers, both in industry and in medicine in the UK. The EA also attempts to standardise and introduce dose reduction procedures both in the medical and industrial arenas. Statutory control can only have meaning when there are strictly quantitative definitions of concepts such as radiation exposure, dose and biological effect and an effective means to measure these quantities. In this chapter we give the standard, internationally agreed definitions, the main ways in which radiation dose is measured and radiation protection implemented. We also describe the biological effects of ionising radiation. Finally we discuss the impact of new imaging technologies on the collective radiation dose to both medical workers and the general public.

3.2 International SI Units of Radiation Exposure and Dose

At the levels generally used in medical diagnosis, humans do not directly sense the presence of x-ray and gamma radiation. Thus the early radiation workers were in much the same situation as early deep miners who descended to their work in fear of odourless and colourless toxic gases that sometimes seeped into the galleries and killed miners without warning. Miners employed caged canaries and then the Davey Lamp to provide a sensitive early warning of the presence of lethal gases. Radiation detection as we saw in chapter 2 depends on photons interacting with matter to produce a physically measurable effect. The earliest radiation detector was the photographic film, this was how the penetrating radiations were discovered in the first place. Although films are not very useful as early warning devices, the film badge is the international method of recording an individual radiation worker's accumulated exposure to radiation.

Exposure

The earliest real time radiation meter and thence the unit of radiation depended on the ionising effect of radiation in air. That is to say the electron-ion pairs, produced when photons eject electrons from air atoms by the photoelectric and Compton effects, see section 2.3. The earliest radiation-measuring instrument was the gold leaf electroscope. In its charged condition the wafer thin gold leaf hangs at an angle to its support post because of electrostatic repulsion between the leaf and its support. When a source of ionisation is brought close to the electroscope, the gold leaf loses some of its initial electric charge when ions of the opposite sign, produced in the air by the radiation, are attracted to its surface and neutralise the initial charge. The amount of ionisation present can be "measured" by the amount of movement of the gold leaf. A modern standard pocket radiation monitor for everyday use employs an electronic version of the same principle. The biological effects of radiation are best quantified in terms of the amount of energy, in any form, that the radiation source actually imparts to the patient or subject rather than the amount of ionisation produced. Ionisation is the earliest stage in this energy conversion process and it is relatively easy to measure. The index of exposure is defined by

$$Exposure = \frac{Q}{M} \qquad \ldots 3.1$$

Where Q is the total electric charge liberated by photoemission in a mass M of air Both older cgs units and the modern SI unified units are still in common use and so we will give the definition in both sets. The old unit of exposure was the Roentgen and is defined by

$$1\ Roentgen(1\ R) = 2.58 \times 10^{-4}\ Coulombs/kg\ of\ air \qquad \ldots 3.2$$

There is no named SI unit of exposure but it is defined as 1 coulomb /Kg of air so that

$$1\ SI\ unit\ of\ Exposure = 3.876 \times 10^3\ R \qquad ...3.3$$

The intensity of a beam of ionising radiation and the strength of a radioactive source can be specified by the amount of ionisation created in a given mass of air at a particular distance from the source. Today, strengths of sources and exposure are specified in terms of the total amount of energy the source deposits in a specified quantity of substance. This reflects the fact that it is the amount of energy deposited by the ionisation in tissue that determines the biological hazard. Thus an air kerma (kerma is an acronym for "Kinetic Energy Released per unit Mass) is often used to quantify source strengths. Air kerma is measured in units of Grays (Gy)

$$1\ Gy = 1\ Joule/kg \qquad ...3.4$$

The approximate conversion between ionisation, R and the equivalent air kerma is

$$0.00873\ Gy \equiv 1\ R\ (Roentgen) \qquad ...3.5$$

For air, the conversion factor is independent of photon energy. For tissue, the energy deposition per unit mass is the dose and the conversion between exposure and dose varies with photon energy and tissue type by about 10%. It is not just total but also the rate of energy deposition that is important. Thus sources are described by air kerma rates in Gy/Hr or Gy/min.

Activity

The numbers of particles emitted by a radioactive source in unit time describes the strength of that source. The original cgs unit was the Curie (Ci)

$$1\ \text{Curie} = 3.7.\ 10^{10}\ \text{disintegrations /s}$$

The SI unit of activity is the Becquerel (Bq)

$$1\ \text{Bq} = 1\ \text{disintegration / s}$$
$$1\ \text{Bq} = 2.703\ .10^{-11}\ \text{Ci}$$
$$10\ \text{mCi} = 370\ \text{MBq} \qquad ...3.6$$

Absorbed Dose

The amount of radiation from a beam or a radioactive source actually absorbed inside the body depends on the radiation intensity at the body surface (exposure) and the atomic density and composition of the absorbing tissue. The tissue related factors and their dependence on the energy of the radiation are described by an empirical dimensionless factor, f (when exposure and dose are measured in the same units) so that

$$\text{Absorbed dose} = \text{Exposure} \cdot f \qquad ...3.7$$

The tissue factor f for x-rays is close to 1, ranging from 1.06 for bone at 20 keV down to 0.79 for bone at 200 keV. The older cgs unit of absorbed dose, the rad was defined as 1 erg /g . The Si unit of absorbed dose is again the Gray

$$1 \text{ Gy} = 100 \text{ rad} \qquad ...3.8$$

Dose Equivalent

In an attempt to quantify the maximum dose levels for different radiations, international protection agencies introduced the concept of relative biological effectiveness (RBE). This is aimed to take account of the long recognised fact that the same level of tissue damage can result from differing amounts of x-ray, electron, proton or neutron radiation. In general the charged electrons, protons an alphas have very short ranges in tissue and thus deposit their total energy in a relatively small volume, hence causing more biological damage. X-rays and γ–rays on the other hand interact very much less strongly with matter, travel further and deposit energy over a much larger volume at a lower energy density.

The dose equivalent H is the product of the absorbed dose D and a quality factor Q determined by the linear energy transfer, LET of the radiation

$$H = D \cdot Q$$

$$Q = 1 \text{ for x-rays, } \gamma\text{–rays, electrons}$$

$$= 10 \text{ for protons}$$

$$= 20 \text{ for } \alpha \text{ particles} \qquad ...3.9$$

The old unit of equivalent dose was the rem (roentgen equivalent man). The modern SI unit is the Sievert (Sv). For x-rays and γ–rays and electrons, 1 mSv $\approx 10^{-3}$ joules/Kg since Q=1

$$1 \text{ rem} = 10^{-2} \text{ Sv}$$
$$1 \text{ Sv} = 100 \text{ rem}$$
$$1\text{mSv} = 100 \text{ mrem} \quad ....3.10$$

Effective Dose

The amount of energy imparted by a photon or particle beam to the body surface can be measured directly and thus there is a simple empirical relationship between exposure and the final absorbed dose. The interior of the body on the other hand is quite clearly inaccessible to extensive direct measurement and thus the amount of energy actually stopped in any particular organ can only be inferred from theoretical models and measurements on phantom objects. To this end radiology makes extensive use of the 70kg Standard Man; a computer model from which it is possible to estimate a weighting factor, W_R, for any particular body compartment or organ. These weighting factors contain information about the relative susceptibility of an organ to radiation damage and thus, when multiplied by the energy absorbed (dose) applied to a particular organ provide an index of relative risk. Thus widespread use is made of the concept of effective dose when comparing the patient dose implications of different diagnostic procedures. Effective dose is then the estimated energy /kg deposited in the particular tissue or organ multiplied by the organ-weighting factor, W_R. The unit is the Sievert. Thus, since gonads have W=0.2 but the breast has only W=0.05 the same energy per unit mass eg 1mGy, absorbed in these two structures results in effective doses of 0.2mSv for the gonads but 0.05mSv for the breast. In the tables shown in section 4 we have made use of effective dose for our comparisons.

3.3 The Biological Effects of Radiation

Between the initial impact of ionising radiation on any animal and the development of a harmful clinical symptom there are a series of about five linked stages. Conceptually the succession of stages may be viewed as a progressive increase in scale and complexity of the objects involved. The various stages discussed below are summarised in figure 3.1. The progression carries us forward in time but backwards in understanding. As soon as we leave the initial atomic stage, our understanding begins to drop extremely rapidly. By the time we reach the very long term stage, basic understanding is very limited indeed, largely as a result of the basic lack of understanding of cell ageing and carcinogenesis. What follows is a very brief summary of the main points in each stage. This account does not do justice to the wealth of work and detailed knowledge in the area of radiation biology. The reader is referred to the bibliography for more complete treatments.

The process starts at the single atom level with the production of energetic photoelectrons according to the principles outlined in chapter 2. This is followed by local chemical changes along the track of the incident photon or particle. Perhaps the most important of these is the production of highly reactive free radicals but we must also include molecular excitations and chemical bond breakages, caused directly by the photoelectrons. We then pass to the bio-molecular stage in which these local chemical changes manifest themselves in damage to their host or neighbouring macroscopic molecules, particularly DNA. Early deterministic symptoms occur on a local scale as a result of cell death or an inability of the affected cells to divide and reproduce successfully. If the extent of the radiation exposure is sufficient then the acute symptoms and finally animal death results from the accumulation of local cell deaths having a harmful effect on a wider population of initially healthy cells.

If the radiation insult is insufficient to cause animal death then a large fraction of the cellular damage can be repaired. Months, years even decades later, a small fraction of the cells damaged in the initial insult may begin to divide in a mutated fashion, leading to malignancy. When ionising radiation is delivered to an individual's gonads, then the long term effect may appear first in the unexposed offspring as a genetic mutation, to be followed, much later by a long term stochastic effect in the exposed individual.

Stage 1 the Atomic Level

The important factor in the initial atomic stage is the density with which the ionisation is deposited along the track of the incident particle. The initial photoelectrons are emitted with energies comparable to those of the incident photons or particles and thus these have more than enough energy to ionise further atoms in their path., see figure 2.4. There results a cascade of short-ranged photoelectrons with progressively smaller kinetic energies. Each emitted photoelectron leaves behind an atomic ion, which is also charged. This process is described quantitatively by the linear energy transfer or LET associated with the type of radiation. In general heavy charged particles such as protons or alphas interact very strongly with any matter and are brought to rest within a few hundred microns. These radiations, along with neutrons, are said to give high LET. The region over which photoelectrons are produced is very small, thus around this track there will be a very high density of short-ranged photoelectrons and thus a high concentration of damage. Cells along or near to this track are unlikely to survive. X-rays and γ–rays interact much less strongly with matter, they have mean paths of many centimetres and thus they are low LET radiations. Roughly speaking the energy lost from a diagnostic x-ray beam, through Compton scattering, is spread over the entire path length through the body. Electrons and positrons although very light are charged, and hence even the 1 MeV positrons used in PET have mean free paths of only a few millimetres. Thus these are also termed high LET radiations. The production of the photoelectron shower takes less than 10^{-12} sec. Once produced, the electron-ion pairs diffuse out of the initial track region over a period of about 10^{-6} sec, finally losing the energy in chemical reactions along their paths.

Stage 2 Chemical Interactions

The dominant source of chemical disruption is the interaction between the shower of many energetic photoelectron-ion pairs of varying energies and the molecules of the exposed tissue. We can distinguish two sources of damage; one direct and, a generally more important, indirect effect. The direct effects arise when a molecule receives a direct hit from a photoelectron. It is likely that a chemical bond will be broken or the molecule will be excited into an unstable state that subsequently leads to chemical bond disruption. If the target molecule is DNA then base changes or strand breaks can lead to significant damage to the cell. The indirect effects begin in the water molecules that make up 70- 90% of all biological tissue. When an electron is ejected from a water molecule, a positively charged molecular ion and a free electron are formed. The free electron, once slowed to thermal energies, becomes trapped by another water molecule to form a negative molecular water ion. Molecular ions are unstable and rapidly disassociate to form an ion and a free radical. A free radical is an uncharged molecular species that has become extremely chemically reactive by virtue of a single unpaired electron in one of its molecular orbits. The hydrogen atom is the simplest free radical, it is uncharged but has a single unpaired electron. Its voracious affinity for other hydrogen atoms and oxygen are well known. Given the chance, any free radical will form a chemical bond with almost all other atoms and molecules found in biological tissue. Once a free radical has bonded to another molecule, the whole is of course a different molecule, sometimes with a completely different chemistry. The formation of free radicals is thought to be an essential part of nearly all chemical reactions. If present in the wrong place in significant quantities, free radicals can set off a chain of molecular events that lead to cell death.

The free radical hypothesis probably explains a fundamental phenomenon of radiobiology called the oxygen effect. All tissue is more sensitive to radiation damage when oxygen is present. Dissolved oxygen in tissue decreases the radiation dose required to elicit any observed cellular effect. In one scenario free radicals are part of this story. Oxygen reacts with free radicals such as H^* to reform water and toxic hydrogen peroxide. Alternatively if an organic molecular free radical is formed, then its reaction with oxygen can set up a chain reaction that can rip through an assembly of molecules, with each individual reaction creating another free radical to seed the next stage. Under these conditions a small initial quantity of radiation induced free radicals becomes amplified, increasing the amount of biological damage that is eventually done. The oxygen effect emphasises the role of complex indirect interactions involved in radiation damage and perhaps provides a partial clue for the explanation of long term delayed effects of very low level radiation exposure. All chemical changes induced by radiation are essentially complete by 10^{-3} sec.

Stage 3 Cellular and Whole Animal Changes

Once molecular changes have been brought about by radiation, three classes of event may follow. The irradiated cells may die as a result of a total dislocation of the internal metabolism. The cells may survive, but with an impaired or destroyed ability to reproduce. The cells may survive and reproduce, but with a slightly altered metabolism, in other words they have become mutants. All three occur to some degree, apparently at all levels of irradiation. The timescale, and the extent of emergent clinical effects in the whole animal depend on the extent of the damage. Even relatively large-scale damage can, and is, repaired by natural metabolic processes. In radiotherapy natural repair is both a hindrance to the effective destruction of some tumours and the means by which the healthy tissue

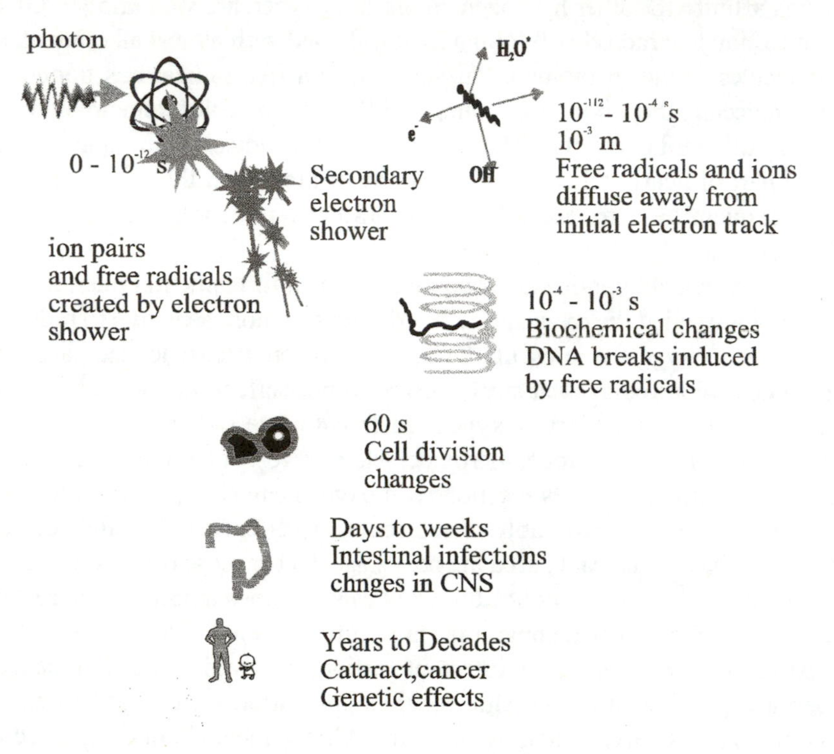

Figure 3.1: The physical changes taking place between the initial radiation exposure and long term biological effects. The primary energetic photoelectron produces a shower of lower energy photoelectrons and associated ions in the immediate vicinity of the primary track. The charged pairs in turn create free radicals and excited molecules, which diffuse away from impact region. After about 1 second biochemical reactions, involving macromolecules including DNA, are complete. After about 1 minute changes in cell division begins. If the exposure is large then the accumulated effects of damaged mutated cells begin to produce short-term acute effects, radiation sickness, within days to weeks in the gut and the central nervous system. The long term stochastic effects of genetic damage and cancer appear after many years.

surrounding a tumour can be kept viable. Both must necessarily be irradiated in order to harm the tumour, but if the dose is administered in parts rather than as a single large dose (fractionation) then the slightly different repair rates of healthy and cancerous tissue can be exploited to keep the former viable while destroying the latter.

Clinical Effects of Ionising Radiation

Ionising radiation causes a mixture of immediate, short term and delayed clinical effects that vary depending on the severity of exposure. For absorbed doses of less than 1 Gy, the immediate clinical picture resembles that of a severe case of sunstroke, with the patient feeling non-specifically unwell with nausea and vomiting. Over the weeks following exposure, blood tests will reveal that the number of white blood cells, initially the lymphocytes but later all white blood cells, and the number of platelets fall. There are usually not any associated complications. In the long term patients exposed to this level of radiation have an increased risk of developing tumours, particularly the tumours of blood cells, the leukaemias. As the absorbed dose increases above 1 Gy the outlook for patients becomes worse. The organs that are most sensitive to the effects of ionising radiation are those, which contain cells that are dividing relatively quickly, the bone marrow, the skin and the gut. Blood cells all have a limited life time, the average life span of a red blood cell being 3-4 months, and the time-expired cells are replaced by newer ones that are produced by the division of a set of primitive bone marrow cells. Cells in the deeper layers of the gut lining, the skin and hair follicles are also dividing continually to replace the cells that are lost from the skin and gut surfaces. Once exposed to sufficient amounts of ionising radiation, the manufacture of new skin, blood or gut cells is halted; however, the cessation of cell production will not be noticed until reserves have run out. The initial clinical effects of the absorption of between 4 and 10 Gy is therefore similar to the non-specific picture described above. In fact, patients may seem to be getting better for the first few days, before the life threatening complications begin. The first problems encountered result from the failure to replace the gut lining and skin cells, leading to intractable, bloody, diarrhoea with a loss of body fluids, the total loss of body hair and skin blistering. Even if the patients are supported through this stage of the illness the worst is yet to come as, in the weeks following exposure, the reserves of blood cells gradually run out. Without white blood cells and platelets, the patient is unable to combat infections or stop bleeding. Many patients will not survive. Of those that do, some will be infertile while others will pass damaged genetic material to their offspring. The survivors will all have an increased risk of tumours and will develop cataracts. The outlook for patients who absorb more than 30 Gy is even worse. These patients soon lapse into a coma and die, within 36 hours, as a result of brain swelling.

3.4 Typical Medical Doses and Dose Limits

The regulation of exposure to ionising radiation has been in operation for most of this century. Continued research has resulted in a strong downward trend in the legal levels of exposure to both the general public and occupationally exposed radiation workers. The overall philosophy is best summarised by the ALARA principle described earlier. In this spirit the general public gains no benefit from exposure to radiation outside the medical clinic and thus levels of incidental exposure are set to be comparable to the very low average level of unavoidable natural background radiation, arising from cosmic rays and terrestrial radioactivity. The statutory limits on radiation exposure to the general public, excluding medical investigations are, for the present anyway, insufficient for medical diagnostic procedures and many orders of magnitude smaller than the requirements for radiotherapy. Thus clinical procedures have their own set of recommendations, but no statutory limits are set on patient exposure. Although all medical investigations using ionising radiation incur patient doses that greatly exceed the natural background rate this is not a cause for alarm. It is estimated that the risk of a cancer arising from radiation exposure of any source is about 0.05 Sv^{-1} averaged over the whole population. Even the largest organ doses following diagnostic procedures amount to tens of milli Sieverts and thus the associated risk for carcinogenesis amounts to only 50 to 100 in a million. Such tiny risks are often put into perspective by comparing them with hypothetical equivalents in everyday life. Thus similar risks arise from smoking 75 cigarettes, rock climbing for 75 mins or travelling for 2500 miles by car. Any state authority has however to multiply these tiny numbers by the total numbers of patients examined each year. In the UK this amounts to 35 million x-ray investigations. Inevitably the medical use of ionising radiation leads to a small increment in the number of cancers that would develop anyway. The increment is small but the very high cost of treating and managing cancer patients ensures that radiation safety and dose reduction will always remain a high priority for the protection agencies.

In the following tables we provide accepted estimates of radiation doses arising from medical procedures and compare them with those arising from cosmic rays and terrestial radioactive radon. It should be noted that the accumulated general population dose arising from all medical uses of ionising radiation amounts to only 12% of the average total of 2.1 mSv per year. Radon contributes 47%, cosmic rays 10%. Manmade sources make up about 25% of the total, roughly equally divided between medical use and sources ingested in our food and drink. Discharges from nuclear installations only contribute 0.1% and

Table 3.1 The estimated averaged radiation dose of the general population arising from both natural and manmade sources

Source of Radiation	**Dose**
Average Annual Dose	
Natural radioactivity in air	0.80 mSv / year
Natural radioactivity in buildings and ground ,	0.40 mSv/year
Cosmic rays	0.30 mSv /year
Food and Drink	0.37 mSv/year
Nuclear Testing	0.01 mSv/year
Nuclear power	0.002mSv/year
Medical sources	0.25 mSv/year
Total from all sources	**2.14 mSv/year**
Legal Dose Limits for Radiation Workers	
Whole body	50 mSv/year
Foetus	1 mSv

fallout from nuclear testing 0.4%. Thus medical uses of ionising radiation do contribute to a significant increase in the averaged exposure of the general population and thus remain a matter, not for concern, but continued vigilance and study. Table 3.2 shows typical radiation doses arising from a variety of medical investigations involving x-rays and nuclear medicine. We have provided for each investigation, after Perkins (1995), a conversion to an equivalent number of chest x-rays and an equivalent period of exposure to the natural background. It should be noted that investigations involving CT and fluoroscopy generally entail a much greater dose. In CT this arises from the large number of exposures required for image reconstruction. In fluoroscopy large doses can arise in surgical procedures that last for an hour or more. The use of the image intensifier greatly reduces the dose required to get a single image. As a rule of thumb 10seconds of screening amounts to one chest x-ray. Fluoroscopy is perhaps the only diagnostic procedure, using ionising radiation, that can routinely lead to short term clinical symptoms such as skin burn and hair loss.

3.5 Practical Radiation Protection and Monitoring

Over the course of this century, the average doses of radiation received by radiation workers both in industry and in medicine have dropped by several orders of magnitude. This is the result of education, research, legal measures, and steady social pressure as well as technological innovation. In fact the power of radiation sources has increased dramatically in this same period, so has the frequency of use of ionising radiation in both industry and medicine. Perhaps the most important measure is that of education. It is practically impossible to eliminate entirely the exposure of staff to radiation in the vicinity of a radiation source. Many fail-safe mechanisms have been invented and installed to reduce or even prevent gross accidental exposure to powerful direct beams, but as soon as an x-ray beam leaves the generating tube, some photons, scattered by apparatus, walls and indeed the personnel present, fill the nearby space with a low level "fog " of ionising radiation. The long term safety of staff then depends on their own knowledge and common-sense and their having an accurate record of their accumulated exposure.

External Protection Measures

The three main factors in effective protection from radiation exposure are distance from the source, the time spent in the radiation field and finally shielding.

Table 3.2 Typical Effective Doses produced by x-ray and nuclear medical investigations related to equivalent numbers of chest x-rays and natural background doses (adapted from Nuclear Medicine : Science and Safety , A. Perkins (1995)

Investigation	Effec Dose	Wr	No of chest x-rays	Period of natural exposure
Radiography				
Dental	0.01	0.01	0.5	1.5 days
Chest	0.02	0.12	1	3 days
Skull	0.1	0.05	5	2 weeks
Dorsal Spine	1.0	0.01	50	6 months
Lumbar Spine	2.4	0.01	120	14 months
Barium Studies				
Oesophagus	2.0	0.05	100	12 months
Large Bowel	9.0	0.12	450	4.5 years
CT				
Chest	8.0	0.12	400	4 years
Brain	2.0	0.01	100	12 months
Thoracic Spine	6.0	0.01	300	3 years
Lumbar Spine	3.5	0.01	175	1..8 years
Nuclear Medicine				
^{99m}Tc Studies				
Bone imaging	3.6	0.12	180	1.8 years
Cerebral perfusion	4.5		225	2.3 years
Myocardial perfusion	5.0		250	2.5 years
Gastric emptying	0.3	0.12	15	2 months
Thyroid imaging	1.0	0.05	50	6 months

A x-ray beam is generally very intense and collimated in a well defined way. Its mean free path in air is many meters and so staff must always avoid intercepting the direct beam. In addition the direct beam must encounter a heavy beam stop to prevent it carrying on through walls, floors and ceilings, into adjacent uncontrolled working areas. Photons are scattered more or less isotropically away from the beam direction when it enters a patient or solid structures. The intensity of this scattered radiation decreases roughly inversely as the square of the lateral distance from the beam. Thus keeping staff a safe distance from activated x-tubes significantly reduces their exposure to radiation. In most radiography suites, the radiographers actually leave the room or retire behind a lead-shielded screen while the beam is on. Where x-rays are used in surgery, (screening) this is of course impractical. Where radioisotopes are being handled, the simple strategy of using tongs to handle sources significantly reduces the dose from the localised source. Practice manipulations, using dummy sources, can significantly reduce the time taken for a particular operation and thus the total exposure.

An individual's occupational exposure to radiation is clearly dependent on his/her working practice and will vary with the type of task involved. It is important that not just the average exposure is controlled but also any peaks contained in that average. This is where personal exposure monitoring plays an important part. Personal occupational exposure is monitored using film badges and pocket ionisation chambers.

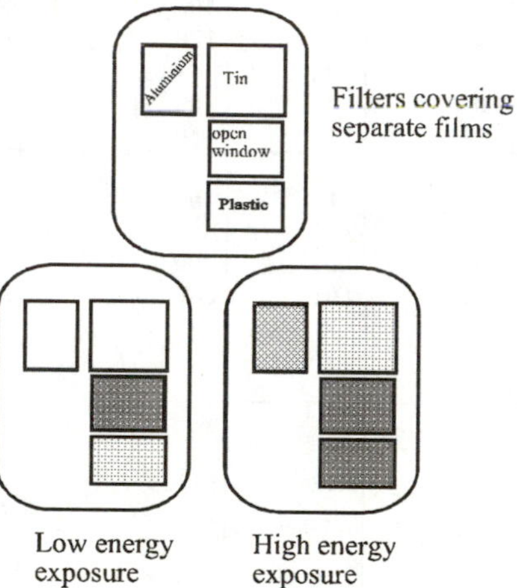

Figure 3.2: The personal radiation film badge monitor. The monitor consists of a piece of photographic film overlaid with small sheets of different materials, which act as radiation filters. Typically these are tin, aluminium, and plastic. After exposure the degree of blackening of the film can be related to the total radiation to which the film was exposed. The degree of differential blackening under the various filters provides a measure of the photon energies in the radiation.

The badges provide the central radiation protection administration with the means to regulate the total exposure of an individual over integrating periods of weeks to months. The pocket meters provide an immediate personal check on the rate of exposure. There are well-defined regulations governing both total occupational exposure and its rate of accumulation. Special regulations govern the young and women of childbearing age and pregnant women.

X-rays and γ–rays are stopped very efficiently, via the photoelectric effect in high Z solids such as lead and tungsten, see section 2.4. Thus thick lead enclosures are used to cloak x-rays tubes, reducing all stray, unwanted radiation to acceptable levels. The low energy, soft x-ray component of the x-ray tube output is removed again by the photoelectric effect in beam filters fitted to the exit aperture of the tube, see figure 4.10. Lead aprons are worn by all radiographers to reduce the amount of scattered radiation reaching the body surface. All modern medical x-ray installations are fitted with adjustable lead diaphragms so that, in each examination, only the region of interest is exposed to the beam. This reduces the exposure and dose to the patient and the exposure of staff to scattered radiation.

Dose Trends in Modern Diagnostic Techniques

The exposure of the general public to ionising radiation, outside the medical clinic is governed by law, as is the occupational exposure of radiation workers both in medicine and industry. The exposure of patients on the other hand is not covered by law, only by a series of recommendations. Whereas it would be impractical to legislate for a particular exposure limit for every possible type of examination the largely unregulated situation did, in the past, lead to an unsatisfactory state of affairs regarding patient doses. It is only in the past twenty years that medical radiation exposure audits have been carried out and some obvious shortcomings revealed. The widespread use of new technologies such as CT and fluoroscopy, while providing very much more diagnostic information, entail greatly increased patient doses which buck the overall trend towards lower radiation exposures to both patients and staff.

In one UK study, carried out by the NRPB and published in 1986, it was reported that there were variations by factors of 20 in the exposure resulting from the same examination in different hospitals and factors up to 100 for different patients, attending the same hospital for the same examination. The reasons for these variations were both technical and procedural. Old equipment, the widespread use of slow intensifying x-ray screens, see section 4.4, together with local preferences for types and number of images, all contributed to the variability. It was estimated that the total amount of medical patient exposure to radiation in the UK could be reduced by 1/3, without prejudicing the medical information gained. More recent surveys have shown that progress has been made in this direction in the traditional radiographic areas.

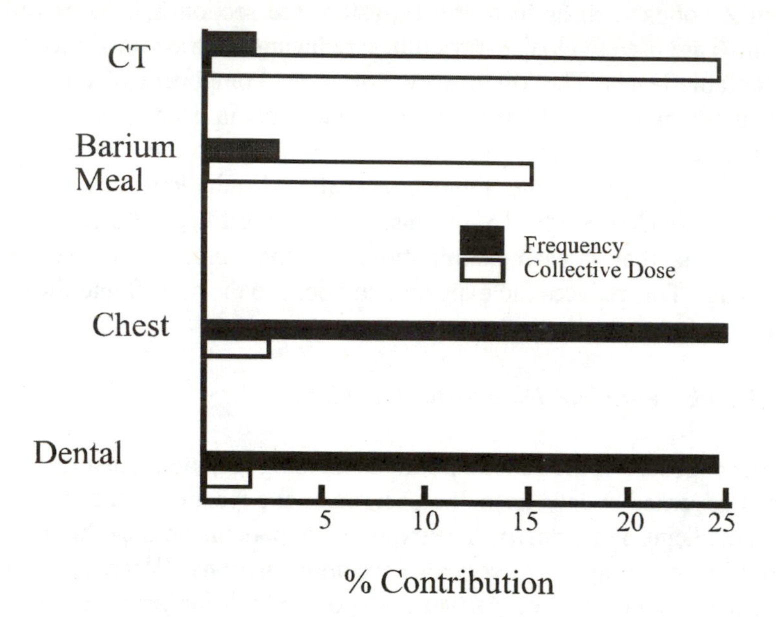

Figure 3.3: Estimated contributions to UK diagnostic radiology practice. The total collective dose from all medical radiological examinations is estimated to be 20,000 man-Sv each year. It is clear that CT is used relatively infrequently in comparison with dental x-rays but makes up 25% of the total diagnostic radiation dose. Adapted from Valetin (1993)

Modern digital developments such as x-ray CT and fluoroscopy both entail significantly higher doses than a single chest or dental x-ray. As we describe in section 4.5, the vastly improved tissue contrast and diagnostic power of CT over projection radiography comes with the penalty of relatively high patient doses. This is inevitable since each tomograph is comprised of the equivalent of at least 100 projection radiographs. This is reflected in the entries in table 3.1. Another aspect of the same issue is the effect of CT on the annual collective medical dose. In the UK this estimated to be about 20,000 man Sv. About 25% of this total is contributed by CT, even though it is used in less than 2% of all x-ray examinations. The radiation dose associated with x-ray CT is one reason why MRI, which entails no ionising radiation, has been developed and introduced into clinical practice so quickly.

A similar scenario results from screening or fluoroscopy. X-rays are used to enable operations to be carried out using small tools inserted into the body through natural orifices or minor incisions. An image intensifier or fluoroscope provides an essential tool for the surgeon, giving her real time views of the positions of instruments. The image intensifier does a remarkable job at reducing the required power of the x-ray beam. However, complex operations last for many minutes if not hours and thus there is significant exposure of both patient and theatre staff. A typical x-ray radiograph uses an exposure time of less than 1 second. Some procedures requiring fluoroscopy can take an hour to complete. In addition although it is recommended that x-ray screening in such operations is done intermittently, in fact, in many cases the screening is nearly continuous. Medical physicists use a rule of thumb that 10 seconds of x-ray screening is equivalent to 1 chest x-ray. Using the values shown in table 3.2 it can be seen that in an hour long operation with x-ray screening, the patient could easily accumulate 360×0.02 mSv = 7.2 mSv, possibly the highest dose in diagnostic radiography.

3.6 Safety in Ultrasound and MRI

Both Ultrasound and MR imaging procedures are achieved without the use of ionising radiation and are thus generally thought of as being "safer" than x-ray or nuclear medicine methods. The effects of ionising radiation are not however the only ways in which biological tissue can be harmed both instantly and in the longer term. Both Ultrasound and MRI impart energy to the patient and thus potential hazards exist. In neither case is there any compelling evidence for long term harmful effects even though in both cases very minor and temporary functional and anatomical changes can be demonstrated at the power levels currently used in imaging. Both types of procedure are covered by well-documented guidelines drawn up by the radiological protection agencies covering recommended maximum power levels and the use of the procedures on the foetus and pregnant mother.

Ultrasound

The relatively high rate of energy absorption of ultrasound energy by biological tissue could spell trouble, since unwanted energy is being deposited in the body. The experience with x-rays and all forms of ionising radiation has encouraged a cautious approach on the part of both clinical medicine and engineering. Although ultrasound produces little or no ionisation at imaging power levels, there are very likely to be small amounts of irreversible mechanical change brought about by imaging. At very much higher power levels, the ultrasound wave can certainly destroy even quite robust tissue. Ultrasound wave energy is transferred to tissue both as random heat motion through induced molecular vibrations and more organised induced motions of cellular structures and changes in fluid flow. In the early days of its use in obstetrics, ultrasound was used quite casually to observe the foetus. Now ultrasound investigations are generally limited to just one scan in a normal pregnancy, in order to check normal development and make a reasonable prediction of the date of birth.

MRI

An MRI investigation imposes three types of magnetic field on patients; a large static field, a switched gradient field and a radio frequency field. Although there are well-documented minor physiological effects associated with exposure to each of these fields it is only the last, the radio frequency field, that appears to be capable of direct damage to tissue. An indirect hazard arises from large static magnetic fields if magnetic objects, such as surgical instruments, are allowed to be pulled into the magnet and thus become high-speed projectiles.

Leaving aside for the moment any hazard arising from magnetic projectiles, the most important safety issue in MRI is the electrical power deposited in tissue by the RF fields. The direct result of RF exposure is tissue heating particularly at the body surface, where the field is most intense. Statutory safety regulations, governing MRI , limit any heating effects arising from RF fields to less than 1°C in any exposed tissue. The RF field intensities required for clinical imaging can approach this level and thus commercial MRI control software automatically vetoes any pulse sequence that would exceed this limit. The MRI radiographer has to ensure that patients and subjects entering the scanner do not have any metallic prostheses inside their bodies since these parts would absorb considerable amounts of RF power and hence rise in temperature by much more than 1°C.

The gradient field if left on continuously, like the much more intense static uniform magnetic field, does not apparently produce any significant biological effects. However in modern imaging sequences the gradients are switched on and off at frequencies of about 1 kHz. The resulting changing magnetic field induces eddy currents throughout the body of the patient (via Faraday's law, see section 2.2). These currents, if allowed to exceed physiological thresholds, can induce muscular contractions and neural excitation. The current regulations limit the maximum permissible rate of change of magnetic field in MRI to 3T s^{-1}. Clinical practice could certainly exceed this limit and so again software vetoes any sequence that would exceed the limit. Short-term effects arising from neural and muscular excitation have been observed but there is no evidence that either the effect persists or leads to other long term complications.

The large static field is potentially a source of hazard as we have already mentioned. Some installations have entry points automatically monitored for the presence of magnetic objects on or in patients using the same type of equipment that is used in airports. In future applications of open magnet MR in the operating theatre, this type of hazard will become very much more important. The magnetic field surrounding a typical MRI magnet decreases rapidly, approximately like a dipole field, but remains considerably above the earth's field of 5.10^{-5} T for a few metres radius from its centre. Strong magnetic fields are known to interfere with heart pacemakers and thus the installations are designed so that the general public cannot unwittingly step inside the $5.\ 10^{-4}$ T contour.

Questions and Problems

1. How are radiation exposure and dose defined and measured ?
2. The old unit of radioactivity, the Curie, is a very large quantity of radioactivity. How many disintegrations per second does it represent? The activity of a typical dose of radioactive tracer used in diagnostic nuclear medicine is a micro Curie. How many disintegrations per second does this represent?
3. A 500 MBq source, with a half life of 10 min, of gamma rays with photon energy of 100 keV is injected into a patient weighing 150 kg. What is the activity 20 min after the injection of the bolus? If 10% the gamma rays are absorbed in the patient what whole body dose after 20 minutes. What is the total radiation dose, assuming that the metabolic clearance rate is of the order of days?
4. Estimate the fractional number of 50 keV x-ray photons penetrating a lead absorber 1 mm thick. What thickness would be required to reduce the amount of transmission to 0.1%
5. An x-ray beam with an intensity of 10 watts cm^{-2} is incident on a human leg weighing 20 Kg. Estimate how much power is deposited in the leg assuming that the absorption coefficient is 0.2 cm^{-1}. If the exposure lasts 0.2 sec estimate the dose to the radiation to the leg.
6. A 1 Curie radioactive source delivers a dose of 10 Gy per minute at a distance of 1 m. Assuming that the emission is isotropic, estimate the distance at which the dose rate from the source is less than the average background radiation dose rate.
7. What is the oxygen effect? How might the concept of free radicals account for this effect?
8. Why is ionising radiation used in the treatment of cancer?
9. Large doses of radiation used in therapeutic applications are frequently delivered in instalments? Why is this necessary
10. Explain why the effects of very low level ionising radiation, have in the past been the subject of controversy and legal argument

4 X-RAY RADIOGRAPHY

100 YEARS AGO

' The Newspaper reports of Prof. Rontgen's experiments have, during the past few days ,excited considerable interest. The discovery does not appear to be entirely novel, as it was noted by Hertz that magnetic films are transparent to the kathode rays from a Crookes or Hittorf tube, and in Lenard's researches, published about two years ago, it is distinctly pointed out that such rays will produce photographic impressions...

Prof. Rontgen has extended the results obtained by Lenard in a manner that has impressed the popular imagination, while perhaps most important of all, he has discovered the exceedingly curious fact that bone is so much less transparent to these radiation than flesh and muscle.'

Nature 23 January 1896

4.1 Introduction

X-ray imaging depends on the partial translucence of biological tissue with respect to x-ray photons. If a beam of x-rays is directed at the human body, a fraction of the photons will pass through without interaction. The bulk of the incident photons on the other hand will interact with the tissue in the ways described in chapter 2. As the beam is moved across the body, the relative proportions of transmission, absorption and scatter change, as the beam encounters more or less bone, blood and soft tissue. Ideally the contrast in an x-ray image would be produced just by the variation in the number of photons that survive the direct journey to the detector without interaction. These are the primary photons. They have travelled in a straight line from the x-ray tube focus and will give rise to sharp edged shadows.

However each point in the image may well receive additional photons that have been scattered by body tissue, see section 2.3. These are the secondary photons. They reach the detector from a wide range of positions in the body and they smear out the sharp shadows formed by the primary component. Medical imaging uses x-rays with energies in the range 20 -150 keV. These are very much higher than most atomic binding energies and thus the inelastic interaction processes of Compton scattering and the photoelectric effect launch charged electrons and ions, with relatively large initial kinetic energies, into the surrounding tissue to produce the patient dose in the manner described in section 3.3. The radiation dose of both the patient and attendant staff

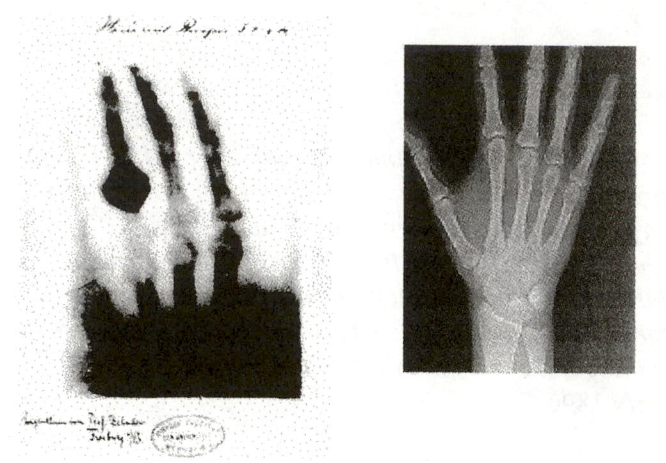

Figure 4.1: A century of x-ray radiography.The left hand panel shows first x-ray radiograph of human bones, Frau Rontgen's hand published in 1896. The large blob is a wedding ring. The right hand panel shows a modern radiograph obtained using x-ray fluoroscopy (Seimens)

must be kept to an absolute minimum and this has a significant impact upon both the design of equipment and the procedures used in x-ray radiography. An ideal imaging scheme would employ photons that are only elastically scattered by biological tissue and never absorbed, but wholly absorbed and never scattered by the chosen photon detector. This ideal is of course impossible. Elastic scattering and absorption mechanisms are significant to differing degrees at all photon energies for any given atom. In general a standard x-ray radiograph provides extremely good contrast between bone and all soft tissue but very little contrast between the different types of soft tissue, see table 4.1. In addition, the standard radiograph is a 2D projection of a 3D patient. This means that shadows formed at different depths within the patient are all superimposed in the final image. Together these two problems impose severe limitations on the amount of diagnostic information that can be obtained. Throughout its hundred year history radiographic techniques have been invented by scientists and clinicians to improve the diagnostic yield. X-ray absorbing dyes can be used to improve the visibility of particular organs or the blood vessels. This technique is called angiography. The development and introduction of x-ray computer assisted

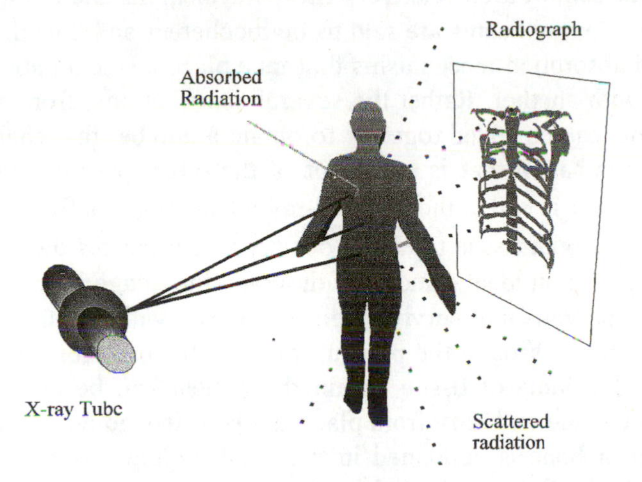

Figure 4.2 : A typical x-ray radiographic geometry. X-ray photons generated by the tube are directed at the patient. A fraction of the photons pass directly through the body to create a 2 dimensional projection of the exposed anatomy.

tomography x-ray CT solved both of the main problems; the CT scan is capable of resolving fine differences in attenuation coefficient and it provides a series of imaged slices through the patient so that the anatomy can be studied in 3D. The major drawback of the CT scanner is that the increased diagnostic information comes at a cost of a considerable increase in patient dose per examination.

All the factors that effect image contrast and sharpness such as exposure time, choice of radiation, and type of projection are very well understood. Manuals for radiologists and radiographers contain very detailed recipes for clinical applications and test procedures that ensure the correct adjustment of the equipment. In this chapter we examine the main features of modern x-ray radiography at an introductory level and explain how the physical interaction mechanisms described in section 2.3 govern the engineering and practical choices that are made in clinical practice.

4.2 The Linear X-ray Attenuation Coefficient

As a result both of their relatively short wavelengths and the manner of their production, x-ray beams are said to be incoherent and thus the various scattering and absorption mechanisms that take place inside a patient do not interfere one with another. Rather the several contributions, from any small region of tissue, can be added together to obtain a number that characterises the amount of radiation that is taken out of the direct primary transmitted beam. This number is called the linear x-ray attenuation coefficient, μ. It is the variation of μ with tissue type and density that produces the contrast in any x-ray image. In an ideal standard radiograph the image is produced just by the primary photons that survive their entire trip, without any scattering, from the x-ray tube, through the patient and onto the detector. We imagine that each small volume of tissue within the patient can be given its own value of μ. The value will vary from place to place depending on how much muscle, blood or bone is contained in the small volume. A representative path through the patient, taken by the x-rays will cross many such volumes in a row as shown in figure 4.4. Each little volume will add to the attenuation produced by its predecessors along the beam according to

$$I_n = I_{n-1}\, e^{-\mu(X,Y,Z)\, dx} \qquad \text{...4.1}$$

where $\mu(X,Y,Z)$ is value of the linear attenuation coefficient of the nth cell

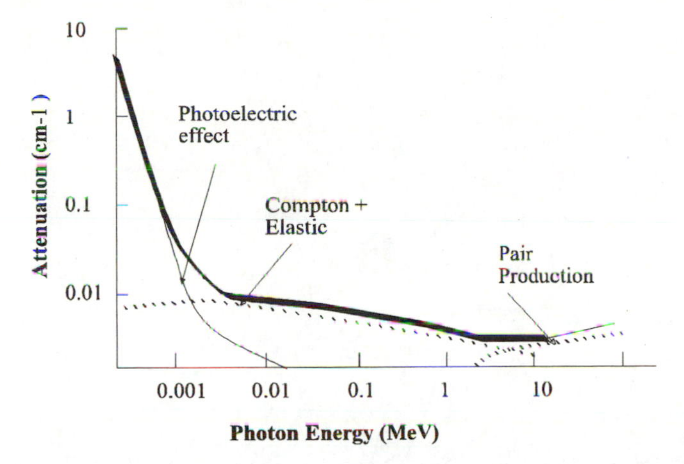

Figure 4.3 The variation of linear attenuation coefficient with photon energy. This curve is produced by the summation of the several mechanisms. We can consider three regimes of energy. At low energies (< 10 keV), the photoelectric effect dominates absorption for all atoms. At very high energies (>1MeV) electron-positron pair production (the inverse of electron-positron annihilation) is the dominant process. In the middle section elastic scattering, Compton, and photoelectric effects all contribute in amounts that vary with atomic composition of the tissue and to some extent with photon energy. Pair production does not enter into x-ray radiology because the photon energy used is too low.

which has a length dx, and , I_{n-1} is the x-ray intensity arriving at this cell, having already survived n-1 cells. The point in the image produced by this particular beam or ray will be determined by just that fraction of the photons which survive the entire trip through the patient. Thus we can write

$$I_{image} = I_o \cdot e^{-\mu_1 \delta x} \cdot e^{-\mu_2 \delta x} \ldots . e^{-\mu_n \delta x}$$

$$I_{image} = I_o e^{-\int_{path} \mu(X,Y,Z)\, dx} \qquad \ldots 4.2$$

Where we have written the integral over the straight-line path taken by our selected x-ray. As we move about in the ideal image, the transmitted x-ray

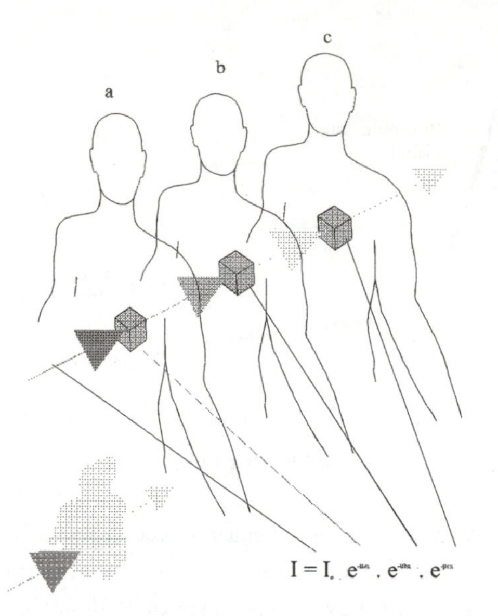

Figure 4.4 : The progressive attenuation of an x-ray beam by successive volumes. We imagine the patient volume to be split into a series of planes perpendicular to the x-ray beams. Here we show just 3 such planes labelled a,b,c for clarity. Along a representative path the beam is attenuated by each volume it passes through. In general each volume will produce a different attenuation. The point in the ideal image produced by this particular beam has an intensity that is determined by the attenuation over the entire path.

intensity varies because $\mu=\mu(X,Y,Z)$ varies throughout the body and the path length changes. The differences in image intensity at two neighbouring points in the image, that is to say, image contrast, depends on the attenuation along two complete lines through the patient rather than two points in the patient. Thus in order to see a very small object, its attenuation coefficient must be big in comparison, not just with a correspondingly small neighbouring point, but rather the integrated attenuation along an entire path. Given the very small differences between attenuation coefficients for soft tissue, small regions of change within soft tissue are very hard to see using a standard projection radiograph. By creating displays of $\mu(X,Y)$ itself rather than a projection, CT resolves this difficulty. Values of x-ray attenuation coefficient, mean atomic number, Z and density, ρ, for some biological materials are given in table 4.1.This table shows that x-rays are particularly good at distinguishing bone from soft tissue but rather poor at differentiating between the different types of soft tissue.

Throughout the previous discussion we have used the phrase "ideal image". This ideal would be achieved if there were no detected scattered photons. The mechanisms, which scatter photons also contribute to the attenuation along the line of sight from the x-ray focus and hence modify the resulting intensity at the particular point in the image. However the photons that leave one line of sight can end up in an unrelated part of the image. The ideal image is degraded in sharpness by these scattered photons since the scattered component of the transmitted radiation does not possess the simple straight line relationship between x-ray tube focus, patient point and image point.

4.3 The Factors Determining Image Quality

All medical anatomical images are taken with the aim of detecting abnormalities of anatomy resulting from injury or disease. The radiologist, whose job it is to examine and report on medical images, has to be able to spot very subtle changes in intensity and make a judgement about whether these changes are real reflections of pathological anatomy or simply artefacts of the imaging process. Given the range of normal anatomical variation, this task would be impossible, except in the most trivial of cases, without a rigid code of radiographic practice and very strict quality controls on the apparatus and materials used to produce the image. Good quality control of both clinical practice and materials requires precise definitions and quantitative measurements of the factors that affect the visibility of small features in the presence of larger scale background variations and noise. In all imaging uses of x-rays there is the additional constraint of radiation dose to both patient and staff that must be known precisely and controlled. All medical images, including x-ray images, are assessed by three quantities, the contrast, the degree of unsharpness (spatial resolution) and the patient dose that is incurred. Apart from the question of radiation dose, which does not apply either to MRI or ultrasound, the definitions and concepts described in this section are common to all medical imaging.

Table 4.1: The linear x-ray attenuation coefficient for some biological tissues

Substance	$\mu(cm^{-1})$	Atomic number <Z>	Density ($gm\ cc^{-1}$)
Air	0.0001	~7	0.0012
Water	0.1687	~7	1
Saline	0.1695	~7	1.0064
Muscle	0.18	~7	1.032
Blood	0.178	~7	1.036
Bone	0.48	~7	1.84
White matter	.1720	~7	1.0274
Grey matter	.1727	~7	1.0355

Contrast

At least qualitatively the idea of contrast is a familiar one. In a high contrast photograph, obtained in good light conditions, objects stand out and their edges are clearly defined. In poor light conditions, or worse in rain or fog, individual objects lose some of their clarity and their edges become softened or blurred. A simple model provides a quantitative description of this important concept. We define contrast in terms of the relative intensity change produced by an object. Referring to figure 4.5 we take the intensity at the middle of our target object to be I_1, and a reference point in the image, just outside the object in the background to be I_2,. Then the contrast, C, is defined to be:-

$$C = \frac{I_1 - I_2}{I_2} \qquad \ldots 4.3$$

We can use this definition to estimate what differences in attenuation coefficient will lead to visible objects in a radiographic image and what this might mean in terms of the dose to the patient. Each of the intensities in our expression for the contrast consists of two terms arising from the primary or line of sight x-rays, P, and the scattered x-rays, S,that reach the detector. Thus we write:-

$$\begin{aligned} I_1 &= P_1 + S_1 \\ I_2 &= P_2 + S_2 \end{aligned} \qquad \ldots 4.4$$

Using the symbols of figure 4.5 we can express the primary contributions in terms of the appropriate attenuation coefficients and thickness of tissue in

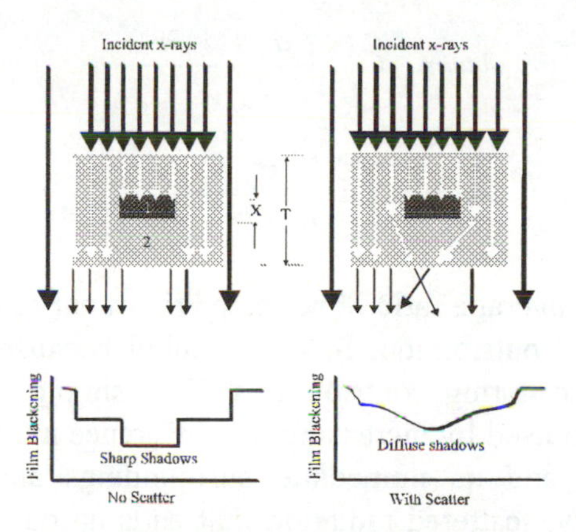

Figure 4.5: A simple model defining contrast in x-ray radiography. A small object of thickness, x, and linear attenuation coefficient, μ_2 is embedded in a larger uniform slab of thickness, T, and attenuation coefficient,μ_1. With no scattering (a) we obtain maximum contrast and very sharp edges in the intensity profile of transmitted radiation. With scatter (b) the contrast decreases and the edges of the profile become softened.(after Webb)

the beam. Thus

$$P_1 = N\ e^{-\mu_1 x}\ x\ e^{-\mu_2 (T - x)} \qquad \ldots 4.5$$
$$P_2 = N\ e^{-\mu_2 T}$$

Where N is the number of photons per unit area, incident on the upper surface of our slab of tissue. When we put the expressions for P into our definition of contrast we have

$$C = \frac{N\,(\,e^{-(\mu_1 - \mu_2)X}\ e^{-\mu_2 T} - e^{-\mu_2 T}\,) + S_1 - S_2}{N\ e^{-\mu_2 T} + S_1 + S_2}$$

$$\textit{using}$$

$$e^{-x} \approx 1 - x \ ;\ \textit{true} \quad \textit{for} \quad x \gg 1 \qquad \ldots 4.6$$

$$\textit{and} \quad \textit{assuming} \qquad S_1 \cong S_2$$

$$\textit{we obtain} \quad C \cong \frac{(\,\mu_1 - \mu_2\,)x}{1 + R}$$

$$\textit{where} \quad R \cong \frac{S_2}{N\ e^{-\mu_2 T}}$$

We have written the average ratio of scattered to primary radiation as R. We have left the scatter contribution, S, as a symbol because it is difficult to calculate accurate scattering contributions. This simple expression shows that contrast is maximised by increasing the difference in attenuation factors between the target and its immediate surroundings and decreasing the amount of secondary, scattered radiation that ends up on the film or in the detectors. Our equation shows that, other factors remaining constant, all x-ray radiographs benefit from a reduction in scatter.

Contrast Enhancement

Table 4.1 shows that, in general, there are only small differences in attenuation coefficient between different types of soft tissue. Thus a small soft tissue target imbedded in a larger mass of soft tissue is generally unobservable unless other measures are taken. Such measures are called contrast enhancement and involve staining a particular region with absorbent dye or, in some cases, by choosing a particular x-ray wavelength. In section 2.3 we saw that the x-ray photoelectric absorption coefficient increased very dramatically with atomic number Z. Table 4.1 shows that the effective atomic number for all soft tissue is about 7. If an organ or region of the body can be stained with a small amount of a heavy element, so that its effective atomic number deviates significantly from 7, then its attenuation coefficient can be increased dramatically as a result of the photoelectric effect. A good example of this is the injection of iodine into the blood stream just before the x-ray image is taken. Iodine increases the attenuation coefficient of the blood with respect to surrounding soft tissue to the extent that small blood vessels less than 1mm across can be imaged. Iodine has an atomic number, Z=53, whereas an average Z for soft tissue and blood is about 7. Combining the numbers in table 4.1 with the simple expression for C we obtain the following;

a) With no contrast agent in the flowing blood, a 1mm blood vessel would give a contrast, C of $(\mu_1 - \mu_2)\,x = 0.002\ \text{cm}^{-1} \times 0.1\ \text{cm} = 2.10^{-4}$ with respect to the surrounding muscle tissue, assuming that scatter can be neglected.

b) Blood containing just 1% of iodine (Z=53) solution, by volume, will produce an increase in C by a factor $\sim (53/7)^5/100 = 250$. This increase is more than sufficient to render fine blood vessels visible, see equation 2.6.

Scatter Reduction

The useful image forming primary radiation from an x-ray tube travels in a straight line from the x-ray tube focus through the patient to the detector. These "rays" form the perfectly sharp shadows that are crucial for optimum contrast. Scattered radiation on the other hand arises from any exposed region of the body and ends up at a position on the detector that bears no simple relationship to the position of the x-ray tube focus. All methods of scatter reduction exploit this geometrical difference between primary and scattered radiation.

Figure 4.6 illustrates some ways in which scatter reduction may be accomplished. Starting at the x-ray tube, diaphragms are used to limit angular spread of the beam and hence the field of view, so that only the region of interest is exposed. This eliminates possible scatter arising from regions of the body that are not part of the examination. In some cases, the region of the body can be deformed to reduce the path length of the x-rays and hence reduce scatter. This is a standard technique in mammography. If the distance between patient and detector is increased (the magnification factor) then scattered radiation is preferentially reduced with respect to the primary radiation. The scattered component tends to be distributed isotropically and thus its intensity falls off approximately as 1/ (patient-detector distance)^2. he primary radiation on the other hand is far less divergent. Finally, by using an anti-scatter screen consisting of thin lead sheets arranged edge on, scattered radiation is preferentially stopped in the lead. Increased patient to detector distance and anti-scatter screens reduce the primary radiation flux density as well as scatter and hence patient exposure dose has to be increased to compensate.

Detector Noise and Patient Dose

Our simple expression for the contrast ,C, is dimensionless and gives no hint of how many photons are required to pass through the patient in order to achieve this level of contrast. Any photon detector, be it a film or an electronic device only records a small fraction of the photons passing through it. In addition any detector introduces an inevitable degree of randomness or statistical noise into the recorded image that can only be reduced to an acceptable level by ensuring that a sufficient number of photons are actually recorded at any place in the image. Statistical noise is sometimes called quantum mottle because of the speckled or mottled appearance on film of featureless regions resulting from the random process of x-ray photon detection. In examinations in which small abnormalities are being sought such false detail in an image is of course unacceptable. In principle increasing the amount of radiation that is used in any examination could always eliminate the statistical noise problem. This is not always possible because of the biological hazard associated with ionising radiation. The radiation dose delivered to the patient is a critical factor in all examinations, using ionising radiation, and thus all examinations should comply with the ALARA principle described in chapter 3. In an ideal world

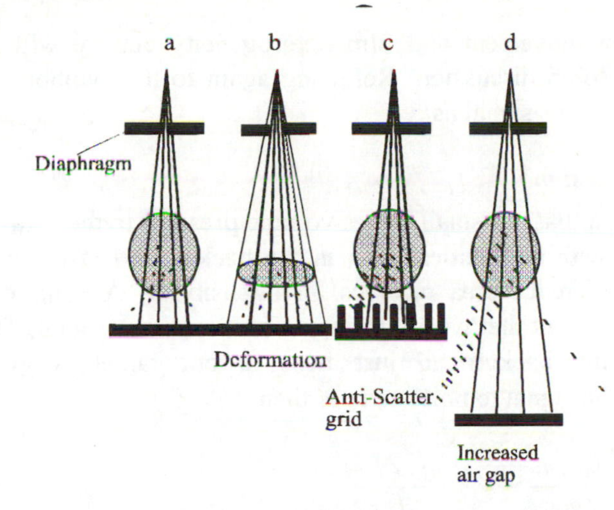

Figure 4.6: Methods of scatter reduction a) Diaphragms are incorporated into the x-ray tube housing to limit the angular extent of the beam and hence the volume of tissue exposed. b) The tissue can be deformed (flattened) to reduce scatter. c) Lead anti-scatter screens are put directly over the film that only let through x-rays travelling directly from the tube focus. d) Increasing the air-gap decreases the flux of scattered radiation relative to the primary beam.

all x-ray examinations would only utilise just enough radiation to resolve the required image feature at a level limited by the statistical noise created by the detection process.

We can use our model to obtain an approximate relationship connecting contrast, detector efficiency, spatial resolution and dose in the regime where statistical noise is the limiting factor. We are going to calculate the number of photons that have to be detected in order to "see" a signal in the image. At first sight it might appear that any number of photons, as long as they are detected might do the job. Unfortunately this is not the case. Photon detection is a random process whether it is achieved using photographic emulsion or electronic detectors. This means that if we detect N_d photons then there will be an uncertainty in the accuracy of N_d of $\pm\sqrt{N_d}$ arising purely from the statistics of random processes. Thus we have an irreducible amount of noise in the image due to these inevitable statistical fluctuations. There may well be other sources of image noise such as beam nonuniformity, patient movement and film heterogeneity but we will neglect these in this simplified discussion. Referring again to the symbols used in figure 4.5 we define our signal as

$$signal = (I_1 - I_2) \times A \qquad \ldots 4.7$$

Where A is the area that our small target volume presents to the x-ray beam. We compare this with an identical area in the background just adjacent to the target. This reference area receives an intensity, $I_2 A$ and it is the fluctuations in this count that we are trying to beat with our signal. Thus the image noise in the background just next to our target is given by noise $= \sqrt{I_2 A}$. The signal to noise ratio is then

$$\frac{Signal}{Noise} = K = \frac{(I_1 - I_2) \times A}{\sqrt{I_2 \times A}} \qquad \ldots 4.8$$

If we now use the expressions that we obtained above for C, I_1 we can write

$$K = \frac{C\, I_2 A}{\sqrt{I_2 A}} = C\sqrt{I_2 A}$$

$$since \quad I_2 = N\, e^{-\mu_2 T} \qquad \ldots 4.9$$

$$and \quad C = \frac{\Delta \mu\, x}{1 + R}$$

$$we\ have \quad K^2 = \frac{(\Delta \mu X)^2 \times A}{Ne^{\mu_2 T}(1 + R)^2}$$

$$if\ we\ chose \quad A = x^2\ and\ solve\ for\ N\ we\ get$$

$$N = \frac{K^2 (1 + R)^2\, e^{\mu 2T}}{(\Delta \mu)^2\, x^4}$$

Although idealised, this expression is extremely important in all photon imaging techniques. It tells us that the number of photons required to be counted for a given contrast and a given signal to noise ratio, K, increases as inverse fourth power of the size of the target object and the inverse square of the difference in attenuation coefficients between target and background. This means that if a target 2mm across is just visible, then the number of detected photons must increase by a factor of 16 to see a similar object that is 1mm across. Our ability to "see" visual signals in the presence of random noise is somewhat subjective, some trained individuals are very much better at it than others. A reasonable value for the signal to noise ratio, above detection threshold, is about K=5.

Spatial Resolution and the Modulation Transfer Function

The question of how big an object can be seen or resolved is a critical issue for all imaging systems. Most of us either have had or will have direct experience of this with our own eyesight. With age, our ability to focus on near objects diminishes, so that reading a newspaper without spectacles becomes increasingly difficult. In order to focus light from the printed page onto our retina, without the aid of spectacles, we have to hold the newspaper further and further away but then the details of the printed letters become smaller and smaller, in comparison with the intrinsic spatial resolving properties of the retina. Thus the letters become increasingly blurred and eventually are illegible.

In general spatial resolution (the size of the smallest clearly visible object) in any x-ray system depends in a complicated way on a number of factors. We have already seen that it depends on the tube focus, degree of scatter, attenuation coefficient and the noise from whatever source. Photographic film, used on its own, does not decrease the resolution, since the size and packing of the AgBr grains is so fine that other factors dominate. This however is not the case when intensifier screens are used or electronic detectors replace film methods. The intensifier screen increases the sensitivity by factors of 100-1000 and hence dramatically reduces the radiation dose to the patient. This comes at the cost of a decrease in spatial resolution, as a result of the spread of light photons within the screen before they activate the film. Electronic detectors used in medical applications are such as digital radiography or CT are composed of many discrete elements each with a finite size. Each element averages together all photon events recorded, over its finite field of view, to produce a single number for that area. Its spatial resolution is then limited approximately to the physical size of the detector element.

Spatial resolution, both in photography and throughout medical imaging, is described quantitatively by the response of the imaging system to a test pattern. This is just like the standard eyesight test card by opticians. The test pattern in x-ray radiography consists of pairs of lines of highly absorbing lead of varying widths and corresponding spacings. The measure of resolution is the difference in contrast between black and white recorded from the test pattern. This is called the Modulation Transfer Function, MTF and it is plotted or quoted with respect to line pairs per millimetre, see figure

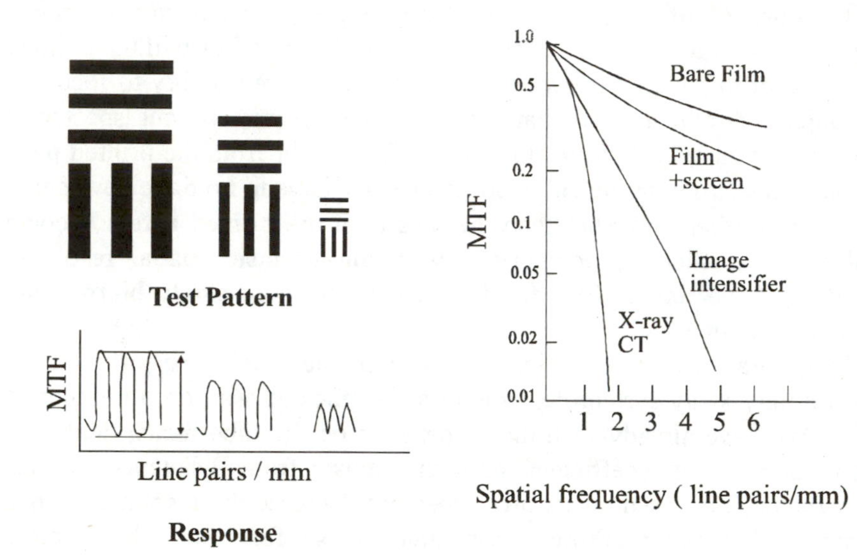

Figure 4.7: a) A typical test pattern used to quantify the spatial resolution of an x-ray imaging system. The pattern would be laid out using lead strips. b) Some typical MTF versus line pairs per millimetre for some x-ray methods. It is generally accepted that an MFT value of 4% corresponds to a threshold of resolution. For CT, spatial resolution is about 1 lp / mm at a contrast level of 1% of water.

4.7a. A plot of MTF versus line pairs is directly analogous to a Bode plot, used in audio engineering, to describe the quality of a sound recording and reproduction system as a function of temporal frequency. Both are in fact the Fourier transforms of the ratio of output to input for different values of test frequency. The composite plot in figure 4.7b compares some typical

MTF's of bare film, film plus screen, an image intensifier and x-ray CT. As we would expect from our discussion, bare film is able to resolve the finest line pairs and thus its MTF stays higher for longer than the other methods. This does not however make bare film the automatic method of choice in radiography. It cannot be used in any digital application and it requires the highest patient radiation dose. The effect of the choice of film/screen speed on patient dose is discussed below

4.4 X-ray Equipment

We have already alluded to the very large literature concerned with all aspects of diagnostic x-rays. Naturally this wealth of knowledge includes large sections on equipment. In the following sections we deal with x-ray generators and detectors. The aim is to highlight the main physics principles and engineering constraints, giving particular emphasis to those issues that impinge directly on the three key points spatial resolution, image contrast and patient dose.

X-ray Tubes

The modern x-ray tube can produce a very large beam intensity, but it uses the same principle in 1999 that Roêntgen employed in 1895. The kinetic energy of accelerated charged particles, electrons, is converted into heat and a broad spectrum of electromagnetic radiation that includes x-rays. The same principle is used to produce visible light photons on the screens of all TV sets, desktop computer monitors and oscilloscopes. They are all cathode ray tubes. In a typical medical x-ray tube, the accelerating voltage or HT is set at 30-150 kV so that electrons acquire a kinetic energy of 30-150 keV just before they hit the target anode. Typically electron currents will be in the range 10-100 milliamperes and thus the electrical power consumed by the tube will be 100.10^3 V x 0.1A =10kW. The electrically charged electrons interact very strongly with the target atoms and are brought to rest within a few mm of its surface. By far the largest part of the electron beam energy is converted into heat and this is deposited within a few mm of the surface of the target. Only about 1% of the available

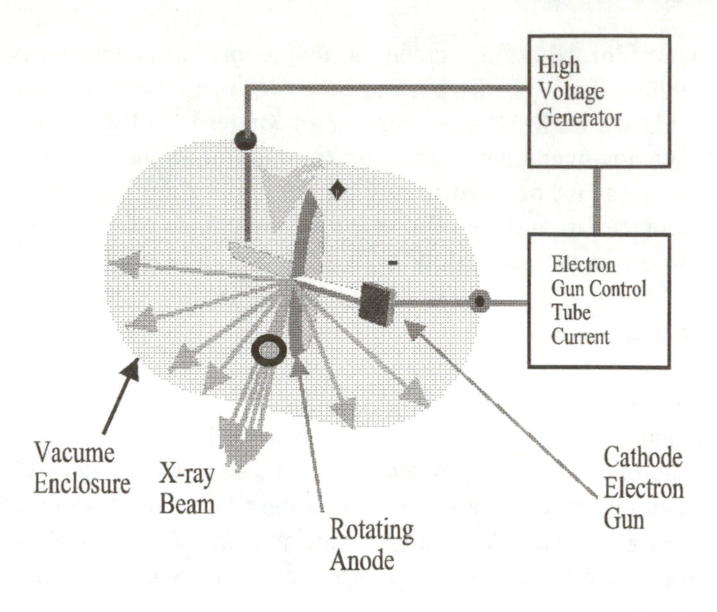

Figure 4.8: A modern rotating anode x-ray tube. Inside the evacuated glass or metal enclosure an electron gun emits a stream of negatively charged electrons that are accelerated by a high tension electric field (30- 150 keV) towards the anode plate. The x-rays are emitted from the anode in all directions. A small window in the enclosure allows a narrow beam of x-rays to emerge from the tube for use in radiology.

electron energy is actually converted into useful x-ray radiation, which escapes from the tube. Without special precautions, the electron beam would rapidly melt the surface of the target, eventually drilling a hole straight through the metal. Thus, the most important problem for the tube designer is to cope with the 10-15 kW of heat energy that the electron beam deposits in a very small volume of the anode.

The heat problem is dealt with in four ways. The electron beam is shaped into a long thin rectangle oriented radially with respect to the anode disc. The anode is made to rotate at high speed (2000-3000 RPM) underneath the beam. Both of these ensure that the beam energy is spread over as large an area of target anode as possible. Even then the anode will get very hot. Very high melting point metals such as tungsten and ruthenium are used as targets. Finally many modern tubes are pulsed and the radiographer is only allowed to take a certain number of exposures before a heat sensor detects the raised temperature of the anode and shuts the tube down until the anode has cooled off sufficiently to start again. Cooling of the anode target can be accomplished by circulating water or oil but many modern tubes just rely on radiative cooling.

High resolution in x-ray radiography demands the smallest possible x-ray source diameter. Although the electron beam is generally spread out along a line, this is oriented edge on with respect to the exit window. Thus x-rays emerge from a small area, limited by the thinner, vertical dimension of the electron beam in one direction and the shallow angle of target in the other. This is called the focus of the x-ray tube. With use, the target anode surface is eroded and thus roughened under the electron bombardment and this roughening leads to an unwanted increase in the size of the focus and deterioration in the uniformity of the emitted beam. The size and stability of the focal spot and the uniformity of the emerging x-ray beam are crucial for high spatial resolution in images. Figure 4.9 shows how, with increasing focal dimension, a penumbra is formed around the image of an absorbing object. This softens the edges of sharp images and reduces image contrast in a very similar way to scatter from the patient. In screening applications, such as mammography, it is important that regular, standardised checks are run on the x-ray tube to maintain a constant focus and a known intensity distribution across x-ray beam to ensure that false detail is not introduced into the resulting image.

X-ray Tube Operation and Rating

The very wide range of applications of x-rays requires a range of tube designs to provide a range of power output, peak x-ray wavelength and operating characteristics. Thus in x-ray CT a relatively high powered, wide angle wedge shaped beam is required from a tube that has to be mechanically rotated at high speed within the scanner gantry, in order to carry out tomography. The mechanical movement effectively rules out the use

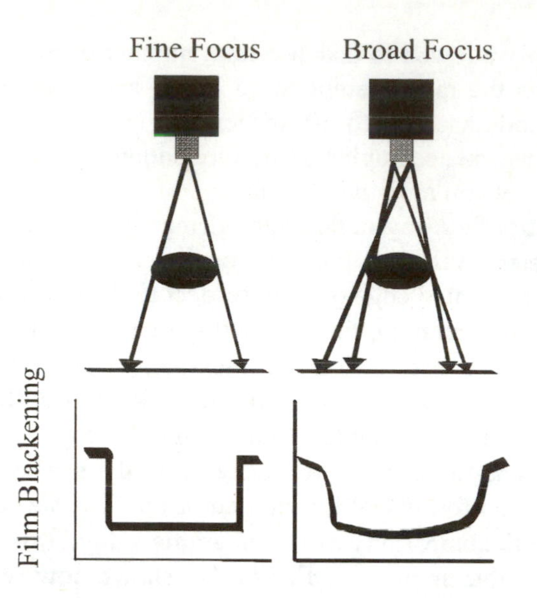

Figure 4.9: The effect of the size of the x-ray tube focus on the sharpness of the radiographic image. The larger focal area gives rise to a penumbra in the transmitted x-ray intensity which softens the edge of the images of attenuating objects, leading to a reduction in spatial resolution in the image.

of cooling water or oil and requires a special slip ring, arrangement to transmit the high tension voltage from the fixed and bulky transformers to the tube itself. The high output power demands the use of a rotating anode, used in pulsed operation. A dental x-ray set on the other hand is relatively low powered, needs only a narrow field of view and the tube is fixed during the exposure. Thus a much simpler water-cooled or air cooled fixed anode can be used.

The radiation output of a x-ray tube clearly depends on the electrical power that it consumes and tubes are described or rated by the manufacturer in terms of electrical power rather than photon flux produced. The electron beam energy is quite capable, when converted to heat, of both destroying the anode and damaging the associated vacuum enclosure. For this reason each tube has an associated rating chart laying out safe operating conditions. The tube is rated in *heat units,* HU obtained from the product kV. mA seconds. Both the anode and the tube are given separate ratings. Thus a standard radiography tube may be given an anode rating of 150,000 HU and the tube 1 million HU. In addition the tube will be described by the time taken for both anode and housing to cool back down to a normal operating temperature after an intensive session of exposures has raised the tube and or anode to its maximum operating temperature .

Different radiographic procedures have been found empirically to be best carried out with particular combinations of focal spot sizes, HT voltages, beam currents and exposure times. The particular choice of settings will be governed by the overall attenuation factor for the anatomy under investigation: how much bone, how much soft tissue, the type of film used, and finally the recommended patient dose for the particular investigation.

The X-ray Spectrum

When any solid target is bombarded by high velocity charged particles such as electrons or protons, the projectile looses energy by glancing encounters with target atoms and by direct atomic 'hits' which eject electron(s) from the target atoms. Most of the projectile energy ends up in random heat energy but about 1% is emitted in the form of x-rays. The spectrum of x-ray photon energies emerging from the target anode shown in figure 4.10 illustrates the two mechanisms of x-ray generation. The broad "hump" results from x-rays produced by a continuum of changes in electron velocity as the electron makes a series of glancing encounters with the target atoms. At each encounter the electron velocity is changed, it therefore decelerates and emits an amount of electromagnetic energy whose frequency depends on the degree of deceleration. This is the bremstrahhlung or "breaking radiation" process. The maximum possible photon energy emerging from the tube is exactly equal to the incident electron kinetic energy. This photon is emitted if the electron is brought to rest in just one encounter. The peak of the spectrum corresponds to the most probable magnitude of electron deceleration. The energy at which the peak occurs varies with applied HT voltage. Thus the mean photon energy in an x-ray beam is governed by the applied voltage. Both attenuation within the patient, and absorption in all photon detectors depend on the mean photon energy, which in turn depends on the HT tube voltage, the kV. The contrast in the final image depends on both photon energy and the radiation exposure. The radiographer has to alter the x-ray tube conditions to be appropriate for the particular investigation and the resulting radiographs are always labelled with a record of the tube voltage, kV and the tube current-

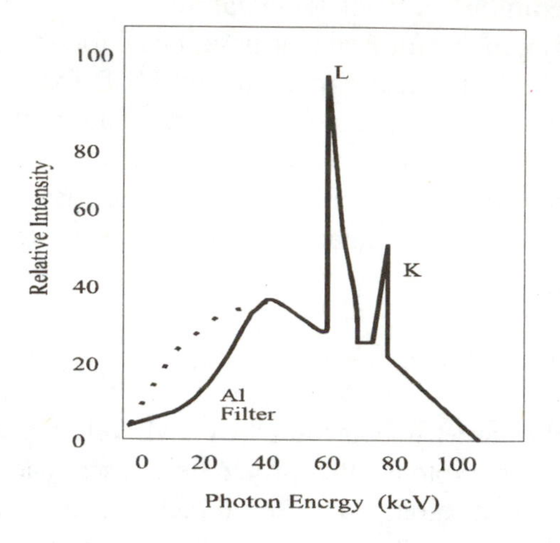

Figure 4.10 : The energy spectrum of x-rays emerging from a typical tube. The broad background arises from bremstrahlung radiation. The photon energy at which the peak of the broad spectrum increases with tube voltage, kV. The sharp peaks are the characteristic line emissions ,K, L atomic transition lines caused by the target atoms relaxing after losing an electron from a "direct hit" from the electron beam.

time product, mAs, used to obtain the exposure. The "kV" provides information about the mean energy of the x-ray photons and the "mAs" provides a measure of the radiation exposure of patient and detector

The low energy, (<10 keV), " soft x-ray" parts of the continuous spectrum are of no value in most imaging applications since this component is heavily absorbed by all tissue via the photoelectric effect. These soft x-rays are removed from the beam by using a thin aluminium or copper metal foil filter as an exit window to the tube, see figure 4.10. The spectrum of the x-ray beam is also modified as it passes through the patient. In general the low energy parts are preferentially absorbed and so the beam progressively “hardens”, that is to say the mean energy shifts upwards with increasing depth into the patient. This effectively changes the attenuation coefficient as a function of depth and can introduce image artefacts, especially in CT.

Photon Detectors

The problems of x-ray detection bring together medical science and particle/nuclear physics. Medical imaging and quantitative measurement has benefited from the technological innovations produced at high-energy physics laboratories such as CERN. For the particle physicist the problem is to detect and localise the results of rare fleeting complex particle interactions. The medical physicist has a similar problem of detecting and localising enough photons to produce an image, but at the same time, minimise the radiation dose administered to the patient. In both cases the aim is to make the best possible use of the photon flux available. Both communities are concerned with the quantum efficiency of a device and its spatial resolution.

The majority of x-ray images are still recorded on photographic film. Digital radiography and x-ray CT both make use of electronic detectors that produce numbers which are stored in a computer and then manipulated to create the required two dimensional image. Gamma imaging similarly depends on the efficient detection of photons whose energies are between 2 and ten times higher than x-rays but the same physical principles and many of the methods used for x-rays are also employed to detect and count γ–rays.

Detection requires that an x-ray hits an atom or atoms of the detector and leaves a permanent, measurable trace of its passage. In order to be able to localise the point of impact of an x-ray, most detectors have to be thin and thus many photons simply pass straight through the detector without leaving any trace whatsoever. The term quantum efficiency describes the number of incident photons required to produce a measurable trace of an impact in the detector. The human eye has the highest known quantum efficiency of unity, in the visible region of the electromagnetic spectrum, since it is capable of registering the impacts of individual visual light photons. The human eye is effectively blind to x-rays and γ-rays. At these very much higher energies practical detectors have quantum efficiencies that are very much smaller, in the region of 1-13%. All detectors exploit the photoelectric effect as the primary mechanism of interaction between the x-ray photon and detector. Many also incorporate energy conversion and amplification whereby a single high energy photon (keV) is made to produce a shower of lower energy photons (eV) that can then be recorded with a higher efficiency.

Photographic Film

The photographic film, although apparently simple and familiar, is actually quite a sophisticated structure that uses about 1/3 of the world's industrial supply of the expensive metal, silver. Both the expense of silver and the incompatibility of the photographic method with digital processing provide strong incentives for research and development in the area of electronic x-ray photon detectors. The well-known photographic emulsion, used in any camera, is sensitive to all energetic photons including x-rays. Silver halide grains, fixed in a thin gelatine film, oxidise when exposed to photons whose energy exceeds the ionisation energy of about 0.5 eV. A typical modern x-ray plate cassette incorporates a number of technologies that maximise the sensitivity, reduce unwanted scatter and preserve spatial resolution. In the end, it is the degree of blackening on the film that determines the quality of the image and this, in a given investigation, depends on a large number of parameters including x-ray tube kV and mAs, the bulk of the patient, the type of film (its speed and contrast type) and the way in which the film is processed. A complete description of all these factors would be inappropriate at the introductory level of this book but we will discuss the main underlying principles.

Film Characteristics

The end point of most x-ray radiographic investigations is the blue green tinged semi-transparent developed negative, which holds the shadow image. This is viewed and assessed by shining a bright light through the negative. The clinician uses this image to look for signs of unusual changes in contrast, brought about by a disease process. Her ability to perceive small contrast changes is critically dependent on how the film was exposed and developed since the photographic process is itself very complicated and non-linear.

Since the image is viewed by transmitted light, image contrast is described quantitatively by the optical density, D produced by the degree of blackening of the exposed film. D is defined, using logarithms, by $D = \log_{10}(I_o/I_t)$, the logarithm of the ratio of the incident to transmitted visible light. Areas of high patient absorption such a bone create pale shadows, little film blackening and small value of D, and areas of low absorption produce more blackening and high values of D. Absolute values are of no use but ratios are; the use of logs provides a convenient format for their graphical display. The amount of blackening in any area of the film is determined by the size and density of the silver halide grains, put into the film by the manufacturer, and what fraction of those grains are turned to silver in the exposure and developing process. During the exposure, the number of affected grains will be determined by the number and energy of x-ray photons passing through the film. Since contrast is the key factor, the response of a film is described quantitatively in terms of ratios of the amounts of radiation actually hitting different parts of the film, called the relative exposure. The absolute difference in the number of photons hitting two regions of film will again depend on the kV and mAs preset used on the x-ray tube and the difference in patient attenuation corresponding to the two regions. The ratio of these two will however only depend on the patient attenuation not the tube settings. Again logarithms aids the display of the ratio. A film *characteristic* is defined as a graph of D versus the logarithm of the relative x-ray exposure. $\log_{10}(E)$. It provides a way of making quantitative predictions of how best to use this particular film for a particular investigation. A typical characteristic is illustrated in figure 4.11.

All film characteristics have the same three important regions. The horizontal position on a common log (E) scale, determines the film *speed.* At low exposures D changes relatively slowly and in a non-linear fashion with log (E), this is followed by an approximately straight region, over which D varies nearly linearly with changes in log(E). Finally at large exposures all the grains in the film are blackened and so D tends to a limit, D_{max}, set by the manufactured composition of the film. The useful region of the characteristic is the central linear part. The slope in this region, called the *film gamma*, determines the contrast produced by small changes in exposure. Ideally gamma would be as high as possible and occupy a large range in log(E), the *latitude* of the film. In practice, these two pull in opposite directions so that a film with a high gamma tends to have a small range in log(E) over which the characteristic is straight and thus is said to have small latitude. The low exposure, toe region, is caused by an inevitable amount of film blackening produced by grains oxidising without intentional exposure to the x-rays of investigation. This is called fogging; it is caused to a small extent by cosmic radiation and background radioactivity penetrating the light protective covering of old film and by the chemical reactions used in the development process itself.

The radiographer aims to make as much use of the linear region as possible by selecting an x-ray exposure that will ensure that all exposed regions of the patient transmit enough photons. The blackening on the film should exceed the intrinsic blackening in the toe region but not stray far beyond the top of the linear region. Figure 4.12 illustrates the appearance of radiographs obtained with too little, the right amount and two much x-ray exposure. In both extreme cases the contrast within the image is insufficient, because small changes in exposure produce relatively small changes in film blackening. The very highest spatial resolution radiographs are obtained using just a fine grain photographic film. As long as scatter from the patient body is minimised and sufficient x-ray photons actually stop in the film, then the spatial resolution in this case depends only on the magnification factor and the graininess of the film itself. The conditions that we have added are of course very important. As the patient to film distance increases the shadow spreads out, enlarging the image dimensions . At the same time however the x-ray photon density is decreased (approximately as the inverse square of the distance from focus to detector). Thus a larger exposure time must be used to obtain a required amount of film blackening. This increases the dose to the patient. In general the absorption efficiency of bare photographic film is insufficient to allow good quality radiographs to be obtained, without giving large radiation doses to the patient.

Intensifier Screens

Intensifier screens are used to increase the overall efficiency of a film /screen combination. An intensifier screen consists of a film of heavy element chemicals such as barium titanate, zinc sulphide, caesium iodide or gadolinium oxisulphide. These act as photon energy converters. When a single x-ray photon is absorbed by one of the heavy atoms (with the ejection of a photoelectron) it is left in an unstable state with one electron missing. These atoms then relax, by emitting a more complex lower energy spectrum of visible photons. If the visible light is emitted in a short time (10^{-8}s) then

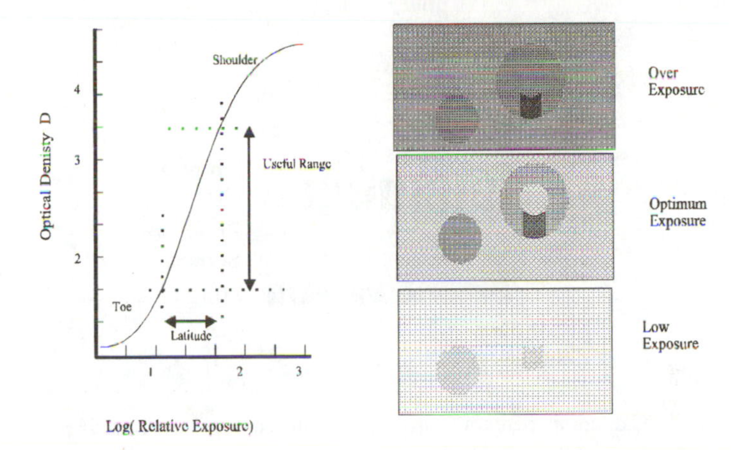

Figure 4.11: A typical film characteristic, D versus log (E), of photographic film to different levels of x-ray radiation exposure. The useful region is the central quasi-linear part of part of the curve.

Figure 4.12: An illustration of the effect of exposure on contrast. The three panels from, starting at the bottom, correspond to the three regions of the film characteristic curve

the substances are said to be fluorescent, if the light emission is delayed then they are said to be phosphorescent. Phosphors, such as zinc sulphide, are used in all computer monitors and television receiver screens to produce a visible and persistent image on the screen that is actually written using an electron beam. The particular heavy element compounds are chosen because they have atomic relaxation mechanisms which involve complicated combinations of low energy, short electron hops, rather than one big jump that would lead to x-ray re-emission. Each of the small hops gives rise to photon emission so that in a well designed screen each captured x-ray produces a shower of about 100 photons of energy 2-3 eV (blue light photons). Each of these photons is capable of producing the chemical change in a nearby photographic silver halide layer and so the quantum efficiency of the intensifier screen/ photographic emulsion combination is greatly increased.

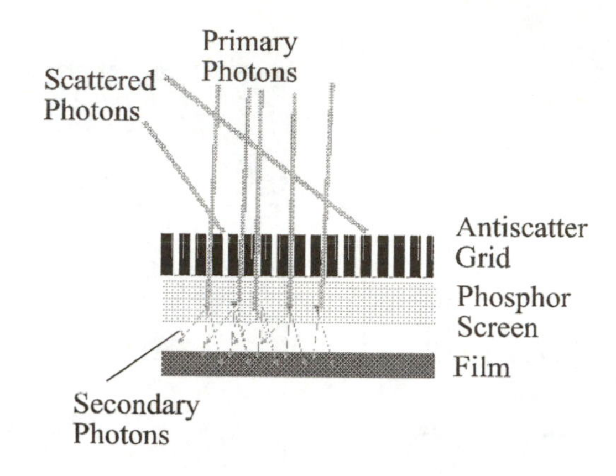

Figure 4.13: The components of an x-ray film cassette. A reusable envelope containing the intensifier screen and the antiscatter screen encloses the photographic film in a light tight package. Each interacting primary x-ray photon produces a shower of secondary, lower energy, photons in the intensifier screen which in turn oxidise the silver grains in the film to produce the shadow image. The intensifier screen is pressed into close, uniform contact with the film to minimise the area of emulsion that is exposed by the secondary photon shower from the intensifier screen.

The secondary photon shower from the intensifier layer is emitted nearly isotropically and thus leads to a degree of blurring even when the intensifier and photographic emulsion are pressed together into close contact.

Film sensitivity or speed depends on how many photons can be made to stop in the thin silver halide/ gelatine layers. The intensifier layer both increases the number of photons produced for each x-ray stopped and, by greatly decreasing the photon energy, increases the chance of photon capture by photoelectric absorption in the photographic layer, see equation 2.6. Spatial resolution depends on the area of photographic emulsion effected by each x-ray. A single x-ray photon will only produce changes in the grain(s) nearest to its impact point since the mean free path of photoelectrons is quite short. A shower of secondary visible photons on the other hand will produce chemical changes in many more grains, covering a larger area. Thus sensitivity or quantum efficiency and spatial resolution pull in opposite directions. A typical film cassette will include an anti-scatter grid , see figure 4.6 in order to reduce the contribution of scattered photons to film blackening. These will necessarily reduce the total photon flux reaching film/screen layers.

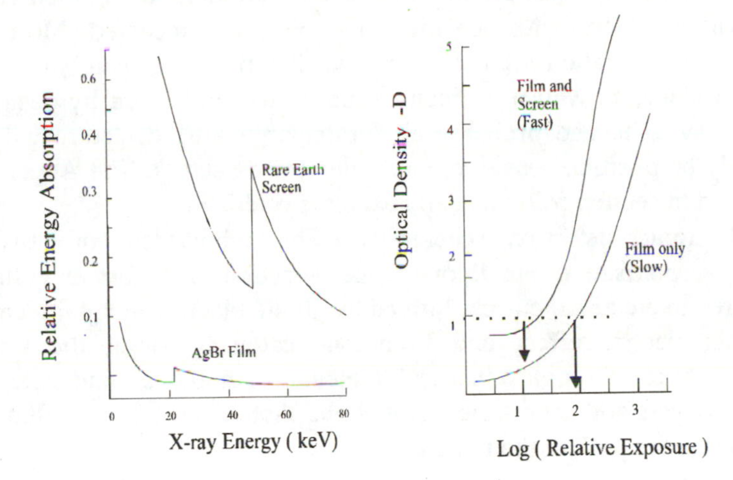

Figure 4.14: The energy absorption curves for silver bromide photographic grains and gadolinium oxysulphide screen.The K absorption edge of the rare earth gadolinium is nearly twice that of silver and hence primary x-ray absorption is higher in screens containing these types of atom.

Figure 4.15: The effect of a phosphor screen on the x-ray film characteristic. For a given degree of blackening, D very much less x-ray exposure and thus a smaller patient dose is required. The rare earth screen produces a factor of 10 increase in film speed as a result of the higher primary absorption in the screen and the production of the secondary visible photon shower, which is more efficiently absorbed by the photographic film.

Electronic Photon Detectors

Many modern radiographic techniques require a digital output from the detector and this essentially rules in favour of electronic devices. Electronic detection schemes must produce two sets of numbers, the photon count and the position co-ordinates at which this count was received. Most systems currently in operation separate these two functions by using small discrete detector elements whose geometrical positions are known by design. Very modern systems and probably all future large area detectors will almost certainly be position sensitive, multi-element detectors. The Anger Camera described in section 5.3 is an early example of this idea.

Electronic detection schemes can be subdivided into two broad categories, ionisation and fluorescence detection. The former collects and measures an electrical current formed by all the electron-ion pairs created by the photoelectric effect and Compton scattering along the track that radiation takes through a particular medium. In the second category the device collects and counts the visible light photons created in a fluorescence layer put in the path of the radiation.

Ionisation Photon Detectors

All x-ray radiation produces electron-ion pairs the along its track through any medium. Both the photoelectric absorption and Compton scattering of x-rays and γ-rays produce relatively energetic primary electrons (1-100 keV). These particles are themselves quite capable of producing more ionisation, electron-ion pairs as they loose energy in the medium. Thus a single interaction of a 100 keV x-ray with an atom in a medium can give rise to a shower of electron-ion pairs involving substantial numbers of pairs. If these secondary charged particles can be collected then the resulting electrical current can be used to register the presence and the intensity of the ionising radiation. The basic principle of the ionisation detector is shown in figure 4.16. A volume of gas is enclosed in a tube, which has a thin window to allow the radiation to pass into the active chamber with minimum attenuation. A constant electric field is set up between a central thin electrode and the metallic wall of the chamber. Electron-ion pairs can be created either in the volume of the gas or on the walls by the passage of a high energy particle or photon. The liberated electrons are accelerated by the electric field towards the positive anode and the positively charged atomic ions are accelerated towards the, negative cathode. The combined motions of these charges causes a measurable pulse of electric current to flow in the external circuitry.

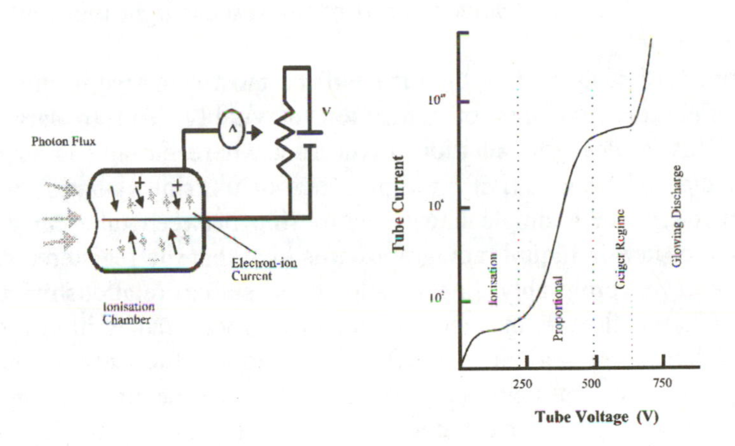

Figure 4.16: The electronic ionisation photon detector. The current voltage characteristic has three important regimes for photon detection, ionisation at low voltages, proportional (gas amplification) at intermediate voltages and the Geiger regime at very high voltages.

The current pulse height - voltage graph in figure 4.16 shows three separate regimes for such a detector. At low voltages, the electron-ion pairs drift, under the action of the electric field sufficiently slowly to have time to recombine with other charges in the gas. This is called the recombination regime. At higher voltages effectively all the pairs reach the collecting electrodes without recombination. In this plateau or saturation region the collected current is insensitive to the exact tube voltage but nearly proportional to the intensity of the incident radiation entering the chamber. At very much higher voltages the electron-ion pairs acquire sufficient energy during their acceleration to produce further ion-pairs through collisions with other gas atoms. A complicated amplification process then takes place that increases the tube current. This regime is called the proportional regime because the gas amplification factor is roughly proportional to the tube voltage. At even higher voltages another plateau region called the Geiger regime is seen. This corresponds to a very complex self-limiting spreading discharge within the tube. Finally at voltages over 1000 V a runaway and continuous gas discharge occurs in which the gas glows, as a result of light emitted from atoms involved in collisions with the accelerated ions. This final stage is used in fluorescent light tubes and neon advertising signs.

Figure 4.16 suggests that the tube will be most sensitive in the Geiger regime since tiny amounts of ionisation inevitably lead to large tube currents. This is ideal for radiation monitoring where the aim is simply to detect, at the highest sensitivity, the presence of ionising radiation without too much regard for a simple linear relationship between tube current and radiation intensity. In digital radiography it is vital that the registered current bears a known, preferably linear, and very stable relationship to the incoming photon flux. If the photon flux to be measured is high then the saturation ionisation regime is preferred. Although the current pulse is relatively low, the simplicity of physics in this regime has an important advantage. At the low current densities and energies of this regime the current pulse is very sharp, lasting only about 10ns and the tube is ready to record another pulse immediately after the passage of the preceding one. This allows higher count rates to be recorded with a linear relationship between current pulse frequency and photon intensity. Such tubes are said to have a short dead time. The dead time of a detector is the time interval during which it is paralysed after the production of one pulse. All detectors have a finite dead time but in some it is considerably longer than others. The effect of dead time is to destroy the linear relationship between photon flux and pulse frequency. As the photon flux increases from very low values, initially the output of the detector will rise in proportion, but when the mean time between electron-ion pair production becomes comparable with the dead time then events are missed and the linear relationship breaks down. In the more complex proportional regime of operation the natural dead time of a gas detector is considerably longer, as a result of the lingering effects of the gas amplification process. The very high charge density around the central electrode and the light photon flux emitted by excited atoms can take several milliseconds to die away. During this time a proportional detector is paralysed and counts are lost. All proportional counters contain small quantities of additive molecular gases which are designed to snuff out or quench the lingering effects of the amplification process. A typical additive to an argon gas chamber might be ethyl alcohol. The quenching process is rather complex, involving the absorption of photons emitted by excited argon atoms and the de-excitation of the primary gas atoms, without photon emission, through atom-ethyl alcohol collisions.

Since these detectors are gas filled, the effective density of the detector is rather low and many incident photons can pass through without interaction. By choosing a gas with a high Z, such as xenon, Z=54 or krypton Z=36, the efficiency is increased through the photoelectric effect, equation 2.6. The gas pressure can only be increased with care since as the pressure increases the effects of after-discharge become more pronounced. It is now common practice to incorporate high Z solid plates into the detector chamber to increase the detection efficiency. These are designed to act as photon converters; the high-energy incident photon ejects a shower of photoelectrons from the plates with sufficient energy to ionise gas. Large area multi-wire proportional gas detectors, originally developed in high-energy physics are increasingly used in medical imaging. Here a matrix of interleaved anode wires, cathode and converter sheets provide position sensitive detection. The ionisation current is kept local so that each anode can act as a separate detector. Such schemes are being developed for line-scan digital radiography, x-ray CT and in PET cameras.

Scintillation Detectors

Our second general category of photon detector, the scintillator-photomultiplier combination has been almost a standard particle and photon detector for 30 years, in both high-energy physics and medical imaging. Figure 4.17 shows the general arrangement of the scintillator crystal backed by a photomultiplier tube. The crystal is chosen for the efficiency with which it can convert the ionisation, produced by high-energy particles or photons, to visible light. A single crystal is generally used so that the light produced by the ionisation travels, largely without scattering, throughout the crystal and can enter the photomultiplier tube whose entrance aperture is optically matched and bonded to the crystal. In general each incoming high energy photon (100 keV) produces many hundreds or thousands of visible photons at 2eV. When these hit the first photocathode of the photomultiplier tube they produce a shower of photoelectrons. The scintillator crystal acts then as a matching device, just like the intensifier screen, which converts the primary high energy photon into visible photons that can interact most efficiently with the first photocathode of the photomultiplier tube, (PM tube). Figure 4.17 illustrates the multiplication process that occurs as the photoelectrons pass down the chain of photocathodes in the PM tube. Each electrode is held more positive than its predecessor and is itself a photocathode. At each stage the electron current is amplified since each incident electron produces about 5 photoelectrons. After 10 stages the current has been amplified by a factor of 5^{10} to give pulse heights on the order of millivolts.

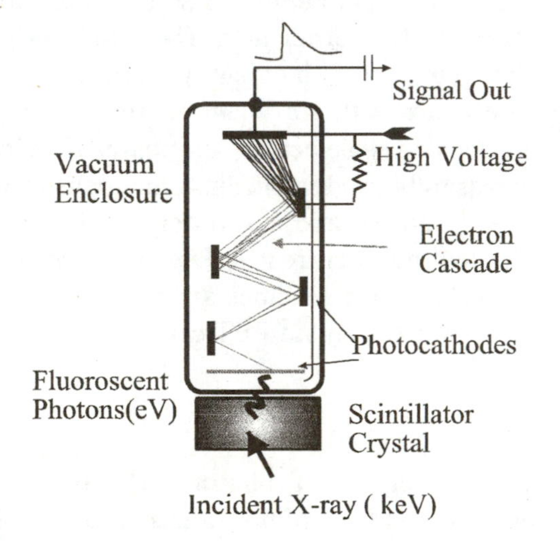

Figure 4.17: The scintillator photomultiplier tube detector. X-ray photons interact with the crystal via the photoelectric effect and Compton scattering to produce many visible light photons. These travel throughout the crystal and some enter the photomultiplier tube where electron current amplification takes place along the chain of dynode photocathodes held at successively higher voltages.

The choice of scintillator crystal depends on the particular application and involves a trade-off between sensitivity and the decay time of the light pulse. Scintillators can be inorganic crystals such as sodium iodide NaI doped with thallium or organic compounds such as anthracene or toluene. In the former electron-hole pairs produced by the initial ionisation are captured by centres created around the thallium dopant atoms. The deactivation of these centres produce the useable visible light. NaI (+ thallium) has a decay time of 250 ns. The light producing mechanism for the organic compounds involves the decay of molecular excitations caused by the primary ionisation. In general the decay time of organic scintillators is rather shorter (30ns) but the light output is a factor of two to five less than that from NaI. In all scintillators there is a definite relationship between the energy of the incoming particle and the amount of visible light produced. In NaI this relationship is almost exactly linear and so these detectors not only count particles or photons but can also provide a measure of their energy. This is put to use in gamma imaging (see chapter 5) where it is important to discriminate in favour of the primary gamma photons, that have travelled from their point of emission within the body to the external detector without interaction, and those photons that have suffered Compton scattering on the way. Since Compton scattering involves photon energy loss, scattered photons have a lower energy and thus can be rejected by energy (pulse height) analysis of the detector tube output.

Semiconducting Solid State Detectors

The rather low density of gas in its chamber and thus a relatively low capture cross-section ultimately limits the gas detector. Many research and development programmes are under way to produce solid state detectors which have higher efficiencies. These devices are generally based on the use of PN junctions produced in silicon or germanium semiconductors. When electron-ion pairs are formed the electrons are accelerated across the PN junction (with amplification) to produce a small current pulse. Apart from their higher intrinsic efficiency such devices would be much cheaper to produce in large numbers, using conventional thin film semiconductor techniques. Finally the very high precision with which such devices are made would lead to much smaller tolerances in the quantum efficiency of individual devices, thus removing a potential source of error in CT, see section 4.6

The Image Intensifier

Quite apart from their use in diagnosis, x-ray systems are also used to image and monitor the positions of tools and instruments used in surgery and in invasive diagnostic procedures. These types of medical imaging procedures are often referred to by the generic terms fluoroscopy or screening . A typical image intensifier is illustrated in figure 4.18. It brings together the methods of photon energy conversion and amplification that we have already discussed. The device has three sections; a large diameter input fluorescent screen/photocathode layer, an electron acceleration and focussing section and finally a smaller diameter output fluorescent

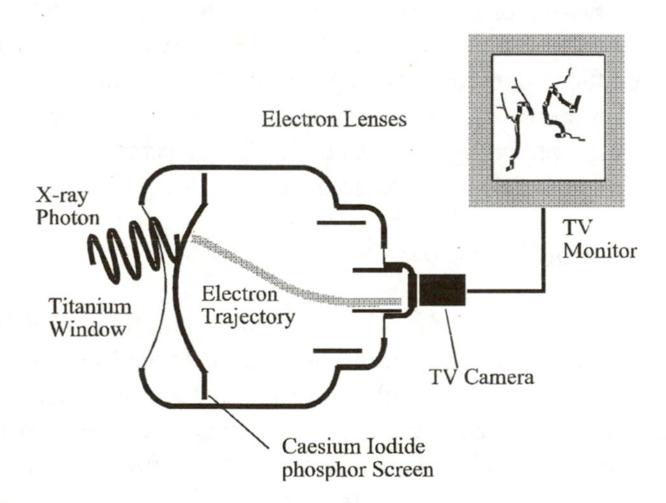

Figure 4.18: The image intensifier or x-ray fluoroscope. X-ray photons enter the device on the left through a very thin titanium window and hit the primary Caesium Iodide, CsI, intensifier screen, where many photoelectrons are created. These are accelerated and focussed onto an output screen that is generally viewed via a TV camera.

screen. The principle is very similar to that used in a TV camera. A large area input window (diameters of up to 50 cm are in clinical use) accepts x-ray photons transmitted through the area of study in the patient. These x-rays enter the evacuated chamber through a thin, high transmission, low scatter titanium metal window. The x-ray photons strike a thin fluorescent screen made from caesium iodide and create a shower of visible light photons. In turn these photons eject electrons which ideally dribble out of the back of the input screen, with negligible velocity, into the electron optics section of the device. The electrons are accelerated by an electric field of about 25kV and focussed onto a second fluorescent with a small diameter (2-3 cm). The image on the output screen can be viewed directly, or more

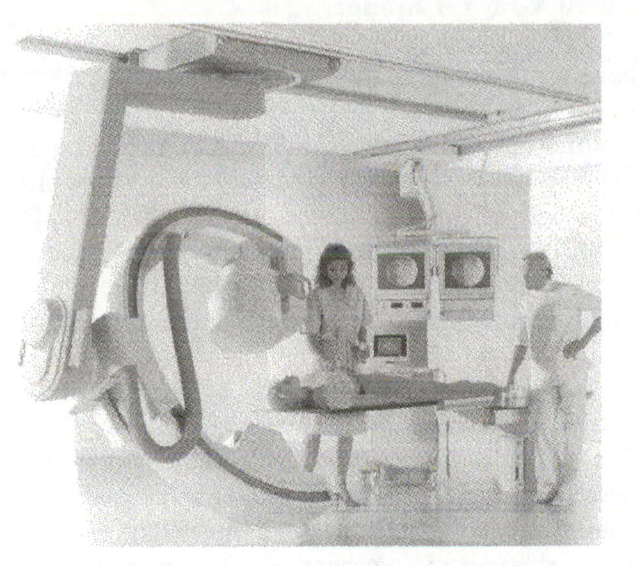

Figure 4.19: A modern fluoroscopy Suite. The image intensifier and x-ray tube are mounted on the C shaped gantry, which allows almost unhindered movement around the patient to gain any chosen angle of view.(Seimens)

commonly by using a TV camera. The electron focussing optics ensures that low velocity photoelectrons emerging from each point on the large aperture input screen have trajectories that end in a unique, corresponding point on the small diameter output screen. The entire device has to operate in a vacuum to eliminate electron scattering and be shielded from electromagnetic interference that would distort the electron trajectories. The overall gain of the intensifier can be as high as 10,000, thus allowing good quality x-ray images to be obtained with low radiation exposure and hence patient dose levels. Some reduction in spatial resolution is an inevitable consequence of the energy conversions that take place. At each stage the emitted visible photons and photoelectrons spread out from the point of primary impact and so smear fine spatial changes in intensity. In electronics terms the intensifier is a low pass spatial filter.

4.5 The Modern X-ray CT Scanner

As we have seen the radiograph is dogged by the fact that the image is a 2D projection of a 3D object and that small differences in x-ray linear absorption coefficient cannot be seen in projection radiology without the aid of contrast enhancement. X-ray computerised tomography or CT goes a long way to solving both of these problems and so diagnosis of many diseases affecting soft tissue can be made a great deal easier. Before Hounsfield's introduction of computerised tomography in 1972 there were in fact several, so called classical tomographic x-ray methods which involved moving x-ray generators and films to selectively highlight particular planes within the body. These methods were generally film based and cumbersome and so have been entirely superseded by the advent of CT.

There was a long time delay between Rontgen's discovery in 1895 and the appearance of the first CT scanner in 1972, (Hounsfield, 1972). In fact Hounsfield never claimed to have invented the method but rather to have put together the first practical scanner. The earliest description of a tomographic method applied incidentally to astronomical observations is due to Radon, (Radon 1917) and crude Russian x-ray tomographic scanners were described by Korenblyum, ([Korenblyum 1959). Perhaps the most obvious reason for the long delay was the need for cheap digital computing facilities. These did not appear until the late 1960's. Without the computer, all modern medical imaging grinds to a halt. The general principles of tomographic reconstruction are described in some detail in chapter 1. It is clear from that discussion that the computer is an essential part of the process since the final images of slices are reconstructed within the computer from a large number (256 or 512) of separate digital x-ray projection measurements.

The First Generation Scanner

The earliest commercial scanner, the EMI CT1000 used precisely the scheme that we used in section 1.2 to illustrate the tomographic process. A single fine pencil beam of x-rays was scanned across the patient and the relative absorption at each point within a single projection was recorded using a single electronic x-ray detector. The necessary number of different projections was then obtained by altering the direction of the x-ray beam. This early first generation scheme dubbed translate-rotate is illustrated in figure 4.20. The original scheme used a matrix of 128.128 points each

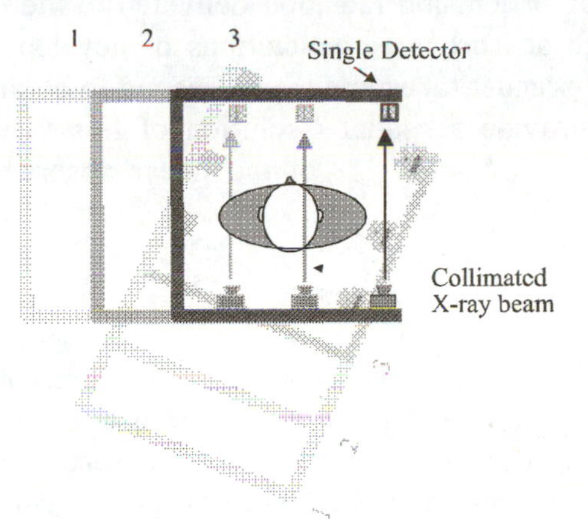

Figure 4.20: The first generation x-ray CT geometry, translate-rotate. The x-ray beam was finely collimated to a pencil beam and a single detector was used. A single projection, the translate phase, was obtained by a linear scan across the patient of the x-ray tube on one side and the detector on the other. Successive projections were obtained by sequentially altering the angle of the x-ray beam. The time taken in data acquisition for one slice was about 300 seconds.

involving mechanical movement of the detector and x-ray tube and took about 5 minutes to obtain data for just one slice. Although the original images obtained by the EMI scanner started the second revolution in medical imaging, it was not long before the limitations of "translate- rotate" became abundantly apparent. Generally many tomographic slices are required to enable diagnosis and so total scan times of an hour or more were quite common. During this time the patient's body had to be immobilised so that gross body movement would not hinder registration of adjacent slices. Even with body restraints the patient had to breathe and her heart had to beat and so considerable internal movement artefact was an inevitable consequence of the very long data acquisition time. All subsequent developments in the mechanics of x-ray CT have been aimed at radically reducing the data acquisition time and more recently with intense competition from MRI, increasing the detection efficiency and thus reducing the dose of ionising radiation delivered to the patient. X-ray CT has gone through at least three generations of development. The present standard clinical scanner takes less than a second to obtain one slice worth of data, it can provide a spatial resolution of less than a millimetre at contrast differences of about 1% of the linear absorption coefficient of water.

Third Generation Modern Scanners

Figure 4.21 illustrates the most common present, third generation geometry. The main changes are in the use of a fan beam, which simultaneously exposes a complete patient cross-section and a multi-element detector bank that collects 600-1200 data points simultaneously. These changes, together with improvements in x-ray power and detector efficiency, have brought the scan time per slice down to less than 1s. The typical fan beam scanner uses a very powerful rotating anode x-ray with an adjustable focal spot size in the region of 1 mm^2 and anode heat capacity in excess of 1MJ (1.5M HU) operating between 100 and 130 kV. The detector generally subtends an angle of about 70° and consists of a bank of 500-1200 separate elements. Different manufacturers use different detection technologies, which trade off the advantages and disadvantages of the electronic detection schemes discussed in section 4.5. Since the advantages of the third generation geometry depend on the simultaneous use of many detector elements, the question of detector stability comes into the equation. High resolution depends on all elements of the detector maintaining, to a high precision, their individual x-ray photon to electric current conversion factors throughout the scan. This is so important that the extreme edges of the x-ray beam are arranged to miss the patient and enter special monitoring elements on either end of the detector bank. The outputs from these elements are then used to keep a running check on x-ray output from the tube and overall variables such as variations in detector high tension that would alter the conversion factor. Element to element stability can only be assured by designing very stable detectors in the first place and then making regular validation measurements on standard phantom objects. The different multi-detector technologies include xenon multi-wire detectors, which have relatively low efficiencies but high stability, banks of mini NaI/PM tube elements which have very high efficiencies but can be plagued by dead time problems and bismuth germanate solid state detectors that are efficient but relatively less stable than xenon. In order to achieve spatial resolutions of less than a millimetre, the detector elements must of course be rather small and the outer reaches of K space must sampled with a sufficient density of measuring points, see figure 1.6. Typically there will about 16 separate elements per degree, each with an aperture of about 1mm. The lowest resolution images are obtained using about 500 separate views or orientations of the tube-detector with respect to the patient. This produces about 400,000 points to be reconstructed into a 512 by 512 reconstruction matrix. Higher resolution is obtained by a method of oversampling, taking as many as 3000 views. Generally the effective width of the reconstructed slice can be varied between 1 and 10 mm. Accurate slice selection is achieved using movement of the patient bed. This is motor driven and its position with respect to the x-ray fan beam can be altered by as little as 0.5mm. Gross positioning over a range of about 1000 mm is obtained from the same mechanism.

Spiral CT

Although the majority of clinical CT machines are the fan beam type, there have been recent innovations in the manner in which data is collected leading to shorter acquisition times and a more versatile data set. Spiral CT is the most important of these developments. Instead of collecting data slice by slice, in spiral CT the patient is moved continuously as the fan beam/detector rotates. This is a 3D, volume acquisition mode and it has two major advantages. Since a complete set of slices is obtained more quickly, movement artefacts are nearly eliminated. In addition the volume acquisition allows a more versatile reconstruction algorithm which effectively eliminates the restriction on slice angle of about 30 ° to the horizontal found in conventional CT machines.

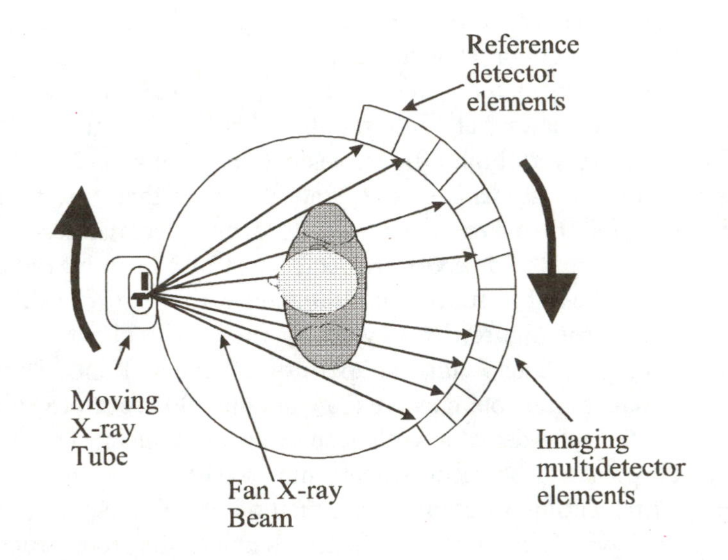

Figure 4.21: The third generation x-ray CT scanner. The x-ray tube produces a narrow "fan beam" that defines the selected slice. Tube and detector bank rotate together about the patient in a series of small angular steps in order to collect all the projections required for 2D reconstruction. At each step the tube produces a pulse of x-rays.

Hounsfield Units

The final image from an x-ray CT scan (the final array of numbers in the box puzzle, figure 1.11) is presented in terms of CT numbers measured in standard Hounsfield units rather than absolute values of linear absorption coefficients. There are two stages in getting from relative absorption values to CT numbers. Each acquired point in each view or projection is a measure of integrated attenuation along one line in one cross section of the patient.

$$I = I_o e^{-\int \mu(x)dx} \qquad \ldots 4.10$$

As it stands this is not very useful and so we take the logarithm of this equation. Thus the logarithm of the relative transmission, I/I_o at any point is simply related to the sum of attenuation factors through the patient.

$$\ln\left(\frac{I}{I_0}\right) = -\int \mu(x')dx' \qquad \ldots 4.11$$

The right hand side now corresponds exactly to the sums along rows and columns shown in figure 1.1, and this constitutes the projection, used for reconstruction. After the computation of the filtered backprojection (see Appendix B) for each slice, the computer holds a 512 by 512 of numbers, each of which represents a scaled value of $\mu(X,Y,Z)$. These are converted to CT numbers by comparing them to what the particular scanner obtains for pure water. Thus we have

$$CT\ number = 1000 \times \frac{\mu(X,Y,Z) - \mu_{water}}{\mu_{water}} \qquad \ldots\ldots 4.12$$

Generally a CT machine will display CT numbers in the range -2000 to +4000. High positive values correspond to compact, highly attenuating bone and high negative numbers approach the low attenuation of air. Fat has a slightly smaller attenuation factor than water and so it has CT number of around -100. A typical scanner specification will quote a noise level of ± 2-5 Hounsfield units.

Digital imaging schemes offer the major advantage of image processing before presentation. Any digital image has an associated colour scale or grey scale that can be thought of as a third dimension in addition to the XY dimensions of the screen. Although a given CT machine would be capable of resolving attenuation differences of 5:6000 in CT number, only a certain number of discrete levels are in fact displayed. A coarse noisy image might only warrant 16 levels whereas a good high-resolution image might require 256 different levels. The relative contrast and visibility of chosen objects within a any digital image can be altered and enhanced by choosing particular bands of CT numbers and either expanding or contracting their range to fit into 256 different levels. To give a precise CT example: suppose that a high resolution image of bony structures was required from a CT scan, then only the high positive numbers eg >500 would be selected for display. All other pixels would be put to zero or some constant neutral value. The total range of selected values would then be scaled to fit into a chosen grey scale of say discrete 256 levels. The use of standardised CT numbers allows a universal grey scale to be established for any image in which each shade of grey corresponds to a particular type of tissue.

Image Artefacts in CT
Beam Hardening

There are two factors that can introduce artifacts into the CT image. These are beam hardening and partial volume effects. As we saw in section 4.4 the mean energy of an x-ray beam increases with its passage through tissue as a result of the preferential absorption and scattering of the lower energy components of the beam. The attenuation coefficients vary with photon energy for the same reasons. Thus, two volumes of the same tissue type, one close to the surface and one deep in side the body will have slightly different attenuation coefficients and thus will produce a contrast artefact in the reconstructed image. The main "hardening" artefact is a progressive reduction in apparent CT number towards the centre of any slice. Similar and more localised effects can occur in the vicinity of bone-tissue boundaries.

Partial Volume

Slice selection together with the reconstruction matrix artificially dissects a particular expanse of anatomy into a set of adjacent boxes. This dissection has no relationship to anatomy and thus a given small box might well contain a bit of bone, a bit of muscle and part of a blood vessel. The resulting reconstructed CT number for this box will not correspond to any one of these, rather it represents the volume average over the three different tissues. This is the partial volume effect. It can introduce false detail into the image and it reduces the utility of an absolute CT number scale. Partial volume effects are present in all the tomographic imaging modalities. The only certain remedy is to increase the spatial resolution so that each reconstructed box contains only a dominant tissue type.

Patient Dose and Spatial Resolution in CT

CT is able to produce very much more information than a single x-ray radiograph simply because it combines information from very many (>300) radiographs. It is thus not too surprising that a typical CT scan produces a very much larger patient dose. A typical chest x-ray might incur a surface dose of 0.02mSv but a single slice, medium resolution CT scan will produce about 8mSv. Although there have been large improvements in detector efficiency and dose utilisation the CT dose must be given approximately by the number of views used for the reconstruction multiplied by a typical radiograph dose. A single CT scan will produce a dose exceeding 4 years of natural background radiation and thus CT cannot be used for mass population screening nor as a real time 3D visual aid to surgery, see table 3.2. Its use is restricted to life threatening and serious illness such as cancer and head trauma. In addition to this, further development of CT to very much higher spatial resolution is effectively blocked by the issue of dose. In section 4.3 we demonstrated on the basis of a simple model that dose is proportional to the inverse fourth power of the spatial resolution. This means that if we wanted to go from a resolution of 1 mm to 0.5 mm then the dose would increase by a factor of 16. This would take patient dose in CT up over 100 mSv not that far from the exposure regime where both immediate and long term radiation damage to tissue is virtually certain, (see chapter 3). Given that MRI is now a clinically established method often providing superior resolution and greater versatility with no real hazard to the patient it is likely that x-ray CT will be phased out of clinical practice in the coming decades. CT has played an enormous part in the second revolution but it might not to continue to be used in the future. The images shown in following pages include several from CT. Although such images are now used casually to illustrate newspaper and TV articles it should remembered that only thirty years ago these were like gold dust. It was CT that started the second revolution in diagnostic imaging.

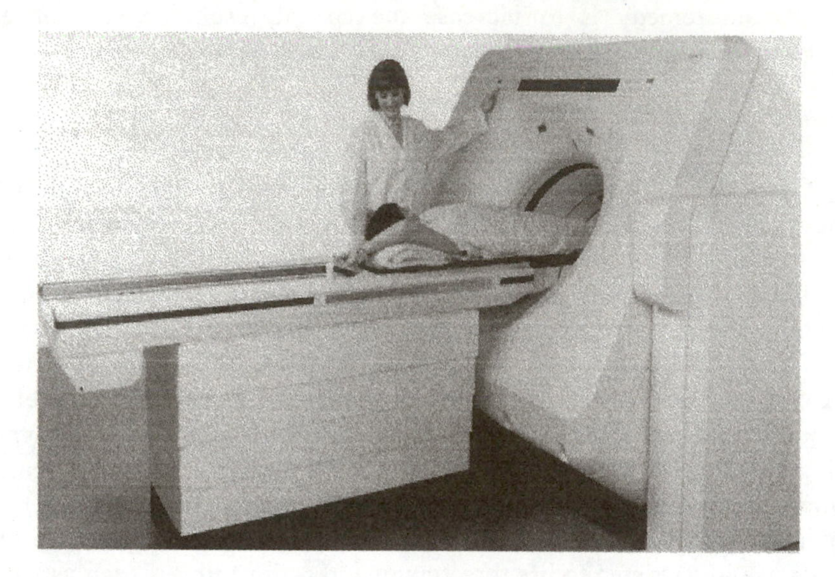

Figure 4.20: A Modern CT scanner: The x-ray tube and detectors rotate within the square enclosure. During the course of the scan the motorised and position encoded patient bed is moved through the large hole. This lateral movement achieves slice selection. Generally the scanner housing can be tilted +-30 with respect to the horizontal to allow slices at different angles to the patient to be obtained. (Courtesy of Siemens Uk Ltd).

4.6 Images from X-rays

Standard Radiographs

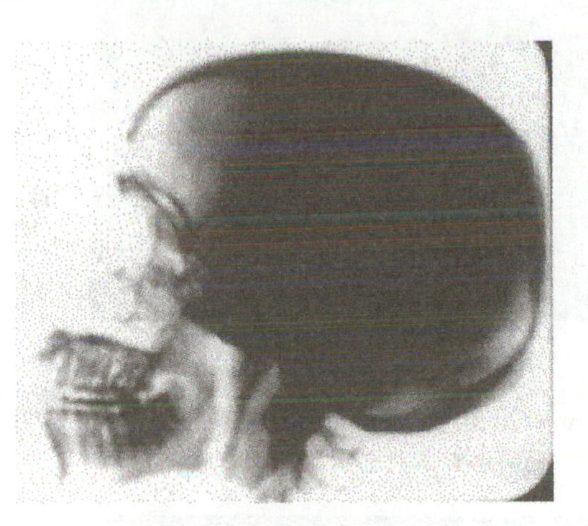

Figure 4I.1:A Standard Skull X-ray. Notice high absorption across the skull and teeth where considerable thickness' of bone intercept the beam. Kodak

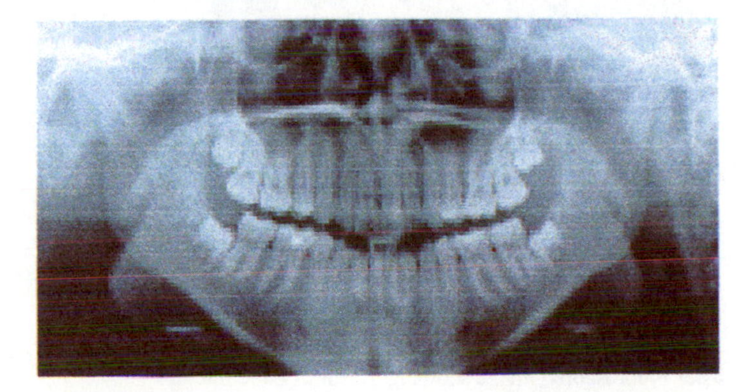

Figure 4I.2 A panoramic dental x-ray

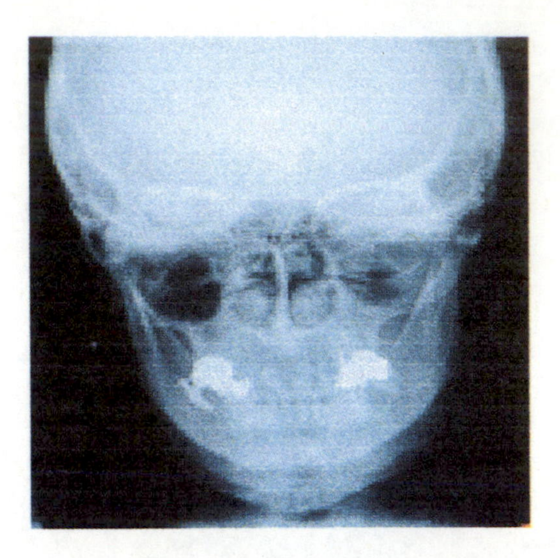

Figure 4L3: X-ray skull and dental details

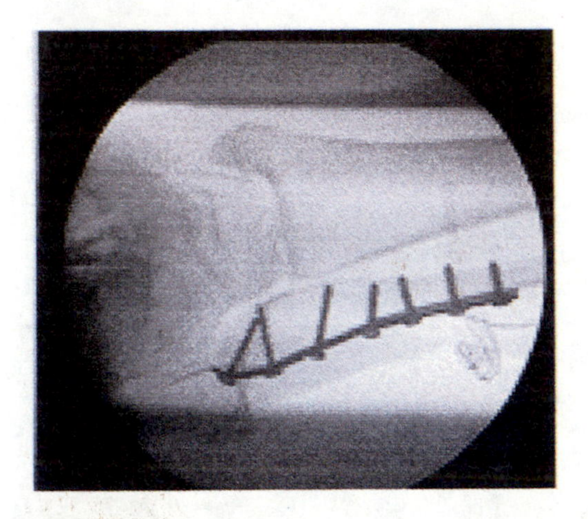

Figure 4L3: Fluoroscopy image showing repair of bone fracture

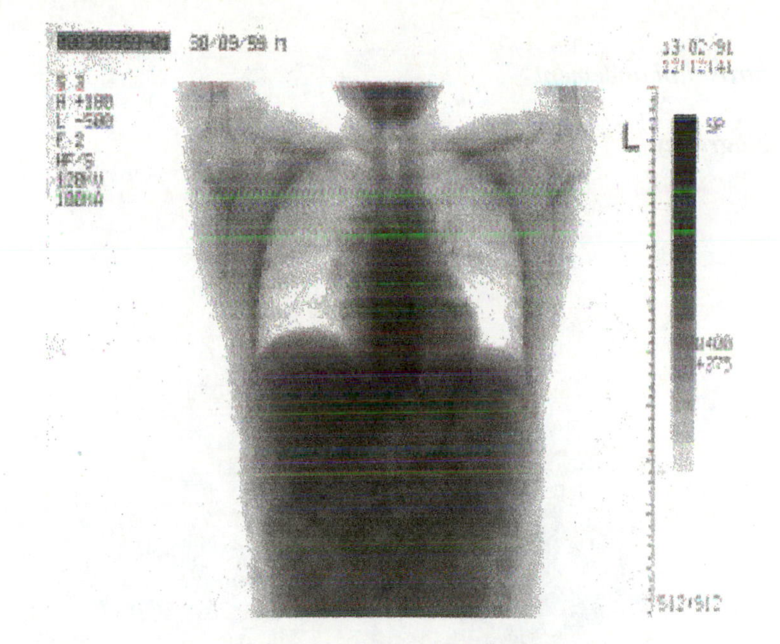

Figure 4L4 Single projection image, taken from a CT scan. This is a typical standard preliminary image used by the radiographer to plan the positions of slices for full CT investigation. Phillips

Contrast Enhancement

Angiography

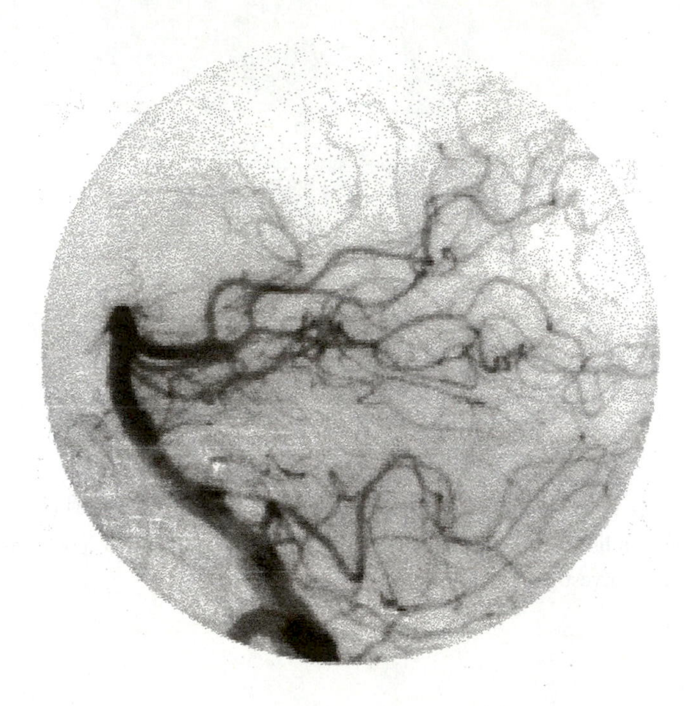

Figure 4I.5 Standard Angiogram of the head using contrast medium, injected into the blodd stream, to reveal the degree of filling of brain arteries. The contrast medium has sufficient absorption for even arteries with diameters of a millimetre to greatly exceed the absorption from the skull (see figure 4I.1) .Seimens

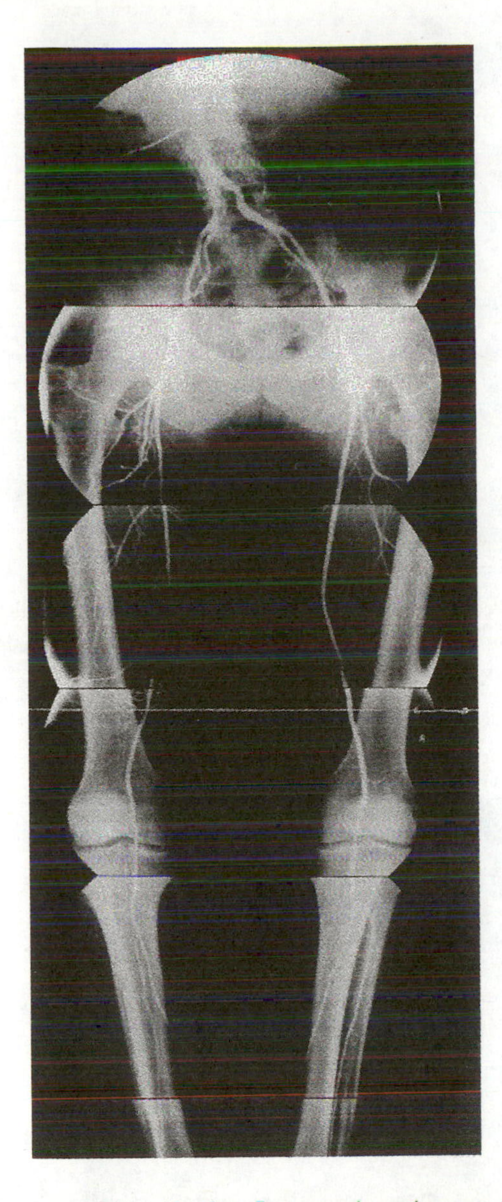

Figure 4I.6 : A multiple wide view fluoroscopic angiogram of peripheral circulation obtained with contrast agent .The extent of each view can be judged by the diameters of the circular limits of each image.

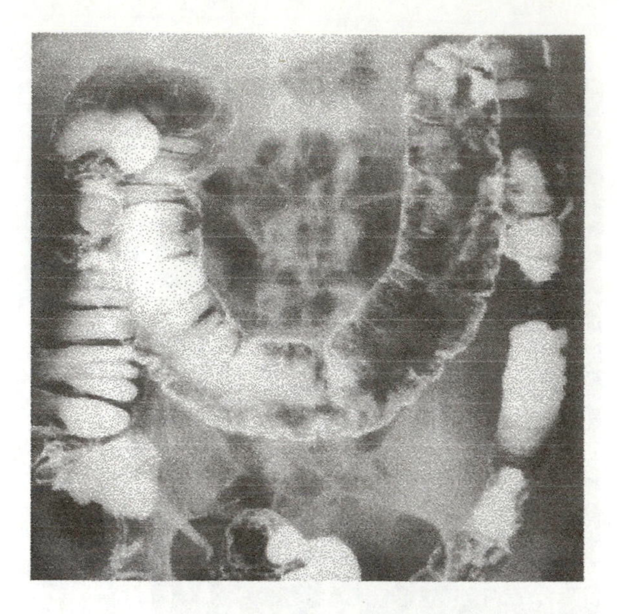

Figure 4I.7: Barium contrast of the lower abdomen showing excellent contrast for the colon .Here the contrast in abdominal soft tissue greatly exceeds that from the skeleton, (spine) .Kodak

X-ray CT Images

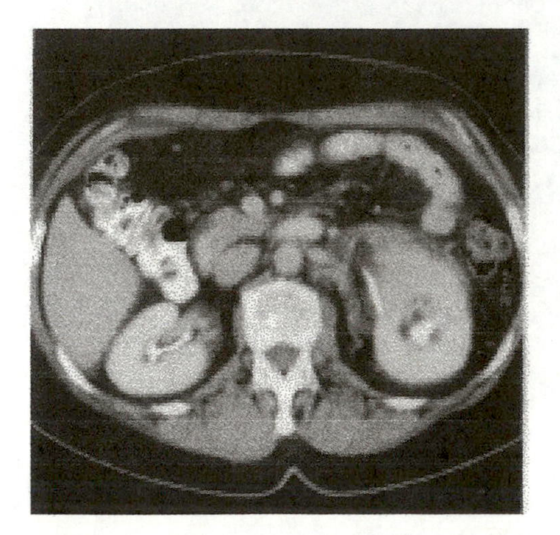

Figure 4I.8: A high resolution CT scan of lower abdomen, showing excellent contrast between soft tissue structures such as the liver , kidney and small intestine. Seimens

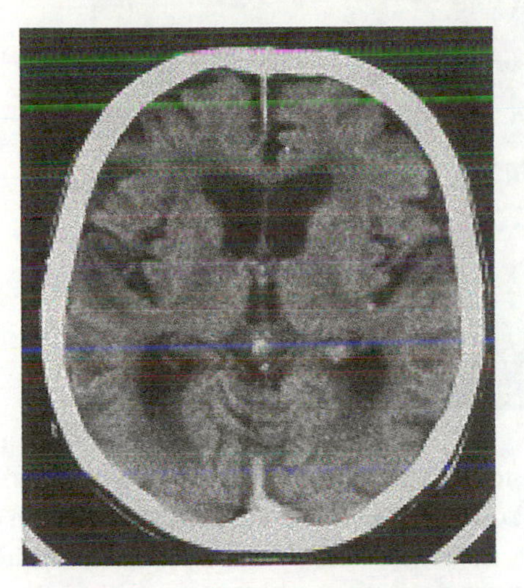

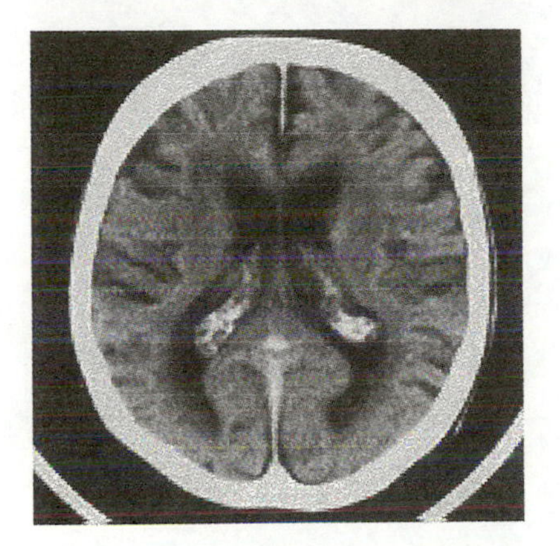

Figure 4I.9: X-ray CT slices at two different levels within the brain. Contrast between grey and white brain tissue is obtained but it is rather less than that possible with MRI

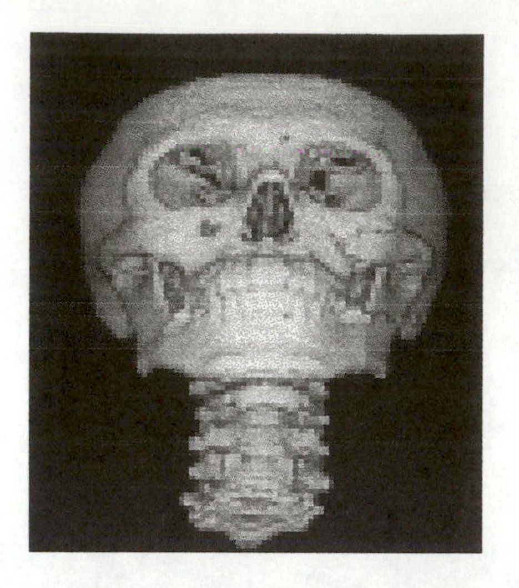

Figure 4I.10: 3D reconstruction of skull from x-ray CT data . This is the result of a computational method called surface rendering. All bone tissue on each slice can be identified by its high Hounsfield number. The 2D "contour" maps of bone on each slice are then combined into a 3D volume image

Questions and Problems

1. Explain the term x-ray linear attenuation coefficient. How does this determine contrast in x-ray radiographs?
2. Describe the main mechanisms responsible for photon absorption/scattering in biological tissue at the following energies a) 10KeV b) 100KeV c) 10 MeV. Which of these mechanisms contribute to patient dose? Why does pair production not figure in normal clinical applications of x-ray radiography and CT ?
3. It is desired to get an image of a blood vessel in the middle of a large region of muscle. If the vessel is 1mm in diameter estimate the value of $\mu\Delta X$ required to produce a contrast of 5% (neglecting any scattering into the detector). Estimate the concentration of iodine required to achieve this contrast assuming that the photoelectric effect dominates the attenuation.
4. Angiographic procedures use contrast media in order to make a fine blood vessels visible. What are the main requirements of the contrast media?
5. Show that the patient dose required to resolve a small feature of size ,X, with an x-ray attenuation coefficient μ, in the centre of a slab of tissue of thickness ,t, with x-ray attenuation coefficient,μ, is proportional to

$$\frac{K^2 e^{-\mu_1 t}}{(\mu_2 - \mu_1)^2 x^4}$$

   Where K is an empirical threshold value of the signal to noise ratio deemed necessary to "see" a feature in a random background.
6. Estimate the statistical "noise" associated with the signal produced by 1000 counts in a counter.
7. A typical X-ray tube is run at voltage of 50 kV and a beam current of 100 ma.

   a) How much electrical power is consumed by the tube?

   b) If the anode has a mass of 1Kg and is constructed entirely from copper, estimate the temperature rise in the anode after 1 second of continuous operation.

c) Discuss the design steps taken to prevent damage to the anode and its bearings.

8. The exit window of an X-ray tube consists of a copper foil of thickness 0.1mm. Assuming that photoelectric absorption dominates estimate the fractional transmission for X-rays at 10keV and 70 keV.
9. What is the increase in patient dose required to double the spatial resolution in a standard x-ray radiograph?
10. X-ray CT allows differentiation between different types of soft tissue whose attenuation coefficients differ by as little as 1%. Explain how this is possible and why x- ray radiography is incapable of such differentiation.
11. Sketch the way in which K space was sampled by first generation CT scanners. What determined the smallest spatial frequency which can be resolved? If the detector width had w, and each projection was sampled at intervals of w/2, show that the number of projections required to reconstruct an object of diameter D with uniform resolution is given by

$$N_{projections} \cong \pi \frac{D}{w}$$

Hint: consider the radial and azimuthal intervals between samples in K space along and between projections in cylindrical coordinates.
12. Explain what changes to present x-ray CT technique would be required to allow microscopic resolution (< 1 micrometre). Could such a device be used on human patients
13. Why does CT inevitably lead to relatively large patient radiation doses?

5 NUCLEAR MEDICINE GAMMA IMAGING

" *The subject which I propose to speak to you about has become, in only a few years, so vast that in order to deal with it in a single lecture I should have to confine myself to listing the main facts following the chronological order of their discovery. But as M. and Mme Curie are to describe to you their fine work on radium, I will summarise the subject and give you some account of my own research......*

...... We are thus faced with a spontaneous phenomenon of a new order. Figure 1 shows the first print, which revealed the spontaneity of the radiation emitted by the uranium salt. The rays passed through both the black paper which enveloped the plate, and a thin sheet of copper in the shape of a cross. "

"On Radioactivity, A New Property of Matter "A.H Bequerel Nobel Lecture 1903

5.1 Introduction

Imaging using radioisotopes dates back to about 1950 and became more widely available in the 1960's with the more general availability of radioactive isotopes, produced as by-products of nuclear energy programmes. The imaging procedure requires the injection or administration of a small volume of a soluble carrier substance, labelled by a radioactive isotope. The blood circulation distributes the injected solution throughout the body. Ideally the carrier substance is designed to concentrate preferentially in a target organ or around a particular disease process. The radioactive tracer is ideally just an emitter of γ–rays whose energies are in the range 60-510 keV which leave the body, to be collimated and counted using a large area electronic photon detector, sometimes called an Anger camera.The image obtained is a map of the distribution of radioactivity throughout the body, brought about by the interplay of a relatively uniform distribution due to the blood circulation and preferential concentration or depletion, due to local metabolic processes. Three types of imaging are used. The simplest is a single projection or planar image, generically similar to a single projection x-ray radiograph. Starting in about 1975, two tomographic techniques have been brought into clinical use which can provide reconstructed 2D gamma images. These are called SPECT, single photon emission computed tomography and PET, positron emission tomography. PET is probably the most successful in terms of spatial resolution but by no means the most widely used clinical gamma tomographic technique.

Gamma imaging, as we shall describe below, fulfils a different role to x-ray radiography and therefore the relative technical merits of the two areas are not strictly comparable. It is important to understand that relatively poor spatial resolution and signal to noise ratio are inherent problems for all gamma imaging methods in comparison with x-ray or MR methods. There are two main reasons for this. First the small concentrations of radioactive substance, that can safely and ethically be administered, produce a relatively small gamma photon flux and thus the techniques are nearly always limited by statistical fluctuations in the numbers of counted photons, see section 4.3. In other words quantum mottle is a serious problem in most gamma images. Secondly on their trip out of the body, gamma photons are subject to absorption and scattering by the same mechanisms that scatter x-rays. These interactions produce significant distortions of the gamma image, especially at low gamma energies. The utility of gamma scanning lies not in very high spatial resolution but rather in its ability to monitor and image metabolic processes with very high sensitivity. Thus gamma imaging is almost exclusively used to produce functional images. This role arises from the fact that many atoms and molecules involved in specific metabolic functions in both disease and health can be tagged by a radionuclide. The earliest example is that of thyroid imaging. The thyroid gland naturally concentrates about 90% of the body's iodine during normal metabolic function. Gamma imaging of the thyroid gland, using radioactive iodine to investigate metabolic disorders or possible tumours, exploits this natural iodine metabolism. Today gamma imaging is used extensively in the diagnosis of most cancers throughout the body and of heart function and disease.

Over the past thirty years there has been a steady development in radiopharmaceuticals, which allow a very wide range of disease processes to be investigated using one of the gamma imaging methods. In recent years PET has established itself, first as a research tool, and now, as a clinical tool in the investigation of both normal and pathological brain function through the measurement of cerebral blood flow and the metabolic uptake of fluorinated glucose. As we shall see in chapters 6 and 8, recent advances in MRI are making rapid inroads into the area of functional imaging, particularly in imaging brain function thus, just like x-ray CT, the long term future of gamma imaging in some applications is uncertain. For the present however, gamma imaging performs a unique role in medical imaging, which no other technique can fulfil.

This chapter begins with a description of the main radionuclides and carrier molecules in widespread use. Section 5.3 describes the physics of the Anger camera and the general constraints that limit all gamma-imaging techniques. The defining characteristics of the generic types of gamma imaging, Planar Imaging, SPECT and PET are described in turn in sections 5.4 to 5.6.

5.2 Radiopharmaceuticals

The Ideal Properties of Radionuclides and Carriers

Gamma imaging depends critically on the design and manufacture of suitable radiopharmaceuticals; substances that can take part in metabolism and are labelled with one or more gamma radioactive elements. Apart from the ability to target a specific organ or disease, the substances must satisfy a number of criteria in order to be successful. Although a few radionuclides such as ^{133}Xe and ^{123}I are used in elemental form, the majority of radiopharmaceuticals consist of two parts, a carrier molecule and a suitable incorporated radionuclide. The choice of each part has its own constraints, the combination of the parts also needs careful consideration.

In general extremely small quantities of the labelled substance are actually administered to the patient, nevertheless great care has to be taken to ensure that the substance is not toxic and does not itself, unintentionally, alter the processes that are being studied. The carrier has to be soluble and eventually cross cell membranes, after oxidation or other metabolic processes. On the other hand, the carrier must be sufficiently stable that it has time to reach a target site and be concentrated there, before its metabolic demise. There are gamma emitting radionuclides of biologically important elements such as iodine, fluorine and oxygen which allow a labelled, naturally occurring, biologically important molecule to be synthesised. More often however either a suitable radionuclide cannot be made or is too expensive to purifiy, then a compromise is attempted in which a practical radionuclide, which has good properties for imaging, is chemically bonded to a synthesised carrier molecule which mimics the properties of a naturally occurring substance.

The radionuclide chosen for labelling must have the right chemical properties which allow it to be incorporated chemically into a carrier, without unintentionally completely altering the designed metabolic function of that carrier. The radionuclide should ideally only emit γ–rays since any accompanying particulate emission would make a large contribution to the patient radiation dose, by its complete absorption within the patient body, but none to the external image. The isotopes used in PET studies, where the emission of a positron is a crucial initial part of the method, cannot meet this ideal . The energy of the γ–ray has to match well with a standard detector, so that it can be detected efficiently but also allow good collimation, using lead or tungsten shielding. In general terms this means radionuclides emitting γ-rays with energies below about 200keV.The half-life of the isotope must match the cycle time of metabolic process under investigation so that a good fraction of the available gamma flux can be utilised in the experiment. Finally whether or not the labelled carrier breaks down inside the body, natural clearance processes should remove all traces of the labelling agent after the investigation has taken place. Given the number and often conflicting nature of these constraints, it is not too surprising that many substances used in practice represent a compromise. We will illustrate the range of applications with a few examples grouped according to radionuclide, rather than possible carrier molecule type or target disease state. This grouping reflects the fact that a few versatile radionuclides with particularly useful engineering properties have come to be used for a wide variety of applications. A very much more complete account of radiopharmaceuticals can be found in Webb's book. A summary table of the properties of common radionuclides is given in table 2.1.

Technetium

By far the most widely used isotope in gamma imaging is ^{99m}Tc. Its production from the longer-lived precursor ^{99}Mo is described in section 2.4. In the hospital, a chemical generator, containing the high activity, but longer lasting (longer half-life) ^{99}Mo, is used to elute or separate the Tc atoms that are constantly being formed by radioactive decay. The Tc is extracted when required in the form of sodium pertechnate ($Na^+TcO_4^-$). This compound is then manipulated chemically to form a wide variety of different labelled carrier substances for different investigations. In general the Tc is incorporated into a chelate structure, that is to say into a chemical ring structure, in order to maintain the right degree of chemical stability within the body. The γ–ray energy of 140keV is very convenient for gamma imaging. It is sufficiently high that photoelectric absorption within the body is quite small, but low enough for lead (again via the photoelectric effect) to be an efficient shield and collimator material. The half-life is just 6 hours and this is an extremely good match to natural clearance times in many metabolic processes. A few examples will illustrate the range of application of Tc.

^{99m}Tc added to Diethylene-triaminepentaacetic acid, Tc-DTPA is used to assess renal (kidney) function since the molecule DTPA is excreted naturally by the kidneys. Abnormal concentration of the tagged molecule, detected by gamma imaging, indicates malfunction. Skeletal imaging can be accomplished with ^{99m}Tc- Phosphonate, brain function with ^{99m}Tc-propylene amine oxime and the heart using ^{99m}Tc- isonitrile. Some Tc compounds are used to actually exploit the changes in the metabolic fate of a molecule when Tc is incorporated. ^{99m}Tc-HMPAO (hexamethylpropylene amine oxime) will cross the blood brain barrier and accumulate in brain tissue in proportion to the amount of blood flow reaching a particular region. Untagged HMPAO on the other hand cannot cross the blood brain barrier.

Iodine and Fluorine

The halogen atoms, chlorine, bromine, iodine and fluorine are of course very reactive and occur naturally in many biological molecules and processes. Common salt, NaCl is the most well known halogen substance that is utilised by the body. Fluoridation of drinking water is thought to help prevent tooth enamel decay. ^{123}I is a gamma emitter, which was the mother of gamma imaging, especially in thyroid investigations. Its gamma energy of 159 keV and half -life of 15 hours are, like the Tc photon, well matched to gamma camera operation. Its major drawback is that it has to be made by particle bombardment and is thus relatively expensive to produce. ^{131}I is produced by fission and thus it is a cheap isotope but its photon energy at 364 keV makes collimation more difficult. Its half life of 8 days is rather long in comparison with metabolic clearance times. In addition the decay process of ^{131}I involves the emission of electrons (beta decay) which limits its use in imaging.

An iodine atom occupies a similar volume to methyl (-CH2) and ethyl (-C_2H_5) groups and has similar chemical affinities to carbon. Thus iodine radionuclides can be made to replace stable iodine atoms in natural iodated molecules or substitute for the methyl and ethyl species with or without useful alteration to the chemistry of the resulting complex. In some substances the exact position of the iodine atom within the molecule alters the chemistry in a useful manner. Iodinated ring structures with ortho and para substitutions can result in different rates of crossing the blood brain barrier.

The fluorine isotope ^{18}F is a positron emitter with a half-life of 110 minutes. This is one of the most commonly used isotopes in PET imaging. Fluorine is incorporated into deoxyglucose, to form fluorodeoxyglucose, FDG, and this compound has been very successfully used in PET studies of regional changes in metabolic rate within the brain, and oncology.

Xenon and Oxygen

^{133}Xe, a gamma emitter, used on its own in perfusion studies was, for some time, a standard method of assessing cerebral blood flow. The gamma energy is quite low at 81keV and the half-life is rather long, 5.3 days but it is a fission product and is thus very cheap to produce. In addition, being an inert gas, Xe crosses the blood brain barrier with ease. It is inhaled or injected into the blood stream until the gas has perfused the whole brain. Once equilibrium is reached, the administration is stopped and the rate at which ^{133}Xe clears from the brain is measured. Regions well supplied with blood will clear more quickly than blood starved regions, so that the measurements provide a measure of regional cerebral blood flow. Absorption and scattering of the 81 keV γ–ray are significant and the long half-life makes specific activity rather low. For these reasons Xe perfusion studies have never been capable of very high spatial resolution.

In recent years PET studies using of ^{15}O has become a standard research method of studying regional cerebral blood flow, rCBF. The radionuclide, ^{15}O is a positron emitter with a half-life of 2minutes. It is produced by particle bombardment in a cyclotron , see section 2.4. Generally ^{15}O is incorporated into water and injected into the blood stream. The emitted positron has a very short lifetime in tissue; it combines and annihilates with a negative electron to form two γ–rays, each with a precise energy of 511 keV. The fact that these two γ–rays travel away from the site of annihilation in almost exactly opposite directions forms the basis of the PET technique.

5.3 Gamma Cameras

All modern applications of gamma imaging use one or more large area multi-detectors, called gamma or Anger cameras, named after the inventor. A standard arrangement is shown in figure 5.1. All gamma cameras have three main parts, a collimator, a NaI crystal scintilllator / PM tube multi-detector and a hardwired logic circuit for position sensitive photon counting, and energy analysis. The gamma camera is used in both planar imaging and SPECT studies. The PET camera utilises scintillation detection but requires a different geometry. This is described in section 5.6.

The Collimator

In any single photon study, the gamma photons are emitted randomly and isotropically, potentially from any part of the body. In order to obtain a projection image, analogous to x-ray radiography, a line of sight has to be imposed and this is achieved using a collimator. Many different "focussing "collimator designs have been tried and some of these are still occasionally used for particular studies. The standard, for both planar imaging and SPECT is the parallel hole collimator which consists of a lead plate about 1 cm thick, pierced by an array of many thousands of small diameter holes, each no more than 1mm in diameter. Each hole is a tube having an aspect ratio of at least 10:1 allowing only those photons travelling close to its axis to traverse the plate. More obliquely angled photons are absorbed by the lead between the holes. The collimator thus defines a line of sight and a projection direction perpendicular to the face of the collimator. γ–rays emitted along the collimation axis from any depth below the face are collected. The collimator is essential but necessarily leads to a drastic reduction in the efficiency of photon detection, since many perfectly acceptable gamma trajectories stop in the lead spaces between the holes, and overall, relatively poor use is made of the available patient dose. Typically the collimator will transmit less than 0.1% of the incident photons and thus it is the main factor limiting the overall detection efficiency of the gamma camera.

The Scintillator Crystal

Once through the collimator, the incident γ–rays produce low energy scintillation photons in the thallium activated sodium iodide , NaI(Tl), crystal and these are detected by the PM tubes . This is described in section 4.4. Standard gamma cameras are now typically 50cms across allowing large areas of the patient to imaged in a single exposure. In modern, multipurpose systems two large area cameras are used simultaneously in order to intercept more of the emitted γ-rays. Each gamma camera has one, large area single crystal of NaI (Tl) backed by many PM tubes. This is illustrated in figure 5.7. The use of a single crystal is essential to ensure that the secondary photon light can travel, without attenuation, throughout the scintillator volume and enter as many of the PM tubes as possible. Typically the NaI crystal is about 10mm thick and this ensures that about 90% of 150keV photons, traversing the collimator, are actually stopped in the scintillator. A fraction of the total scintillation light produced by each gamma event enters the PM tubes.

Count Rate

The overall sensitivity of a gamma camera is quoted as a count rate for a given source strength , cps/ MBq . Typically a modern gamma camera will achieve of order 5-10 cps/MBq. Thus for every million γ–rays emitted by a radioactive source in each second only about 10 are actually counted. This reduction is caused by attenuation in the body tissue between source and detector (10-50% transmission depending on the depth), by the collimator (0.1% transmission) and finally the area covered by the detectors. In some investigations the total count rate can be quite large >100kcps, and this causes problems associated with the deadtime of the detector. If γ–rays arrive in too quick a succession, then a proportion of the counts is missed, because the

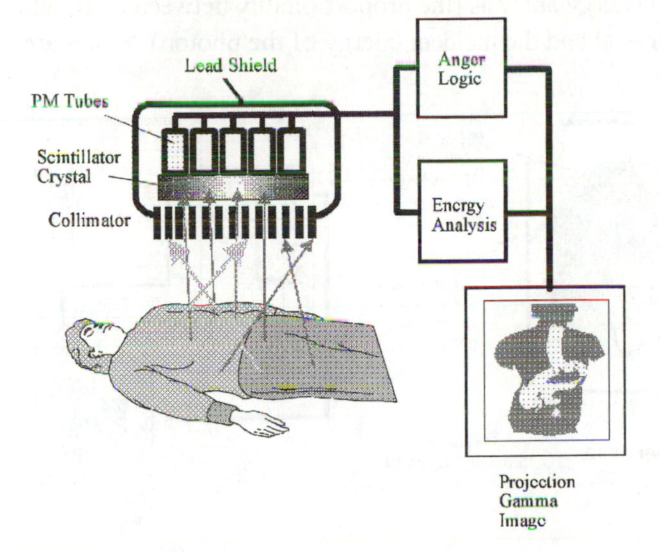

Figure 5.1 Large area gamma projection imaging using an Anger or gamma camera. Incident γ–rays are collimated, converted to visible photons then counted by the PM tubes. The XY positions of initial gamma impacts in the crystal are estimated using the Anger logic described in figure 5.2. The pulse height analyser separates primary from lower energy compton scattered γ–rays by energy analysis

electronics is still dealing with one event when another arrives. A typical camera will experience a 20% drop in apparent count rate when the total true rate exceeds about 150 kcps. The error in recorded count rate increases steadily with the incident photon flux. Typically the maximum possible recorded count rate is about 250 kcps. Beyond this level, increases in incident flux produce no changes in recorded rate and so the detector is saturated.

XY Position and Energy Analysis

Two more features have to be added to the gamma camera in order for it to produce useful 2D image information. The XY position of the initial gamma interaction within the crystal is required and there has to be discrimination between primary γ–rays and lower energy Compton scattered gammas. The first uses an electronic circuit sometimes called the Anger logic and the second makes use of photon energy analysis (the proportionality between total light output of the NaI(Tl) crystal and the incident energy of the photon). γ-rays are

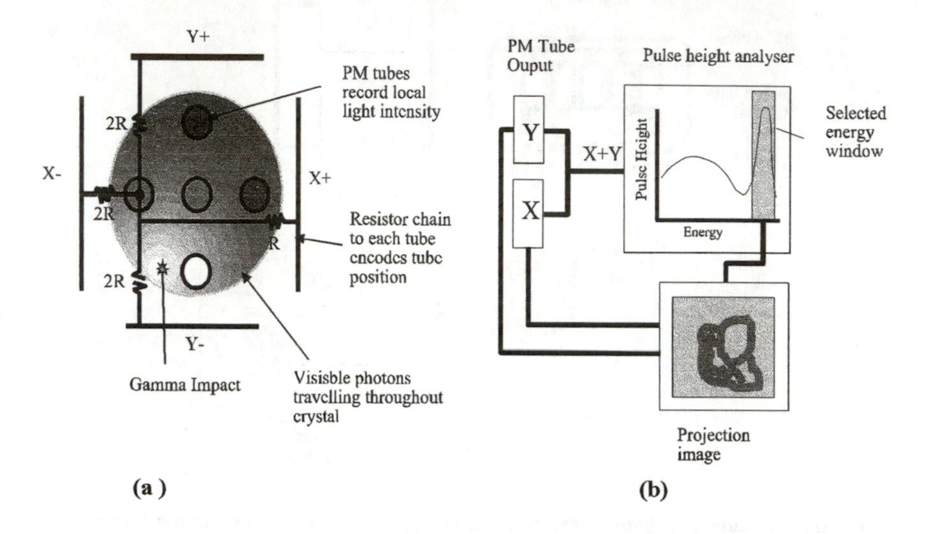

Figure 5.2a The Anger Logic. Light photons created by the initial gamma impact travel throughout the NaI crystal. Some of the light enters the array of PM tubes, so that each tube records a local light intensity proportional to the solid angle subtended by the tube aperture at the point of gamma impact. Each tube output is connected via scaling resistors to the four summing amplifiers, two for each axis, X,Y. b)The difference signals provide estimates of the X,Y positions of the initial gamma impact. The sum of all signals is fed to the pulse height analyser to discriminate against lower energy Compton scattered photons

detected first by converting their energy (~140 keV) to lower energy, visible photons (~2-4 eV), that are more easily stopped in subsequent parts of the detector. Thus, when the NaI stops a γ–ray, a shower of visible photons is produced in the crystal which travels throughout the crystal, ideally without internal re-absorption. A modern gamma camera will have up to 60-90 PM tubes arranged behind the crystal to detect these photons. Some visible light will enter just about every PM tube but those closest to the initial event will receive most light. A crude XY position could be obtained simply by noting the tube with the largest current signal for each event. This however would limit spatial resolution to the size of the PM tube window, which is about 2-3 cm in diameter. Finer resolution is obtained using Anger's circuit, figure 5.2a. Each tube output is connected via electrical resistors to four summing amplifiers X^+, X^-, Y^+, Y^-. The ratios of the resistor connections to the X and Y amplifiers determine the tube's position within the array, by dividing the current signal from each tube in these known proportions. Summing the outputs from all tubes, scaled by their resistors according to tube position, allows the X, Y position of the gamma impact to be estimated to about 5mm. The position co-ordinates of the event X_{event}, Y_{event} are determined by

$$x_{event} = \frac{k(X^+ - X^-)}{\Sigma}$$

$$y_{event} = \frac{k(Y^+ - Y^-)}{\Sigma} \qquad \ldots 5.1$$

$$\Sigma = X^+ + X^- + Y^+ + Y^-$$

where k is an empirical constant of the particular camera. The sum of all four amplifier outputs is a measure of the total visible light and thus the initial energy of the incident γ-ray. The summed signal is sent to a pulse height analyser circuit, which performs the energy analysis. This is illustrated in figure 5.2b. Both XY position and energy analysis are carried out by dedicated hardwired circuits to increase overall counting rate and reduce dead time effects, caused by detector paralysis after a gamma impact. Those photon counts that are not vetoed by the energy analyser are stored, according to their XY position, and a single projection image is accumulated as the prescribed counting period progresses.

Gamma Camera Performance

The spatial resolution and sensitivity of the gamma camera are determined by the geometry of the collimator, gamma photon scatter and absorption, both within the patient and within the collimator, and finally, the statistical uncertainty associated with a relatively small number of counts per image pixel. The following sections describe the main factors determining performance.

The Inherent Low Count Rate in Gamma Imaging

Although there are many points of equivalence in the methodologies of x-ray radiography and gamma imaging there are some crucial differences which together are ultimately responsible for the relatively poor practical spatial resolution of gamma imaging. Figure 5.3 summarises the main factors that reduce the count rate in typical gamma imaging schemes. All of x-ray radiography takes place in transmission, using a source of radiation whose strength and position are known very accurately. Gamma imaging, on the other hand, is seeking to determine where and with what strength the sources of radiation are located. Generally the source is injected into the body as a few cc bolus of solution which is then distributed by the circulation to nearly all parts of the body. Differential uptake of typical carrier substances is generally quite low; a good figure may be a 20% difference in contrast between the target and the uniform background. Iodine is of course an extreme counter example since 90% of the body's iodine is contained within the thyroid gland. In addition, radioactive sources emit their radiation not in a single direction, like an x-ray beam, but isotropically, that is to say with equal strength in all directions. In order to restore a definite line of sight into the geometry, the standard gamma camera has to make use of a lead collimator that selects just those photons travelling in the right direction to form a useful projection image. Unfortunately in doing this the collimator has to reject over 99.8% of all incident photons. Finally, all gamma cameras, because of their finite physical size, can only intercept a fraction of the possible solid angle of photons. All these factors together contribute to the fact that even PET, which has the most favourable geometry, is only about 0.5% efficient in terms of its utilisation of the available dose of radiation. This in itself would not be so serious if the available gamma flux were as high as the x-ray flux used in radiography. It is illuminating to compare the available photon fluxes per cc of target tissue for x-ray radiography and gamma imaging. A typical diagnostic x-ray beam would have a photon flux of about 10^{10} photons s^{-1} cm^{-2} at the patient body surface. Since the attenuation through uniform tissue varies as $I = I_0 e^{-\mu x}$, the change in intensity due to a cube of side dx at a depth of x cm is given by $dI = -\mu I_0 e^{-\mu x} dx$. If we assume a constant linear attenuation coefficient of 0.2, and a cube size of 1 cm, at a depth of 10 cm, then each cc of tissue at this depth can be said to remove about 2.7×10^8 photons per second

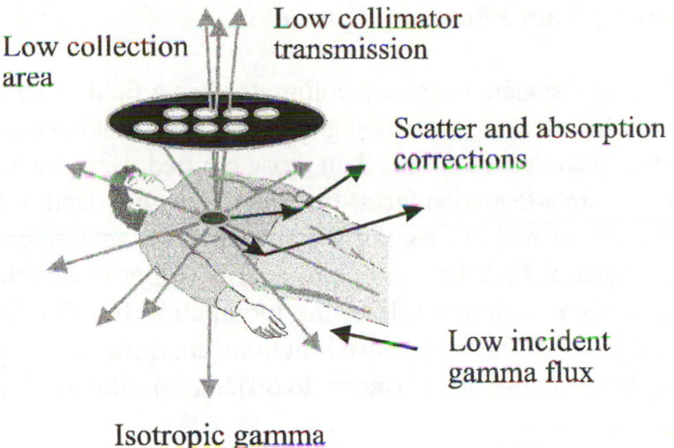

Figure 5.3 The several factors that reduce the gamma photon count rate in gamma imaging as compared with x-ray radiography

from the beam and thus contribute this amount to the radiograph. A typical gamma imaging session will start with the injection of a radioactive source with a strength 500 MBq, (see chapter 3). If we assume that this activity is distributed instantaneously throughout the blood pool of about 4 litres then, very approximately, each cc of blood has an initial activity of about 1.25×10^5 MBq $cc^{-1} = 3.3\mu Ci\ cc^{-1}$. This is already three orders of magnitude below the x-ray case. This radioactivity is subsequently emitted over a sphere, so that at a distance of 10 cm we have a flux of $1.25 \times 10^5 / 4\pi 10^2 = 100$ photons /second /cc blood / cm^2 detector. Again we have ignored attenuation of the γ–rays on their way out of the target volume. It is clear that the available flux in gamma imaging is several orders of magnitude below that of the x-ray case and, when the other common factors such as overall detector size, detector efficiency are taken into account the actual count rate is relatively very small. In a typical single photon imaging procedure, a total of one million counts might be required to obtain statistical significance. This total number of counts can take many minutes to collect. During this time a certain amount of random patient movement is likely and this degrades the ultimate spatial resolution. By comparison, modern x-ray CT images are obtained in less than 1 second.

Gamma Imaging Source Response Functions

Each hole of the gamma camera collimator has a field of view that widens with depth into the patient. In addition gamma photons that emanate from deeper positions have to cross more tissue than those emitted from shallower locations and thus there is an attenuation factor that increases with depth into the patient. The overall performance of imaging systems like the gamma camera is often described in terms of their line and point source response functions, LSF and PSF. These perform a similar role to the modulation transfer function, MFT described in chapter 4.These response functions are quite simple to understand by imagining the response of the camera to different positions of a point source

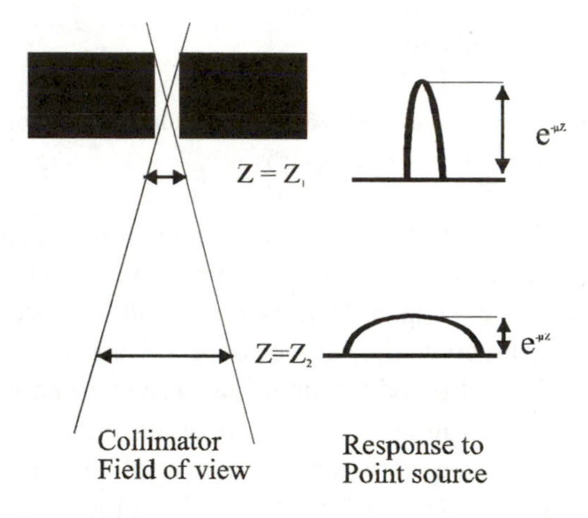

Figure 5.4: The schematic effect of collimator hole field of view and tissue absorption on the gamma camera response to a point gamma source source increases, so the amplitude of the camera's response decreases exponentially, as a result of gamma photon attenuation in the body. Close to the face of the collimator, the source will be imaged as a sharp point. As depth increases, the test source can be moved sideways over a sizeable distance in the XY plane

of gamma photons. An ideal camera would produce tight, sharp responses of equal height at all positions and depths within the field of view. Figure 5.4 shows that the typical gamma camera does not give this ideal response. As the depth of the source increases it can be moved laterally over the collimator hole field of view without making very much change to the image; more off axis scattered photons are accepted inside the cones defined by the collimator holes. Thus the sensitivity of a gamma camera decreases exponentially through attenuation and its lateral spatial resolution decreases with source depth into the patient as result of collimation. This is why all single photon gamma imaging systems are inherently less successful than PET, in terms of spatial resolution.

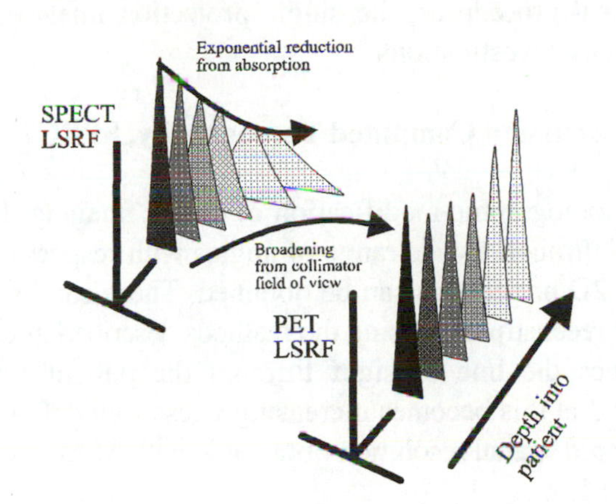

Figure 5.5 Quantifying the spatial resolution and response of a gamma camera using a known radioactive source geometry. **SPECT:** As the source to collimator depth increases so the gamma count rate goes down as result of photon absorption and scattering within the patient. At the same time the field of view of the collimator increases which smears the line response by an increasing amount in a lateral direction. **PET:** The PET camera has no front collimator, rather the two ends of the line of sight are relatively well localised using coincidence detection. The absorption correction for coincidence events depends on the total thickness of tissue traversed by both photons not the depth into the patient. Together these lead to a tighter response function throughout the depth of the patient

5.4 Planar Imaging

The gamma scan analogue of the x-ray radiograph is called Planar Imaging. It involves counting gamma photons, keeping the Anger camera in a fixed position and orientation with respect to the patient. The collimator on the front of the Anger camera defines a definite line of sight for photons reaching the detector crystal. The variation of intensity with position in the camera is then a 2D image (1 projection) of the variation of radioactive density immediately below the camera. Planar imaging with a parallel hole collimator, like the x-ray analogue, provides no depth information, rather the activity over the entire patient thickness below the collimator is integrated to form the image. Planar gamma imaging, like its x-ray analogue, has the merit of simplicity and relative speed and thus, in many clinical procedures, the single projection image is the chosen method for preliminary investigations

5.5 Single Photon Emission Computed Tomography, SPECT

SPECT is one tomographic modification of planar imaging. If the gamma camera can be tilted through a wide range of angles with respect to the patient, then a collection of 2D projections can be obtained. These can be combined to make a tomographic reconstruction using the methods described in chapter 1. The collimator determines the line of sight through the patient for any given projection. The fact that this becomes increasingly less well defined with depth contributes to a reduced spatial resolution obtainable with SPECT in comparison with PET. see figure 5.5.

Improvement in the resolution of SPECT can be made by reducing the effect of the inevitable absorption and scatter through the use of empirical correction factors. One way to do this is to take the geometric mean of counts obtained in anterior and posterior views, 180° apart. The data for each projection can then be made up from the geometric mean of the two counts. Figure 5.6 illustrates how this results in an absorption correction, for a given view, that varies with overall patient dimensions, L rather than the depth of the source into the patient, x. The geometric mean of the two counts results in an absorption correction that varies as $\exp(-\mu L)$ rather than $\exp(-\mu x)$.

In both SPECT and PET the final image is obtained by reconstruction. This in itself amplifies the effects of low signal to noise ratio. The standard filtered backprojection scheme involves the ramp filter discussed in appendix B, its effect is to amplify high spatial frequency noise since its amplitude increases with spatial frequency. Random statistical errors from projection element to element look like high spatial frequency noise in the data. In addition each reconstructed pixel in the image takes contributions from the entire set of projections and this amplifies the effects of overall signal to noise ratio The variance or root mean square statistical fluctuation from pixel to pixel within an image of a disc of uniform activity is given by

$$RMS\% = \frac{120 \times (number\ of\ pixels)^{3/4}}{(Total\ Count)^{1/2}} \qquad \ldots 5.2$$

Thus if an image is to be reconstructed into $64 \times 64 = 4096$ pixels from a total of 1 million counts, then RMS% = 61%. Naively we might have imagined that there are effectively $10^6 / 4096 = 244$ counts per pixel and thus a statistical uncertainty of 6% on each pixel value. Both SPECT and PET reconstruction software

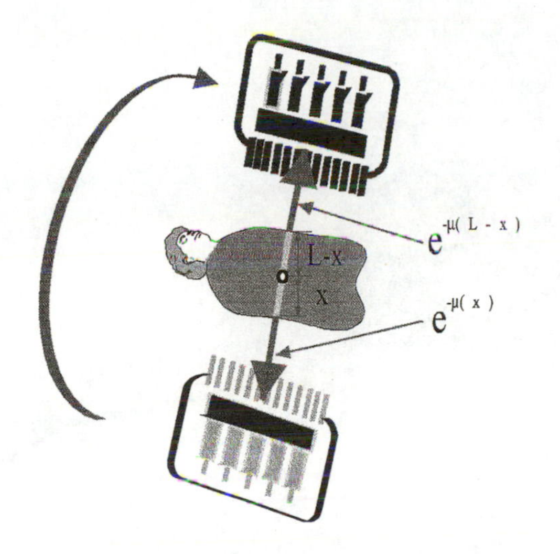

Figure 5.6: Reducing the effect of absorption corrections in SPECT and Planar imaging by combining counts from above and below the patient. The combined (multiplied) counts vary roughly with the relatively constant total path length ,L, through the patient rather than the variable depth into the patient. The product of the two counts only depends on L not x

methods have partial remedies. Clearly the number of reconstruction pixels has to be chosen carefully and spatial filtering is used. Thus instead of using the simple Radon filter, a ramp, modified by one of several so-called window functions such as , Hanning and Parzen are used, after the manner described in appendix B. In effect all these schemes are trading a reduced spatial resolution performance for a reduction in statistical fluctuations. In other words these are sophisticated low pass filter or spatial averaging schemes. These problems limit the spatial resolution obtainable by SPECT to 8- 15 mm . This is an order of magnitude worse than that of x-ray CT and MRI but both SPECT and Planar imaging provide invaluable functional information not obtainable at all by x-ray CT. PET achieves a better spatial resolution because the signal to noise ratio is more favourable and more accurate attenuation corrections can be obtained.

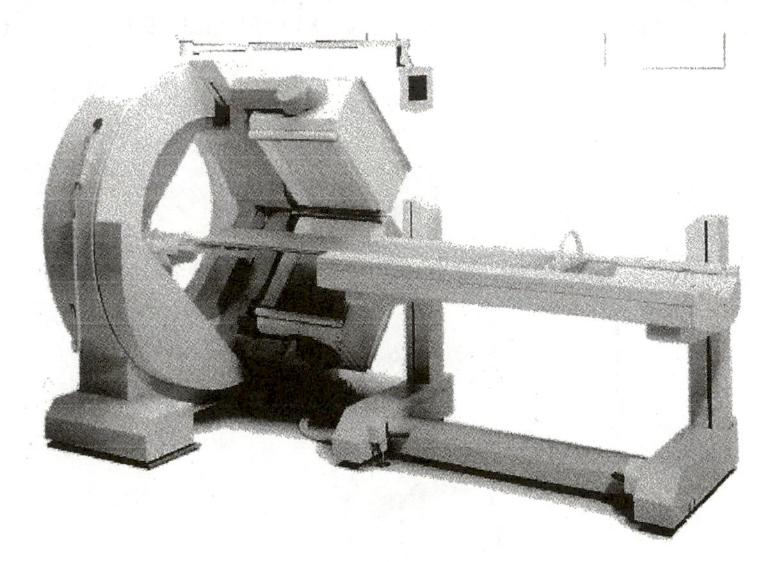

Figure 5.7: A modern multipurpose gamma imaging installation. There are two square gamma cameras which can be rotated independently around the patient to produce SPECT data, or held fixed to produce a planar scan.(Seimens)

5.6 Positron Emission Tomography — PET

PET or positron emission tomography is a more recent clinical modification of gamma imaging that makes use of the two γ–rays, emitted simultaneously, when a positron annihilates with an electron. The tracer, introduced into the patient, is a positron emitter such as $^{15}O, ^{11}C, ^{18}F$, bound to a suitable carrier. The radioactive decay produces a positron, e+ with an initial kinetic energy of ~ MeV. Although it has a high initial kinetic energy, the charged positron has very strong interactions with the surrounding tissue. These interactions transfer the positron kinetic energy to the tissue in a series of scattering events that produce a broad spectrum of x-ray energies (bremmstrahlung radiation) and a shower of photoelectrons. Typically the positron travels less than 5 mm in biological tissue from its point of emission. The high electron density of biological tissue ensures

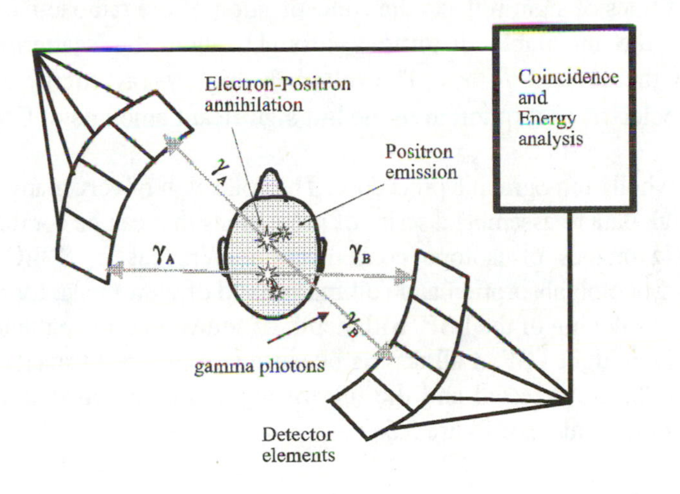

Figure 5.8: Positron Annihilation. The injected tracer is a positron emitter. The emittedpositron travels about 5mm before annihilating with an electron to form two γ–rays which travel away from the annihilation site in opposite directions. Coincidence detection is used to discriminate against spurious background counts and define a line of sight

frequent electron/positron encounters; one of these will result in the disappearance or annihilation of the two particles, replacing them with two γ–rays. The conservation of energy demands that the energy of the two γ–rays is supplied by the total energy of the positron and the electron. By the time the annihilation takes place, nearly all the initial kinetic energy of the positron has been dissipated. The annihilation event results in the conversion of the combined electron/positron rest mass, 1.024 MeV into photon energy. Two γ–rays, each with an energy of nearly 512 keV are produced to conserve energy. In addition since the electron/ positron pair is essentially at rest, when annihilation takes place, the two γ–rays have to leave the annihilation site, travelling in nearly opposite directions in order to conserve linear momentum. The two γ–rays together define a line that passes close to the point of emission of the original positron. Coincidence detection is employed with a circular position sensitive detector, encircling the patient. A single coincidence event, with its two detected ends, defines a line of sight through the patient, somewhere along which, the annihilation took place. Since the positron mean free path is very short the distribution of lines of sight reflects the concentration of the radioactive tracer, leaving aside the inevitable distortions brought about by scattering and absorption of the emitted γ–rays. The relatively high γ–ray energy ensures minimal photoelectric absorption in tissue but significant amounts of Compton scattering.

PET is a wholly tomographic procedure. The collection of very many events provides enough data to assemble a series of projections that can be combined to reconstruct 2D images of isotope concentration. Whereas in SPECT, the combination of photon absorption and collimator field of view make for a rapid decrease and broadening of the LSF, with depth of source into the patient, PET retains a relatively tight LSF at all depths because two photons traversing the entire body width are detected and the line of sight is determined from two external detection points, see figure 5.5.

The PET Camera

The key difference between single photon and PET imaging arises from how the two schemes define a line of sight. In SPECT, as we have seen, this is done using a collimator, which is inherently inefficient, since many "good" photons are stopped by the lead along with those travelling on oblique trajectories. In PET the two γ–rays, travelling in nearly opposite directions, themselves define a line and thus a collimator is not required, as long as position sensitive detection is sufficiently precise. The absence of a collimator in PET makes a major contribution to an overall increase in its efficiency and spatial resolution with respect to SPECT.

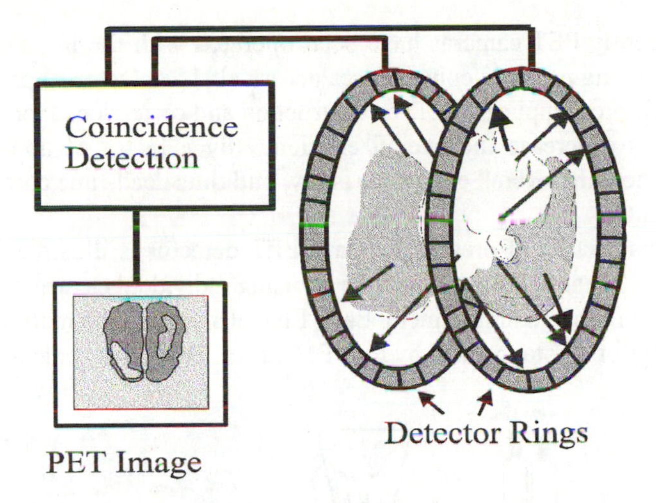

Figure 5.9: A schematic PET Camera. A typical camera has 16 rings each of diameter 760mm each containing 384 detector blocks. As originally conceived coincidences were allowed in 3 adjacent rings between one detector block and the 5 opposite blocks across the axis of the camera. The absorbing septa between rings reduce scatter from adjacent slices but severely reduces the count rate for each slice

The PET Detector

The basic PET camera is shown schematically in figure 5.9. It consists of a number of segmented 360 ° ring detectors with a common axis, arranged at intervals of about one centimetre apart. The majority of PET cameras are optimised for brain studies and thus the counter ring is of a considerably smaller diameter than its equivalent in an x-ray CT machine. Slices, for reconstruction , are chosen by using particular detector rings along the axis. Each segment within a ring is a position sensitive detector. Counts from all the segments in a number of rings are collected and those counts, arising from a given segment and a bank of five segments across the diameter of the camera, are examined for coincidences. In many cameras, data for a given slice is obtained by examining just three rings for coincidence. Absorbing septa placed between rings effectively stop all γ–rays, from one slice, contaminating neighbouring slices as accidental and single photon events in the detectors. This increases the ratio of true to false coincidence counts but significantly reduces the overall counting efficiency of the camera. Recently PET cameras have been operated with the interleaved septa retracted allowing possible coincidences across all 16 detector rings. Although this leads to more complicated 3D reconstruction and correction algorithms it has been shown to increase the overall efficiency by a factor of about four, in situations where the overall count rate is low and thus dead-time corrections are less important.

The construction of present standard PET detector is illustrated in figure 5.10. Position Analysis is achieved here by using individual elements rather than area detection in the gamma camera. Each PET element is a bismuth germanate, BGO scintillator crystal backed by four PM tubes. BGO has some advantages

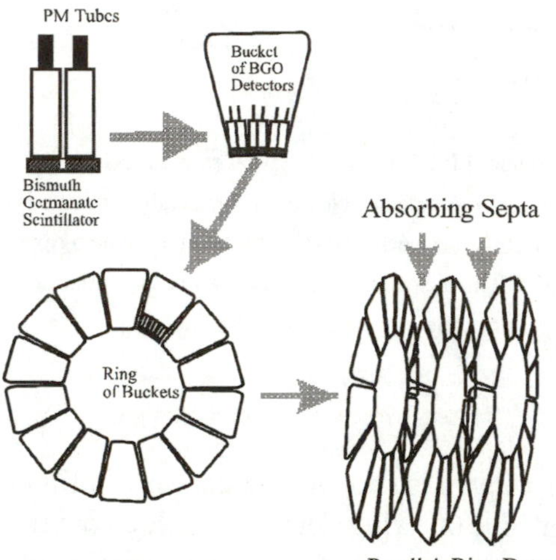

Figure 5.10: The construction of the PET camera detector. A Bismuth Germanate, BGO, scintillator backed by four PM tubes forms the basic element. Four or more of these are grouped into a "bucket" or cassette. Many cassettes are arranged to form one ring. Several rings are arranged along a single axis, with retractable absorbing septa between the rings.

and some disadvantages as a scintillator with respect to NaI(Tl), the standard crystal used in SPECT. BGO has a high density and relatively large atomic number and hence an increased stopping efficiency for 510keV gamma photons. Unlike NaI (Tl), BGO is not hygroscopic and thus the detector is simpler to construct, thus reducing the cost of producing many elements. Its bad features are that the visible light output is 8 times less than that of NaI(Tl) and the decay time of light pulses is 300ns (80 ns for Na(Tl)). This reduces the maximum count rate obtainable without dead-time corrections. Each BGO crystal is segmented into an 8 by 8 matrix to improve the spatial resolution achievable from the four PM tube outputs. A single BGO crystal along with its PM tubes and associated electronics is referred to as a block. In most systems a number of blocks are grouped into a bucket or cassette that can be replaced as a single unit. Many cassettes are assembled form one ring and then the complete PET camera consists of many axial rings, generally interleaved with retractable absorbing septa.

True and False Coincidence Events in PET

PET depends critically on the validity of the events captured under coincidence and there are three classes of false events that contaminate the true coincidence count tally. These, if left uncorrected, significantly degrade the final image. Figure 5.11 illustrates the four types. Accidental coincidences can occur simply because two completely unrelated photons happen to arrive within the electronic coincidence time window of about 20ns. Scattered coincidences arise from a single annihilation, but scattering of one or both photons destroys the correct line of sight. The main mechanism is Compton scattering, involving a change in a photon energy, but the energy resolution of the scintillators is insufficient to discriminate against small angle Compton scattering. The inevitable detection of single events by the scintillator crystal/PM detectors does not directly alter the final image but does set an upper limit of the overall count rate and effective dead-time for true coincidences. If the detectors are flooded with counts then the dead-time, during which subsequent counts are missed, becomes important. Scattered coincidences contribute between 10 and 30 % of the trues rate and accidental coincidences contribute another 15%. A typical detector will "see" 50 -100 times as many single events as coincident events. This places a well defined limit on the maximum amount of activity allowed in the FOV.

PET Correction Factors

As we have already seen, the coincidence count contains accidental and scattered events, for which the line of sight has been destroyed. Considerable

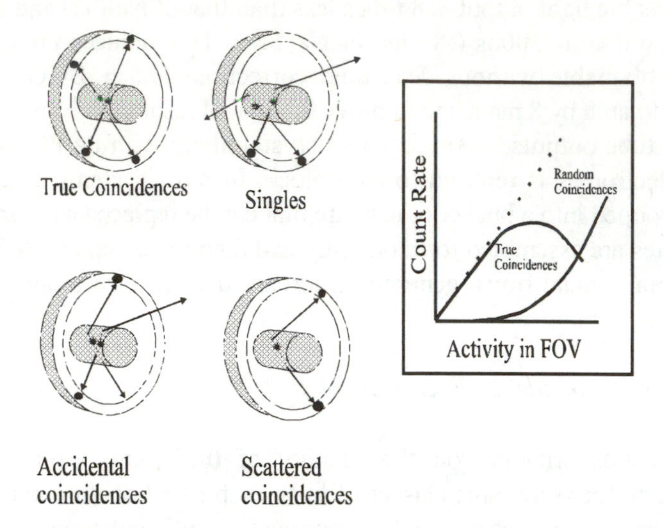

Figure 5.11: The four classes of detected event that make up the total counts in a PET camera. The singles rate affects the maximum count obtainable by limiting the effective dead time for coincidence detection. Accidental and scattered coincidences contribute to random noise to the image as detected events with incorrect positions. The schematic graph shows a count rate loss curve in which the true rate saturates slightly before true the true coincidence rate equals the accidental rate

effort is put into estimating correction factors to reduce the effects of these counts and to reduce the effects of attenuation and scatter. The composition of the coincidence count rate for a given instrument is assessed using a variety of phantom objects containing radioactive material, placed in the FOV. This provides a calibration and a set of correction factors, which can be applied to each patient investigation. Corrections for attenuation within the patient, on the other hand, have to be carried out for each and every investigation.

Accidental coincidences can be estimated experimentally by deliberately offsetting the time window. These delayed coincidence counts could not possibly arise from true annihilation pairs and thus must be spurious. Since these events are randomly distributed in time they provide an estimate for the accidental contribution to the total count obtained with no time delay. There is no comparable experimental method for estimating the scattered coincidence rate and thus software corrections are applied. These are based on models of how the probability that a photon will be scattered, varies with the position of the initial annihilation event within the patient.

Patient attenuation corrections are obtained experimentally, using data obtained in transmission scans. Typically, before the real PET experiment takes place, three or more radioactive line sources are rotated around the patient and those γ-rays that have entirely crossed the patient are counted. During a period of several minutes, enough different trajectories can be accumulated to calculate an empirical attenuation factor for every real gamma pair trajectory obtained in the subsequent PET scan. It should be noted that this empirical correction method could not be used in SPECT because the attenuation of single photons varies with the depth of the source rather than the overall cross section of the patient. Finally dead time corrections, obtained from standard phantom experiments are used to correct the true coincidence count rate for variations in dead time arising from differences in count rate with slice position and angle.

The theoretical resolution of PET is 2-5 mm (set by the positron mean free path); this is a factor of about two better than SPECT. The patient dose implications of PET are more complex than those of SPECT. All the initial positron energy is deposited in the tissue and this adds to the energy deposition, caused by Compton scattering and photoelectric absorption of the outgoing gamma photons. The separate transmission scans used to determine attenuation correction factors add a further dose burden to the final image. Typically the amount of radioactivity administered for a single PET study is similar to that used in SPECT, with typical administered doses of 500MBq.

PET has not yet found widespread use as a whole body gamma imaging scheme. In fact PET is only now beginning to be used as a clinical tool. This is mainly due to the cost of the PET camera and the need for a nearby cyclotron to produce suitable short half-life positron emitting isotopes. Most clinical PET studies have traditionally been performed in oncology and cardiology but more recently PET has been used with considerable success to localise and define the extent of epileptogenic brain regions in patients suffering drug refractory epilepsy. Current clinical use of nuclear diagnostic medicine is described in section 8.5.

In recent years, with the widespread introduction of general purpose dual headed Gamma cameras, there have been attempts to perform PET studies using this arrangement, see figure 5.7. This is attractive because a single instrument can then perform a wider range of investigations and because NaI(Tl) is an intrinsically better detector than BGO. PET-like studies can be accomplished with a collimator designed for 511keV rather than 140keV photons or by the addition of coincidence electronics. In the former the collimator significantly reduces spatial resolution and through Compton scattering in the collimator itself, diminishes the quantitative information available from the dedicated PET system.

5.7 Images from Gamma Imaging

SPECT

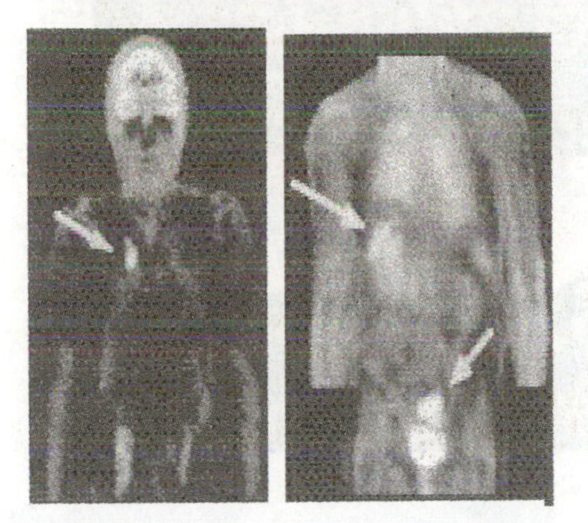

Figure 5I.1: Two whole body scans showing *** take up around cancers in the Lung and the colon (Seimens)

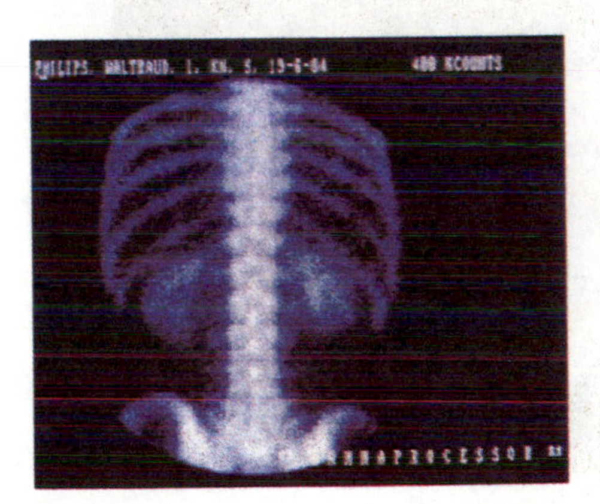

Figure 5I.2: SPECT Bone study (Philips Medical Systems)

Figure 5I.3: A SPECT study of blood Perfusion around heart in normal and abnormal cases (Philips Medical systems)

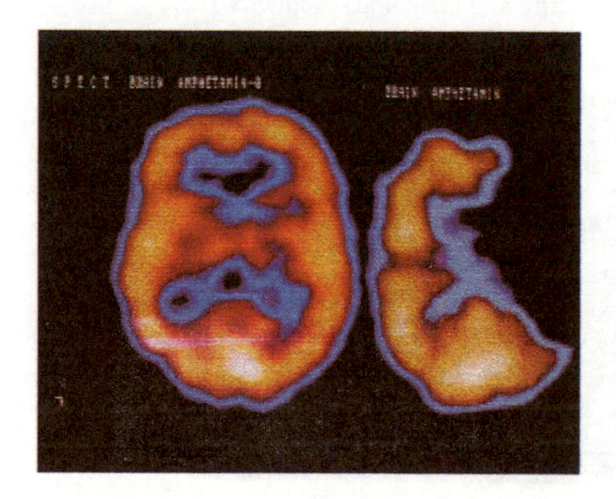

Figure 5I.4: SPECT study of amphetamine uptake in normal and abnormal brains (Philips Medical systems)

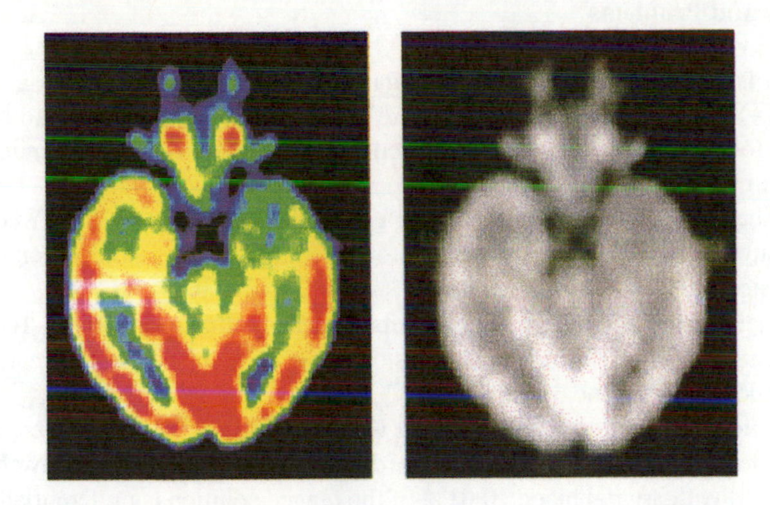

Figure 5I.5 : A PET study of *** take up in an epileptic brain.
St Thomas PET centre

Questions and Problems

1. What is the source of contrast in any gamma image ?.
2. The 141 keV gamma photon produced by technetium is considered to be ideal for imaging. Why is this preferred to very much lower or very much higher gamma energies?
3. Describe the main features of the Anger gamma camera. How are the XY co-ordinates of the detected gamma photon encoded ? Why is pulse height analysis used?
4. Calculate the acceptance angle for a tubular lead collimator of length 1cm and diameter 1mm. Calculate the areas of tissue is "seen" by the collimator at the depths of 10 and 30 cm.
5. A solution containing a gamma emitter with a half life of 30 min is injected into the bloodstream of a patient in order to image an organ that is well supplied with arterial blood. 0.01% of the tagged solution is preferentially taken up by the organ during each heart cycle.
   a)Sketch the time evolution of the count rate recorded by a gamma detector
   b)Estimate the time after the injection at which the count rate would be a maximum.
   c)What other processes might have to be taken into account in determining this time?
6. A typical Anger camera parallel hole collimator consists of a lead plate 1cm thick and 40 cm in diameter with $3x10^4$ circular holes of diameter 1mm drilled perpendicular to the large faces of the plate. Make an estimate of the attenuation coefficient for 140 keV photons normally incident on the septa of the collimator. Estimate the fraction of normally incident photons that will get through the collimator. The linear absorption coefficient of lead is obtained from the following observation: 1 mm of lead reduces the intensity of a 140keV gamma beam by a factor of 1000.

6 MAGNETIC RESONANCE - MRI

[In a liquid], ".. the nucleus rides out the storm like a well balanced gyroscope on perfect gymbals" , **[Purcell (1948)]**

6.1 Introduction

The imaging plower and versatility of MRI arises from the variety of contrast mechanisms that the resonance process provides. Whereas x-ray CT is stuck with just one parameter, the linear x-ray attenuation coefficient, in conventional water proton MRI there are three primary parameters or flavours that contribute to contrast in a typical image. These are the " free water density", longitudinal relaxation time, T1 and transverse relaxation time, T2. The water proton resonance can also be made sensitive to fluid flow and tissue magnetic susceptibility. The nuclear magnetic resonance technique can be applied to nuclei of biological importance other than water protons. Finally at high frequency resolution all NMR nuclei can provide information about molecular structure. It is likely that in the coming decades, applications of spatially localised NMR spectroscopy will become a clinical reality and be incorporated into existing MR facilities.

Our aim in this chapter is to provide a semi-quantitative account of the main physical principles of NMR and MRI applied to water protons. We start with the main principles of pulse NMR, namely RF excitation and transverse and longitudinal relaxation time. This is followed by a qualitative description of the microscopic processes responsible for magnetic relaxation. The section on NMR is completed with a description of the main pulse sequences, used originally in NMR, but adapted for use in MRI. Spatially localised NMR or MRI using Fourier reconstruction or spin warp imaging is described in section 6.4. This is the most versatile and commonly used protocol for anatomical imaging. The last three sections provide an introduction to fast imaging methods, flow imaging and finally we give a brief treatment of the main types of artefact that can corrupt MRI images.

We have already introduced the microscopic principles of nuclear magnetic resonance in section 2.5. There we described a quantum picture of nuclear magnetic energy levels being split by an amount proportional to the applied magnetic field, and the resonance process in terms of the emission and absorption of radio frequency quanta. The aim there was to emphasise the fact that although MRI, CT and PET seem very different, in fact they all rely on similar processes of emission and absorption of photons from atoms or nuclei. In this chapter we describe both NMR and MRI using an alternative approximate macroscopic classical picture that is the standard conceptual framework for quantitative discussions of MRI. It turns out that classical physics in this context provides accurate predictions using equations and concepts that are a good deal more transparent than the underlying quantum theory. The mathematics of the classical treatment of NMR is outlined in appendix C.A classical treatment is possible for three reasons. First, all NMR and MRI signals arise from the coherent motions of a very large number, 10^{14}, of nuclear spins and thus we are always dealing with macroscopic averaged behaviour rather than individual nuclei. Secondly the quantum energies of RF waves are so small that they can be treated as though they were continuous. Finally the interactions between nuclear spins and atomic chemistry, which form the basis of contrast in MRI, are sufficiently weak to allow very large numbers of nuclear spins to remain in a coherent state of motion for relatively long periods of time. The coherent nuclear spin motion allows the myriad of individual spin motions to be described, as if they were a single quantity, the bulk nuclear magnetic moment, M. Thus, in the standard description of NMR, the classical macroscopic notions of torque and friction are used, together with Newton's laws, to predict the motions of M, following RF excitation, as if it were a compass needle.

Both NMR and MRI depend on dynamic interactions between nuclear magnetic moments, the applied fields, and the environmental fields arising from surrounding atoms. An intense static magnetic field B creates the nuclear magnetic polarisation. Each cc of tissue contains about 10^{20} water protons and, as we saw in section 2.5, the equivalent of about $1:10^6$ of these are pointing along the direction of B. Each small volume of tissue thus contains about 10^{14} polarised nuclei and it is this collection that we describe by the symbol, M, the bulk nuclear magnetic moment per unit volume of tissue. A pulse of RF field applied at the Larmor frequency is used to tip M away from its initial equilibrium direction; it then starts precessing at the Larmor frequency, about the direction of B, just like a disturbed compass needle. The NMR signal is the alternating voltage induced, again at the Larmor frequency according to Faraday's law, by this rotating nuclear compass needle in a nearby loop of wire or pick-up coil,. The rotation of M does not continue for ever, instead there are the equivalents of frictional forces, arising from the electric and magnetic fields of the surrounding atomic chemistry, that damp the motion of M in two ways. These frictional forces are incorporated into Newton's equation of motion for M in a similar way to that in which air friction is introduced into the equation of motion for a bullet or a car travelling at high speed. First there is the longitudinal relaxation process which can be interpreted as a frictional force damping the precessional motion and dragging M back towards B. The time it takes for M to return to equilibrium is described by the time T1. Second, the transverse relaxation process is interpreted as the breakup of the single rotating vector M into a myriad of individual moment vectors, all rotating at slightly different rates as result of slight variations in the static magnetic field experienced by each nucleus. The time taken for the transverse process to destroy the coherent rotation of M is described by the time T2. Spatial variations in T1 and T2 provide two of the contrast mechanisms for conventional MRI.

The third contrast mechanism, described by the term "free water density", is in fact also related to T1 and T2. It turns out that both relaxation times are very much shorter, 10^{-5} s in solids than in liquids, 10^{-2}-1 s. A typical RF excitation effectively sets all water protons in coherent motion, irrespective of whether they reside in cellular water or bone. After excitation however, the signals arising from protons in bone and the more treacly components of soft tissue disappear so rapidly that they give no contribution to a conventional MRI signal. All the water molecules in soft tissue can be roughly classified as either "free" or "bound". It is the "free" molecules in the low viscosity liquid components of tissue and fluids that produce the bulk of MRI signal. The fraction of "free" water molecules varies from tissue to tissue and with some disease processes and thus it is the third flavour of MRI contrast.

6.2 Pulsed Nuclear Magnetic Resonance

RF Excitation
Free Induction Decay

The basic signal element in pulsed NMR and all of MRI is called the free induction decay, FID. It is the electric voltage, induced in a nearby coil, by the rotating nuclear magnetic moment set in motion by an RF pulse. The frequency of the FID is the Larmor frequency, f_L, set by the static field B so that

$$f_L = \frac{\gamma B}{2\pi} \qquad \text{... 6.1}$$

For water protons, the most common subject of clinical MR, $\gamma / 2\pi = 42.573$ MHz/T. Throughout this chapter we will, for convenience, drop the factor of 2π and use the symbol γ to refer to the ratio of frequency to magnetic field. T1 and T2 determine the envelope of the FID. In this section we describe how the FID is created. A more formal mathematical treatment of the nuclear spin equations of motion is given by a set of equations called the Bloch equations, these are described in appendix C. The key conceptual point arising from this description is that of the "rotating frame". It is imagined that the nuclear spin motion is like that of a roundabout, which has a definite rotational frequency equal to the Larmor frequency. By choosing to describe the nuclear motion from the viewpoint of somebody sitting on the roundabout, the equations of motion become much simpler.

In order to gain information about M, we have to disturb it, away from its equilibrium direction. This is done using a short pulse of radiofrequency, RF waves with a frequency tuned to the local Larmor precession frequency. Once the RF pulse has finished, there are no man-made external torques acting on the nuclear magnetisation and the subsequent motion is then described as free. The rotating nuclear magnetisation, although free of externally applied torques, is still subject to the randomising effects of the internal molecular fields that give rise to the T1 and T2 relaxation processes. The free motion of the nuclear magnetisation is thus damped and the signal decays in time, creating the FID signal envelope.

Figure 6.1 illustrates the geometry that we will use in this and subsequent sections to describe pulsed NMR. Purely as a matter of convention, the direction of the large static magnetic field, B is taken to be the Z axis. In a MRI whole body machine this is the head to foot axis of the patient. The X and Y axes are of course perpendicular to Z. We have arbitrarily assigned the X axis to be the direction of radio frequency field B_{RF}. This choice causes the nuclear magnetisation to precess, under the influence of B_{RF} in the YZ plane. We imagine that a continuous RF field, B_{RF} is applied along the X direction, while the nuclear magnetisation, M is initially polarised along Z, the direction of B. For an arbitrary choice of RF frequency, f, there will be very little disturbance to the direction of M, so long as $B_{RF} << B$. However when f matches the Larmor frequency, $f_L = \gamma B$, the effect of even very small RF field amplitudes becomes greatly magnified, as a result of the resonance process. Exactly at $f = f_L$ the nuclear magnetisation behaves as though it is influenced only by the small field, B_{RF}. M begins to precess about the X direction at a frequency $f_{flip} = \gamma B_{RF}$. After a time

$$T_{flip} = 1/\gamma B_{RF} \qquad \text{....6.2}$$

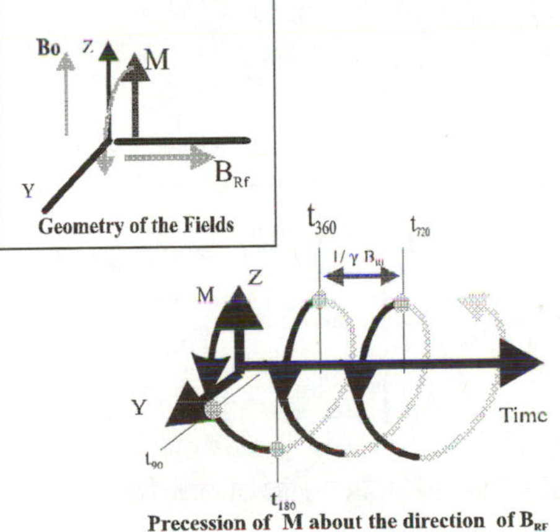

Figure 6.1: The geometry used in the discussion of NMR /MRI. In the resonant condition M precesses in the rotating frame, about the X axis with a period $T_{flip} = 1/f_{flip} = 1/\gamma B_{RF}$. If the RF field is switched off after a time $T_{flip}/4$, M is tipped by 90°. Doubling either the time or RF amplitude will produce a tipping angle of 180°. In the laboratory frame the spins are also spinning at f_L about the Z axis, see figure C.2.

M will complete one complete revolution about the X axis. After $T_{flip}/4$, M will have been rotated by 90 ° and will be pointing along the Y direction. .Similarly after a time $T_{flip}/2$, M will have been rotated by 180° and will be upside down, pointing along the -Z direction. If the duration of the applied RF field, the pulse length, is limited to $T_{flip}/4$, then the net result will be that M has been tipped by 90° and is then free to precess and relax in the applied static field B. An RF pulse, which has this effect, is called a 90° pulse. Doubling the duration or the amplitude of the RF field pulse will cause M to be flipped by 180° and this is called a 180° pulse. These are the two standard manoeuvres of both NMR and MRI used to create the initial signal. Special fast imaging schemes, such as FLASH and FISP often use a smaller, variable tipping angle. Unless otherwise stated, we will restrict our discussion to the use of the standard tipping manoeuvres.

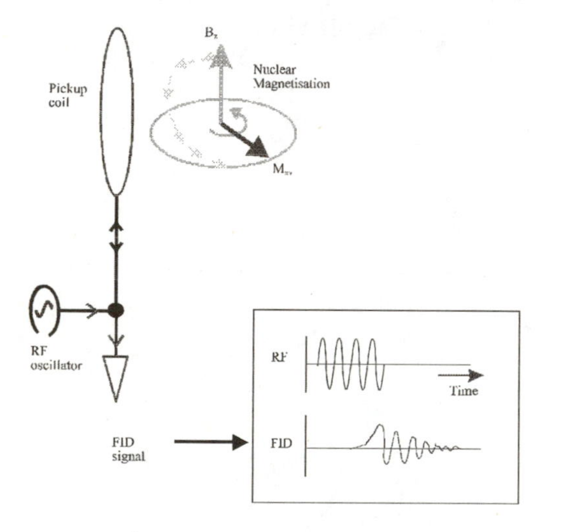

Figure 6.2: The Production of the NMR Signal. A radio frequency pulse tips M in to the XY plane, M precesses in the XY plane at the Larmor frequency f_L, inducing a sinusoidal voltage in the nearby detector coil. In a magnetic field of 1 T the Larmor frequency of water protons is 42.54 MHz. The envelope of the resulting induced FID is described approximately by an exponential decay with a time constant which is the shorter of T1 or T2.

After the application of a tipping RF pulse, initially M precesses, describing a circle in the XY plane at $f_L = \gamma B$. A nearby pick-up coil, placed so as to be sensitive to the rotating component M_{XY}, will produce an ac voltage at the frequency f_L, due to the rapid rotation of M. This is the FID signal, illustrated in figure 6.2.

Spin Relaxation, T1 and T2

Now we must include the frictional forces that damp the motion of M after the tipping pulse. Friction, in this classical model, is caused by the interaction with the atomic fields characterised by T1 , T2.. The T1 process can be thought of as forcing M back into the Z direction, the T2 process effectively breaks up M into its constituent parts. Both cause the oscillating signal to decay with time.

T1 Relaxation

After excitation, the nuclear spins are precessing in the presence of small fluctuating magnetic and electric fields, set up by electronic motions in the surrounding atoms. Depending on the temperature and the nature of the material, these fluctuating environmental fields give rise to a spectrum of photons; those with just the right energy can induce transitions between the nuclear spin states and restore equilibrium. Classically this is equivalent to a frictional force tending to realign M with the Z axis. In the classical picture the vector M spirals in 3D back towards the direction of B. This is called longitudinal spin relaxation since it involves the restoration of the equilibrium value of M_z, the component of magnetisation along the main field. The relaxation proceeds approximately exponentially with a time constant T1. The T1 process in NMR/MRI provides a classic example of time's arrow. The 90° flipped M_{XY} has a higher magnetic energy and is more organised and has a lower entropy than when it is in equilibrium, aligned along B,see section2.2, 2.5 and Appendix C. All the energy exchanges between the rotating M and the surrounding atoms tend to take this extra organised energy out of the motion of M_{XY} and transfer it back to the atoms in the form of random heat energy. The distinguishing feature of the T1 process is that it must involve a transfer of energy between the nuclear spins and the atoms. Sometimes, especially in physics, this process is called the spin-lattice interaction.

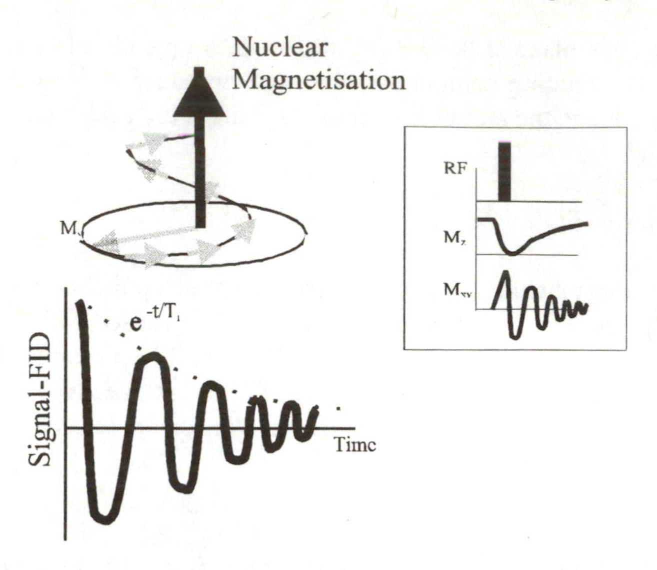

Figure 6.3: Longitudinal Spin Relaxation. Immediately after a 90° RF pulse M_Z is brought to zero and is rotating at f_L in the XY plane. The exchange of photons between the surrounding atoms and the nuclear magnetisation causes the equilibrium state to be re-established in a characteristic time T1. Classically we may picture this as M spiralling back towards the direction of B, causing the signal amplitude, which is proportional to M_{XY} to decrease exponentially with time.

T2 Relaxation

Although NMR and MRI are observed ideally in a large, spatially uniform, static magnetic field, B, we have to consider the imperfections in the large polarising magnetic field, the small spatial variations in static field caused by the weak, generally diamagnetic, magnetic susceptibility of the tissue and some very weak nuclear-atomic interactions. The last and smallest of these is the chemical shift, which is effectively a small variation (a few Hz) in the resonance frequency from individual hydrogen atom to atom resulting from local atomic and molecular fields. Next the quasi-static magnetic fields, set up by neighbouring atoms, causes the Larmor frequency to vary from place to place in a piece of tissue with a molecular length scale. These frequency dispersions are considered to be intrinsic to a particular type of tissue and are characterised by the intrinsic T2. Finally the man-made field is not perfect, but its strength varies by a small amount in space, introducing a frequency variation of a few tens of Hertz from one side of the patient to the other. As a result each nuclear spin is actually precessing in a slightly different static field and each has a slightly different Larmor frequency. The dephasing processes are together described by an effective transverse relaxation time T2*.

Although the RF tipping pulse starts all spins rotating in phase in the XY plane, they do not remain so. Rather, as time progresses, M_{XY} breaks up, as some spins speed ahead and others lag behind the mean. The voltage induced in the receiver coil again decreases, but this time because the phases of the precessing spins are spread out in the XY plane. They are still precessing out of equilibrium, but their net induced signal diminishes as time progresses.

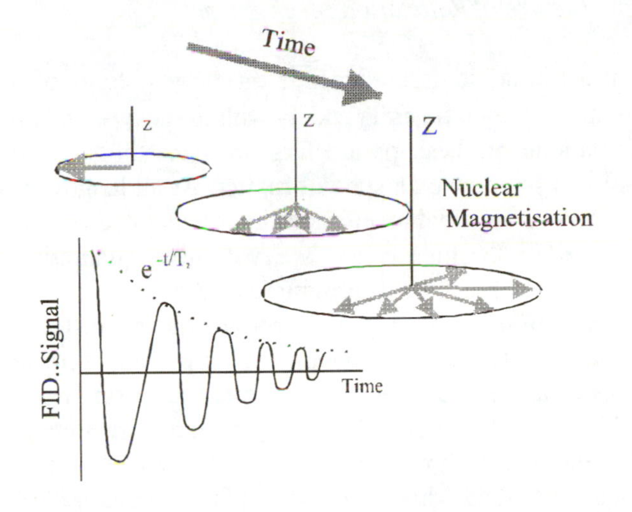

Figure 6.4: Transverse Spin Relaxation leading to T2. Immediately after the RF tipping pulse, all the spins comprising M are in phase. As result of the spread of static fields there is a spread in Larmor frequencies and so M_{XY} dephases, leading to a decay of the NMR signal characterised by a time constant T2. This process occurs at the same time as T1 decay (figure 6.3) and thus we can picture the motion of M as a complex dephasing together with a gradual spiralling back towards B.

Eventually for every positive increment in receiver voltage, produced by one spin, there is a negative contribution from another spin with an opposite precessional phase and so the induced signal is then zero. The FID amplitude created by the RF pulse decreases simultaneously as a result of both the T1 and T2 processes. In general the intrinsic values of T2 are less than T1 for free water protons and thus the signal dephases in the XY plane rather more rapidly than M returns to equilibrium. In addition, the effect of the man-made field generally dominates and so T2* is generally a good deal shorter than T2. The T2 process takes place without altering either the energy or the entropy of the rotating nuclear spins, and in consequence the ideal process is reversible. That is to say any dephasing, caused by a strictly static magnetic field variation, can actually be reversed and the rotating nuclear spin state refocussed. This is the spin echo process that is described below.

A Microscopic Picture of Relaxation Mechanisms

MRI contrast in static tissue depends on the variations of the free water density ρ and the relaxation times T_1 and T_2 with tissue type and disease process. Detailed calculations of these parameters are exceedingly difficult because biological tissue is so complex a state of matter. At all length scales there is a complex network of semi-rigid membranes interpenetrated by fluids of varying degrees of viscosity. In this section we will give qualitative microscopic description of the processes that determine relaxation.

In all biological tissue, a given water proton is never further than 2.10^{-10} m away from another proton. If a collection of water protons were held rigid, as in ice, then the static magnetic field, at a given proton site arising from neighbouring protons, would amount to about 10^{-4} T. In general the orientation of the proton spins would be distributed at random and so, in the collection there would be a small static magnetic field whose direction changed randomly from proton to proton with this magnitude. In such a field, the initially coherent rotating nuclear magnetisation would dephase in a time $1/f_{rand} = 1/(\gamma \times 10^{-4}) = 2 \times 10^{-4}$ sec. This is about the right value for T2 in pure ice at very low temperatures. Water protons are subject to magnetic fields from other protons and, depending on the chemical environment, magnetic fields set up by atomic electrons. Both will cause the coherent rotating state to break up. These microscopic interactions, called the dipole-dipole interactions, form the basis for all intrinsic relaxation in matter but we have to add one more twist to the story. At any finite temperature all atoms and molecules are in constant motion, brownian motion, arising from the random energy transfers or heat. Once we allow the neighbouring protons and atoms to move and rotate, then the resonating proton spins experience both static and dynamic magnetic fields. Dynamic environmental fields that are changing at the same rate as the Larmor frequency have a special importance in NMR and MRI. These fields, although random in direction, have exactly the right frequencies (quantum energies) to induce transitions between the nuclear spin states or classically, to return the rotating magnetisation to equilibrium. This is the microscopic picture of the T1 process. All other frequencies, including the zero frequency static field, cause some degree of dephasing, the T2 process. The degree of T1 and T2 relaxation depends on the proportion of the environmental field that is changing at a rate f_L, for T1 and how long in comparison with $1/f_L$ that field remains constant for T2. In a solid, at room temperature, atomic motions are constrained by chemical bonding and so the environmental fields are relatively constant in time, with correlation times of the order of 10^{-5} sec. This is very much slower than the proton Larmor period of 2×10^{-8} sec in 1T, and thus in a solid ,T2 dephasing is very effective; the rotating nuclear magnetisation will only survive for about $T2 = 10^{-4}$ sec.

When we consider a low viscosity liquid, the dynamic situation changes dramatically. Here the environmental fields only stay constant for times of about 10^{-12} s, as the molecules tumble and move. These times are very much shorter than $1/f_L$. Now, since the atomic fields are changing in a random fashion, the rotating nuclear magnetisation experiences a series of very short, randomly directed pulses of field. These tend to average to nothing over the relatively long Larmor period. Thus in a low viscosity liquid the T2 process becomes ineffective and T2 increases to about 1 sec for pure water. This is called motional narrowing; a progressive reduction in fluid viscosity leads to a progressive reduction in resonance linewidth and a consequent increase in T1, T2. Our opening quotation from Purcell describes this process rather eloquently.

In biological tissue the states of matter range from solid bone to nearly free water. Solid bone plays no role in conventional MRI, simply because its T2 value is so short. A typical RF pulse excites most protons in bone but the resulting signal has dephased to nothing before MRI measurement begins. In soft tissue there is a dynamic equilibrium between those water molecules that are essentially free and those bound to large molecules such as proteins. The free component dominates the MRI signal, since it has the longest values of T1 and T2, but the bound component and the dynamic division between the two components alters the relaxation processes from tissue to tissue, hence providing the three contrast mechanisms, free water density, T1, T2. It should be noted that the division into bound and free water is a dynamic, not a static process. In reality individual water molecules hop between the two at variable rates that depend on the details of the tissue type. Rather complex scenarios can occur, for example, a group of initially free water molecules are set in resonance by an RF pulse. These molecules then get attached to a protein molecule, carrying their nuclear precession with them. Once attached, the nuclear relaxation process suddenly becomes more efficient and both dephasing and spin-lattice relaxation is hastened with respect to the free water situation. Given this type of process and the complexity of biological tissue it is not too surprising that it is very difficult to get accurate quantitative predictions of the magnitudes of T1 and T2 for a given tissue type. Furthermore nominally the same tissue type appears to give rise to a range of values when measured on different individuals. When NMR is used in chemistry or physics on small homogeneous samples, the intrinsic relaxation times are of fundamental interest; they are susceptible to first-principles calculation and their values can provide very useful quantitative insights into the physical and chemical state of the sample. To this end both T2 and T1 are measured very precisely, using the pulse sequences that we describe below. In MRI, T2 and T1 have a slightly different significance because of the variations that we have just described. The variability means that tissues from the same organ or disease process, taken from different individuals, apparently give rise to a range of values of T2 and T1 rather than single values, characteristic of the particular tissue. In addition the time constraints of MRI imaging are such that neither T2 nor T1 are ever measured with high precision. Thus in MRI we will talk of images being weighted by T2 or T1. Seldom, if ever, does an exact value of the relaxation time enter a description of the image. This is why we have dubbed T2 and T1 contrast flavours, in order to emphasise the fact that approximate, rather than definite values, are being used as image descriptions.

In fact the variability of T2 and T1 is something of a disappointment for the clinical application of MRI especially in the context of cancer diagnosis. It was originally claimed that T1 and T2 could be used to obtain a completely non-invasive biopsy of tissue suspected to be malignant. If it had turned out that a given type of carcinoma or metastasis gave rise to a specific value of relaxation time, then this would have been extremely useful in diagnosis. It is not the case but nevertheless particular combinations of image flavours have been found that give a very high diagnostic yield for particular conditions. Table 6.1 shows some representative values of relaxation times reported for different tissue types and disease processes. It can be seen that the values are all considerably greater than 10 ms and that in general T1 is larger than T2. In a practical MRI machine, the static field, B cannot be perfectly uniform over the whole body Typically in superconducting systems, the field varies by $1{:}10^6$ over a spherical volume of diameter 50 cm. In a 1 Tesla magnet this means that

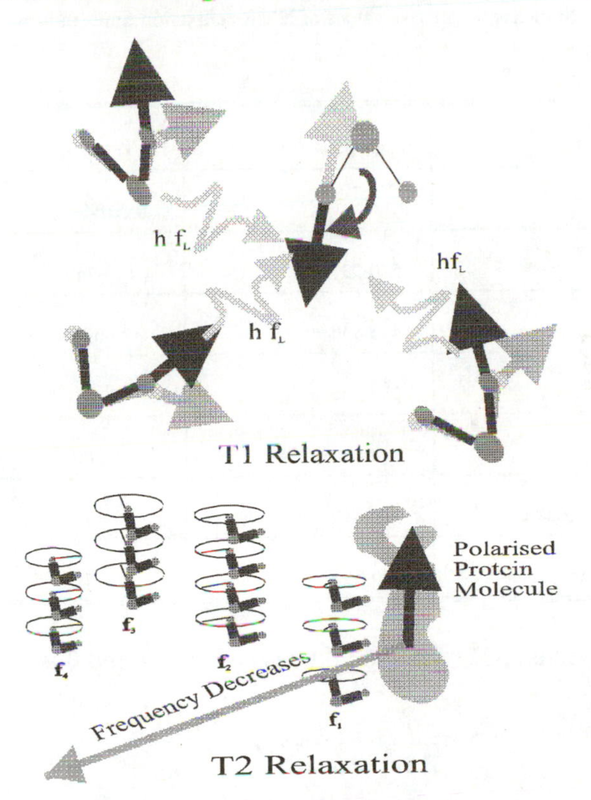

Figure 6.5: T1 relaxation is illustrated by vibrating water molecules emitting photons with the Larmor frequency, f_L, which flip the nuclear spin of neighbouring proton. T2 relaxation arises primarily from more static magnetic field variations caused by polarised molecules whose moments remain relatively fixed in orientation.

there is a difference in Larmor frequency of 42 Hz from one side of the sphere to the other and this, on its own, will cause the coherent nuclear spin state to break up in a time of about 12 ms. In this time the FID signal arising from all the nuclear spins, on one side of the sphere, will be out of phase with those on the other side. Since the spin dephasing has two contributions, one arising from intrinsic atomic variations in magnetic field from place to place in the sample, the other from the variations in the applied static field the two are separated

Table 6.1: Some representative values of NMR relaxation times in human tissue.

Tissue Type	T_1 (ms.)	T_2 (ms.)
Bone	0.001- 1	0.001-1
Glioma	540-930	120-240
Muscle	460-650	26-34
Fat	180-300	47-63
Body Fluid	1000- 2000	150-480
Brain Grey Matter	340-610	90-130
Brain White Matter	220-350	80-120

The effective transverse relaxation time is called T^*_2. and this is written as

$$\frac{1}{T2^*} = \frac{1}{T2} + \gamma \, \Delta B \qquad \ldots 6.3$$

Where ΔB is the small, inadvertent change in applied magnetic field from one side of the sphere to the other, caused by imperfections in the main magnetic field..

6.3 Pulse Sequences

Both NMR and MRI make extensive use of pulses of RF field and, as we shall see below, slower changes in magnetic field gradient, in order to highlight different aspects of the three flavours of contrast and to achieve spatial localisation. All three contrast flavours are related to the longevity of the coherent nuclear magnetic state. Thus, examining the coherent nuclear magnetic state after different time delays, following the initial excitation, can make different contrast choices. The examination process generally involves the use of one or more RF or magnetic field gradient pulses which together, in the jargon of NMR/ MRI, are called a pulse sequence. As a general rule T2* is very much shorter than either T2 or T1 and thus, unless special measures are taken, the effect of magnetic field homogeneity would always dominate the relaxation process. Most of the standard pulse sequences that we describe below are different ways of getting around this problem and allowing estimates of T2 and T1 to be made. The early MRI systems made most use of three standard methods inherited from NMR, Saturation recovery, Inversion recovery and Spin Echo, all of which use sequences of RF pulses and a fixed magnetic field. More recent schemes also involve the use of pulses of magnetic field gradient in order to create what are called Gradient Echoes.

T1 Measurement
Saturation Recovery

The simplest method of examining the relaxation of the coherent nuclear state uses sequences of 90° selection pulses to test the recovery of M_Z. Figure 6.6 illustrates the principle. The signal FID, following the first RF pulse, decreases in amplitude according to exp(- t / T2$^{'}$) and thus becomes too small to measure in about 100ms (about 5 T2*). As time goes on, the T1 process will restore the equilibrium state and the magnitude of M_Z will increase from zero back towards its equilibrium value, M_o. If a second, 90°, RF pulse is applied after a time interval T_R <5T1, it will be only be able to flip the recovered portion of M_Z back into the XY plane as a coherent state. Those spins that are still precessing in the XY plane are flipped by the second RF pulse but do not form a coherent state. The initial value of the FID signal created by the second pulse is given by

$$S(t=T_R)=M_o(1-e^{-\frac{T_R}{T1}}) \qquad \ldots 6.4$$

and this is dependent on the ratio of T1 to T_R. If an accurate value for T1 were required then a sequence of such measurements, taken at different values of T_R, would provide data, which could be numerically fitted to the exponential form and a value of T1 estimated.

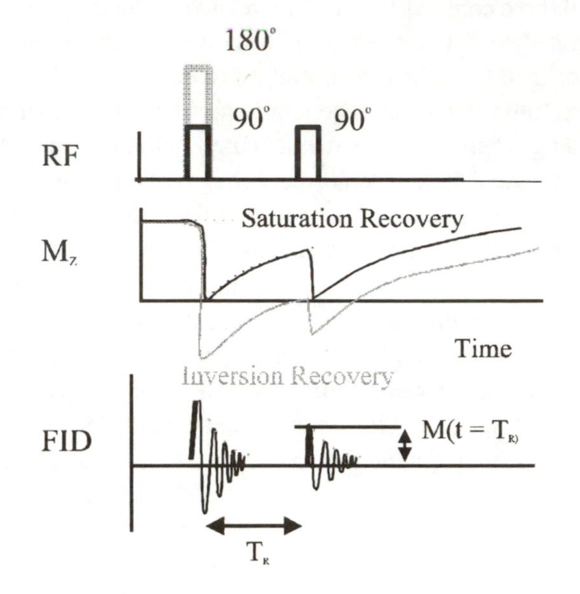

Figure 6.6: Saturation and inversion recovery. The equilibrium nuclear magnetisation is excited using an appropriate RF pulse. After a variable delay, T_R , a second RF pulse is used to measure the amount of longitudinal magnetisation that is restored in time T_R. The second, interrogating. pulse can only set in motion those spins that have relaxed to their equilibrium orientatrions. In saturation recovery the excitation uses a 90^0 while inversion recovery begins with a 180^0 pulse. Both use 90^0 pulses to interrogate the relaxing longitudinal magnetisation.

Inversion Recovery

Inversion recovery is essentially the same process as saturation recovery but the initial selection is made with a 180° inverting RF pulse. By completely inverting the nuclear magnetisation, a longer period of observation is obtained which allows more significant data points to be collected and thus better values of T1 estimated.

T2 Measurement
Spin Echo

Very early in the development of NMR, Hahn discovered that some of the dephasing of the precessing nuclear magnetisation can be reversed to produce echo signals. An idealised T2 process does not involve any change in spin population; the spins remain in the XY plane,

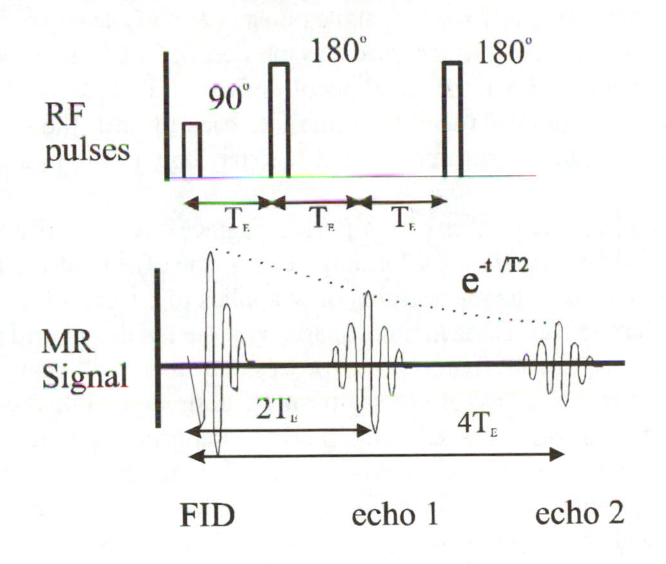

Figure 6.7: Spin Echo and the measurement of T2. A 180° RF pulse applied Te seconds after the initial 90 ° pulse, causes a spin inversion. The spins refocus in the XY plane and the signal returns at $2T_e$ as an echo. By adding more delayed 180 ° pulses, a sequence of echoes can be obtained. The initial, maximum signal amplitude of the nth successive echo has decreased according to exp(-2nTe/T2) measurements to be made of T2 alone, provided that T1 >> T2. In general echo delay times, T_e, are very much shorter than repeat times ,T_R..

but are simply spread in precessional phase, making them invisible after a time of about 5T2. Since there have been no changes in the spin populations, this dephasing is reversible. No energy has been exchanged with the lattice and there is no change in entropy within the spin system. Spin echo exploits this reversibility in the following way. At a time Te, after the initial 90 ° RF pulse, a second pulse either of double amplitude or of double duration, a 180° pulse, is applied to the freely precessing spins. This second pulse turns the entire spin system through 180° and reverses the sense of rotation of the M_{XY} spin components. After a time of 2Te, M_{XY} has rephased and the FID signal returns as a spin echo. Hahn drew the following analogy between a spin echo and a false start in a running race like the London Marathon. Initially all the runners are grouped in a relatively thin phalanx just behind the start line. On the starting pistol, this group surges forward so that the phalanx as a whole moves away from the start line. At the same time it spreads, as the fleet of foot surge ahead and old and lame lag behind. When the recall pistol is fired at Te seconds the runners immediately about turn and run at the same rate back towards the starting line. At 2Te the entire phalanx has reformed at the start line, facing in the opposite direction.

Spin echo pulse sequences with differing degrees of complexity are widely used in both NMR and MRI. By forming echoes, the effects of the rapid $T2^*$ decay can be reduced. The re-focussing only applies to the contributions to the $T2^*$ process that are truly static in time. Spatial variation in the applied polarising field, B is the largest contributor to this process but there will also be a small contribution from the intrinsic T2 of any tissue. The strength of this contribution is determined by intrinsic static atomic field inhomogeneities that survive the constant thermal agitation. Thus regions with a high concentration of ferrous iron atoms would produce a static local field anomaly that would contribute to the intrinsic T2 and will create image contrast if T_e is suitably chosen. MRI Image weighting, using variations in T_R and T_e, is discussed below. In conventional MRI many methods include some spin echo re-focussing in order to reduce the effects of magnet field variations. As we saw in the previous section a good magnet with a spatial homogeneity of $1:10^6$ will produce a $T2^*$ of about 20ms and this is clearly very much shorter than the intrinsic T2 values listed in table 6.1

NMR Spectroscopy

When an NMR experiment is performed at very high resolution on a chemical specimen, structure is observed in the resonance line. It is found that the resonance signal, from say water protons, is in fact composed of a number of discrete sub resonances separated by a small amount in frequency from the resonance frequency expected from the value of the known applied magnetic field. The separation of the components is quite small, 1- 1000 ppm, and increases with increasing field strength. This is called the chemical shift and it can be pictured as a small atomic magnetic field produced at the nucleus , as a result of the alignment of the orbits of the atomic electrons. The same atom, when incorporated into different molecules or in fact different positions within the same large molecule, can undergo a change in the configuration and the motions of its outer, valence electrons, see figure 2.4. In NMR these changes produce different atomic magnetic fields at the nucleus and thus different chemical shifts. This is the basis of NMR spectroscopy. A study of the distribution and relative intensities of the component spectra from a given atom in a chemical sample allows chemists to deduce the structure(s) of the molecules containing that atom. NMR spectroscopy is now one of the most important analytical tools in chemistry and biology. It is even possible to gain considerable amounts of structural information about very large bio-molecules using a technique called Fourier transform spectroscopy.

Whole organ NMR spectroscopy has had some success in-vivo human and animal studies, especially using the phosphorous resonance. The phosphorous atom plays a pivotal role in the chemistry of ATP – adenosine triphosphate a major link in the chain of energy metabolism. NMR can follow the changes that take place in these molecules, as they change during the metabolic cycle, by revealing the changes in chemical shift spectra of phosphorous and thus the structures in which this atom is found. Clearly this type of investigation may have potential for the diagnosis of metabolic disorders throughout the human body.

Spatially localised NMR spectroscopy has yet to be implemented to the same extent as anatomical and now functional MR. There are many reasons for this but we will highlight two of these. Since the chemical shift is very small, the applied magnetic field in an NMR spectroscopy experiment has to be much more uniform over the region of interest, in order to resolve the separate spectra. For hydrogen the maximum observed shift in any biological molecule is 10 ppm, about the same as the spatial homogeneity of the field in commercial systems. Thus very accurate spectra can only be obtained over a limited spatial region in a whole body magnet. The difficulties of imaging the spectra of nuclei, other than hydrogen, are compounded by the reduced imaging sensitivity of all nuclei with respect to hydrogen. Table 2.2 shows that ^{31}P has a relative imaging sensitivity of only $6.\ 10^{-2}$

6.4 Spatially Localised NMR: MRI

The key to spatial localisation in NMR is the linear relationship between Larmor frequency, f_L and magnetic field strength, equation 6.1. Beginning in 1973, several different localising schemes have been proposed that utilise spatial and temporal variations in the applied magnetic strength to perform various types of imaging. These schemes range from the point by point methods used originally by Damadian, to the modern fully 3D data acquisition schemes still under development. All present day MRI systems make use of the Fourier technique, first described by Kumar. The standard tomographic recipe of carving 3D into 2D slices and then strips is implemented in a particularly elegant fashion in this scheme. Much of the discussion of modern MRI makes extensive use of Fourier Methods and the K space ideas used in chapter 1 and described in appendix A. In conventional MRI, the 3D volume is sliced first by superimposing a static magnetic field gradient,

$$G_{ZZ} = d\,B_Z / d\,z \qquad \ldots 6.5$$

parallel to the Z axis with a strength of about 10^{-3} T/m = 1mT/m. For water protons this means that for every 1mm shift in position along Z, the Larmor frequency changes by about 40 Hz. When the magnetic field gradient, G_{ZZ}, is applied each slice through the subject perpendicular to the Z axis has effectively been labelled by a definite frequency. Any plane perpendicular to Z can be chosen by using a selective RF pulse whose centre frequency matches that of the selected position on Z and whose bandwidth determines the selected slice width along Z. After a selective RF pulse, an entire plane of spins perpendicular to the Z axis will have been tipped into the XY direction, the subsequent precession of these spins creates the FID signal arising from the entire selected slice. This isolates the chosen slice of patient for imaging, since all other parts of the body cannot give rise to any NMR signal.

MRI can be accomplished by means of backprojection, using a series of projections selected by imposing magnetic field gradients in a sequence of different directions. This however is not the scheme used in modern MRI, rather the Fourier scheme is used. In this method the spins, within a slice, are further encoded in Larmor frequency and phase by the application of two more gradient fields, Gxz, Gyz. These magnetic fields are directed along the Z axis but their magnitudes vary parallel to the X and Y axes respectively, see figure 6.12. The gradient fields are switched on and off in a particular sequence to perform the encoding trick. The magnitudes of all three gradient fields are small in comparison with B so that the

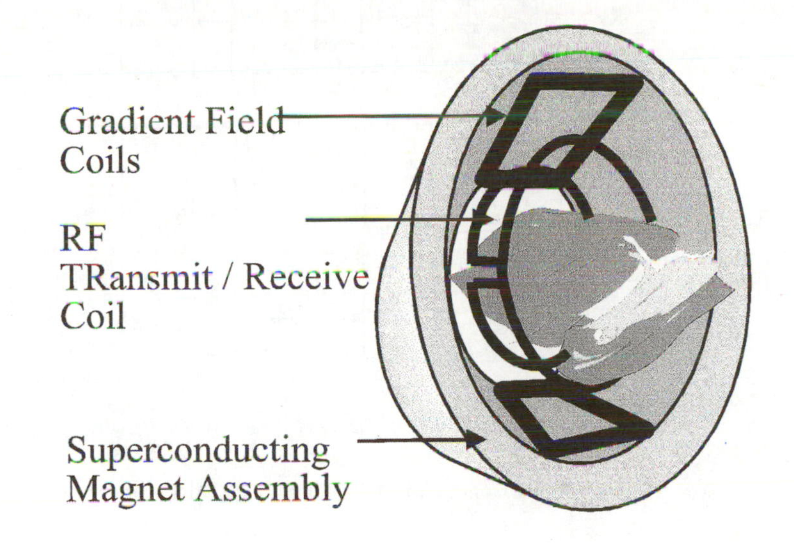

Figure 6.8: An illustration of the bore of a whole body MRI magnet. The internal coils in the illustration are hidden from the patient who simply sees a straight smooth "tunnel".

total variation in Larmor frequency across or along a human subject is only a few percent of the centre frequency, f_L, set by the value of the large static field, B. This choice is determined by the radio detection schemes used and the engineering constraints imposed by the need to switch these fields on and off very rapidly. By keeping the signal frequency band narrow, frequency selective (high Q) receiver coils and phase sensitive detection schemes can be used to improve the signal to noise ratio

The Main Components of the Modern MRI Machine

MRI requires three types of magnetic field generation, static, pulsed gradient and RF and a radio detection and digitisation scheme. Here we will give a very brief description of these component areas

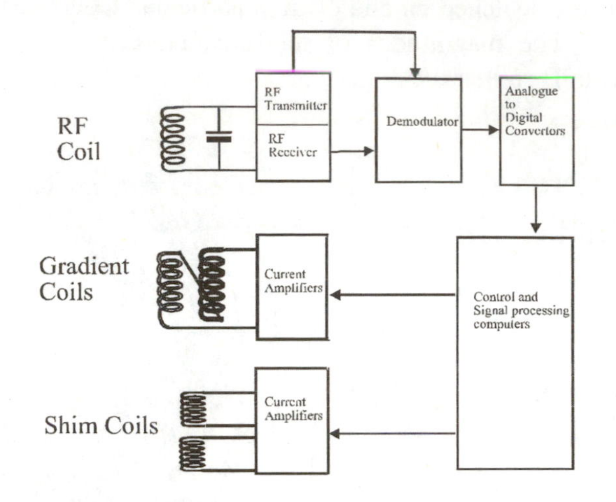

Figure 6.9: The Components of a Whole Body MRI Scanner.

The Main Static Field

The main polarising field in a typical clinical system has a strength of 1.5T and most clinical systems achieve this using a superconducting magnet. This has two crucial advantages. First, a persistent electric flowing in the windings of the solenoid generates the field. Persistence of the current is guaranteed by connecting the ends of the coil together using a superconducting short circuit. Once the magnet has been charged at commissioning time, it is never intentionally allowed to warm up. Typically the decay of the persistent current will be on the order of a few parts in 1million over a month, ensuring that the centre Larmor frequency hardly changes. The second advantage is an economic factor; the persistent current flows almost entirely without any dissipation of energy and thus there are no electricity costs, after installation, and no need for large and expensive high voltage, high current generators to be installed. The running costs of the magnet are dominated by the cost of keeping the magnet at the boiling point of liquid helium using liquefied helium gas. Typically the helium reservoir has be recharged a few times each year. In a typical system the magnet is 2.2m in length with a room temperature patient access bore diameter of 0.8m. The superconducting magnet/helium dewar assembly weighs 5-6 tonnes. With a set of additional static shim field coils, the spatial homogeneity is 5 ppm in a sphere 50cms in diameter.

Spatial uniformity of the main field is of paramount importance in MRI since it can dominate the spin dephasing rate, T2. Given a finite length solenoid, the Maxwell equations show that it is impossible to design a field that is perfectly uniform everywhere within the volume of the magnet. A compromise is sought in which the field is guaranteed to be uniform to about 5 ppm over a spherical volume, 50cms in diameter. Three separate stages of compensation for unwanted gradients can be identified. During magnet construction, the distribution of the superconducting windings is calculated to satisfy the design specification. The final field of the installed magnet will initially fall short of this ideal as a result of small manufacturing errors and spatial variations in the local terrestrial field at the MRI site. One can get a feel for the size of this problem by noting that for a 1T magnet the manufacturer is aiming to guarantee the field to a level of about 10^{-6} T, while the earth's field is typically 5.10^{-4}T. Worse than that, building reinforcing steel, when magnetised, can easily produce a local field anomaly that is at least this big and this necessarily will vary rapidly in both strength and direction with position. Thus, at commissioning time, the precise local variations in field have to be measured (using an NMR probe!) and eliminated, either by the use of supplementary shim coils built into the magnet or iron plates fixed rigidly to the magnet housing. Finally, since the local field can vary slowly over time and a patient body will disturb field uniformity, small temporary field adjustments using shim coils are made prior to a scanning session.

The magnet must necessarily produce a field whose lines of force encircle the coil windings, giving rise to a significant magnetic field, external to the bore of the magnet, extending for several metres. A soft iron shell that acts like a magnet "keeper", reducing the spatial spread of the external field, surrounds many magnets. The external field produces one of the potential safety hazards discussed in chapter 3. Untethered ferromagnetic objects are strongly attracted into the magnet bore from a distance of about a metre. Much weaker fields extending out to several metres can upset the operation of heart pacemakers.

Open Magnet Systems

In the last five years a new generation of MR systems have appeared. These embody a new design philosophy and are aimed at a new range of clinical applications. These are the open magnet systems that use much simpler permanent magnets or electromagnets, rather than superconducting coils to produce the main polarising field. At present the field strength is generally quite low at 0.2T, but this is perfectly sufficient to obtain very high quality images of clinical utility. Two advantages follow from the open design. Most importantly it allows MR imaging to be used in an operating theatre as a completely safe alternative to x-ray fluoroscopy for some procedures. Second the open design is a much less hostile environment for the nervous or critically ill patients and for

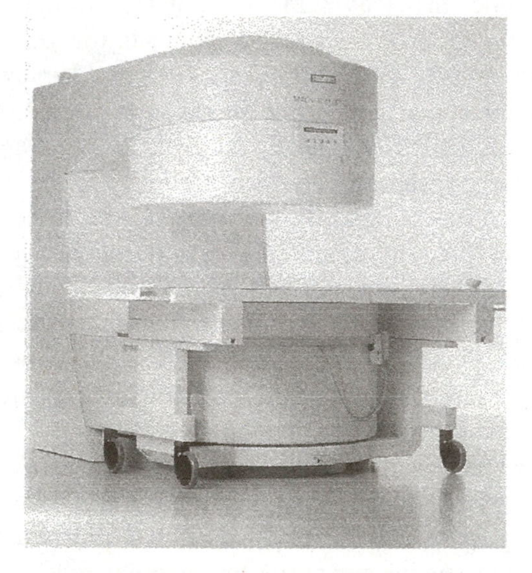

Figure 6.10 : An example of an open magnet system. The main field is produced by a large diameter electromagnet / iron pole with the pole gap arranged vertically across the patient table. The field in this case is 0.2T (Seimens)

small children. The open aspect and a very much reduced level of acoustic noise from the gradient coil switching (see below) both contribute to a more friendly imaging system.

Very much smaller systems MRI systems exist that use permanent magnets to produce fields of up to 0.1T. These are designed for non-medical applications and have been used for limb imaging. These developments might well herald the beginning of a second phase of MR development in which the large "main frame" MR system gives way in some areas to very much smaller dedicated imaging systems that are to an extent portable.

Gradient Coils

Standard clinical MR imaging requires the application of short pulses of gradient fields with strengths of 1-5mT/ m, lasting tens of milliseconds. The geometry of the coils has to be considered very carefully in order to optimise the region of space over which the gradient is uniform in strength and direction. In this case uniform means varying exactly linearly with distance,thus the gradient coils such as those shown in figure 6.11 would ideally give rise to a field $B_Z = G_{ZZ} Z$. Any deviation from gradient uniformity introduces spatial distortion into the final image. Again Maxwell equations ensure that gradient fields cannot be perfectly uniform in space and thus an engineering compromise is sought. A large gradient field strength, applied for a short time interval, also requires an engineering compromise. The field magnitude requires a large value for the product of current and number of turns on the generating coil. The rapid switching requires a very low value for the ratio of the coil's inductance to its resistance. The latter effectively rules out the use of superconducting coils since these would make it impossible to switch the field rapidly. Thus copper coils with a finite electrical resistance have to be used and a trade off, between current amplitude and number of turns, achieved. Typically the MR gradient coils comprising a few turns, transiently carry very large currents on the order of 100A. This minimises the coil inductance and minimises the rise time for field changes. Even with highly optimised coils there is still a finite gradient field rise time that has to be taken into account in designing the imaging sequence. An unwanted consequence of the large switched gradient coil currents is the ferocity of the transient magnetic forces that act on the windings. These cause the coil wires and their supporting structures to flex, which in turn produces very significant

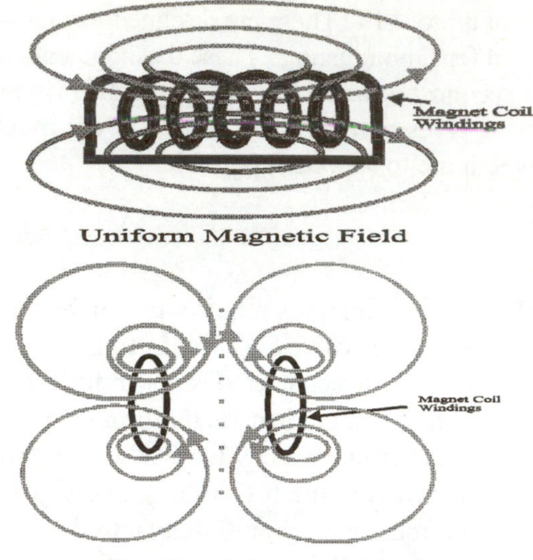

Figure 6.11: The schematic pattern of field lines around the main, B field (superconducting magnet) and gradient coil windings. The gradient field is most simply produced by a pair of coils carrying opposed currents. The field is precisely zero midway between the coils. The sign of the magnetic force alters across this central plane and the strength of the field (the separation of the lines of force) varies along the axis joining the two coils. The figure resembles the scheme for a Z gradient X and Y gradient fields are generated using more complex arrays of straight wires.

acoustic noise levels at the pulse switching frequencies, everywhere within the imaging suite, but particularly within the bore of superconducting magnets. The noise level is quite disturbing to some patients and thus earplugs or earphones transmitting music are routinely used to mask this noise

The Radio Frequency Circuit

Most modern MR systems use the same coil(s) for both the transmission of RF energy to the subject and the reception of the MR signal. The circuits used are very similar in concept to an ordinary radio receiver with a tuned input coil connected to an amplifier and a RF demodulator. The large amplitude applied RF transmission pulses must not reach the high gain amplifier chain and so signal detection in both pulsed NMR and MRI takes place between, not during RF pulses. During the pulses, diodes short circuit the amplifiers to protect them from destruction. This has to be designed very carefully in order to ensure that, as soon as possible after the RF pulse has finished, the detection circuit is ready to receive the MR signal. Speedy recovery in these circuits is not favoured by a high degree of tuning since this tends to make any transient last for a long time. Most clinical MR systems employ a range of different RF coil geometries in order to optimise the signal to noise ratio for different investigations. Thus there are rigid " bird cage whole body and head coils", " flat head coils", and semi-flexible "spine coils", "ankle coils", "knee coils". Each one is carefully engineered to get the best, most uniform coverage of the anatomical region of interest with the best signal to noise ratio. All MRI RF coils are constructed from room temperature wire or tube to allow easy substitution of different coils for different investigations.

The radio frequency electronics has to accomplish three main tasks. First, as we have just described, the system has to switch the amplifiers off during RF pulse transmission in order to protect them from destruction. Second it must amplify the MR signal with the addition of minimal electronic noise.

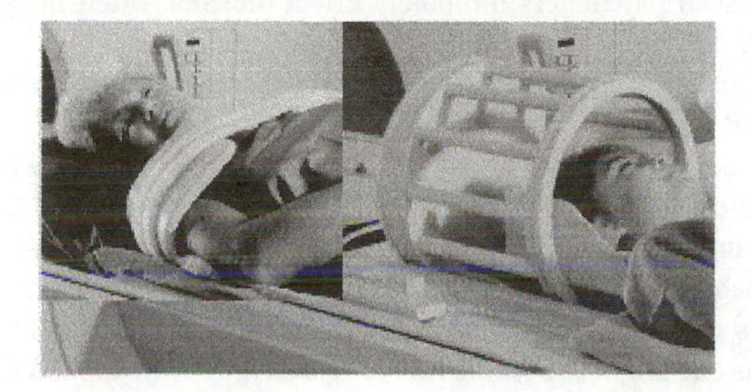

Figure 6.12: Two types of Transmit /Receive coil used frequently in MRI. On the left is a dedicated shoulder coil, on the right a more general purpose "bird cage " head coil. The multiple windings in the latter ensure a more uniform signal intensity both in RF pulses and in reception . A similar but enlarged arrangement is used for whole body coils.

Finally the signal has to be demodulated and filtered and then converted from analogue to digital form, before storage on the computer. The MR signal can be thought of as a frequency modulated radio signal, the information being contained in the modulation not in the carrier. Thus a typical MR signal will have a centre frequency of 42MHz but a bandwidth, around this, of 40kHz as a result of the localising schemes described in the next section. The carrier or centre frequency is removed by synchronous demodulation, leaving just the audio frequency signal envelope for further amplification and conversion to digital form.

Signal to Noise Ratio

NMR, at room temperature, is an intrinsically weak physical effect and so the MR signals arriving at the input coil can be close in magnitude to electrical noise components arising from the electronics, fluctuations in patient body conductivity and external sources of electrical interference. Thus, detailed consideration has to be given to the question of just how big the signal is, in comparison with accompanying noise. This is a very complex, even vexed, subject, well beyond the scope of this book, but one that is invariably treated at length in dedicated MR texts. Clearly an infinite signal to noise ratio, S/N would be desirable but MRI has to make do with S/N figures of between 10 and 20. In standard MR imaging nearly all the scan parameters ultimately effect the S/N, often in quite subtle ways. We will describe a few of the main factors involved.

The MR image is composed of a collection of small volumes or voxels from which an NMR signal is obtained. The signals from all these small volumes are initially necessarily combined into one frequency modulated signal because the whole anatomical field of view is excited and thus contributes to the overall signal picked up by the receiver coil. Right at the start of the process, the signal strength in each voxel is determined by the size of the nuclear magnetisation it contains. This increases linearly with the main field strength and the volume enclosed. Thus one might expect the S/N to rise linearly with applied field. In fact it rises very much more slowly than this, closer to $B^{0.5}$. Thus if one compares early MR images, obtained at about 0.15T, with modern equivalents obtained at 1.5T there is certainly not a factor of ten improvement in S/N between the two. Since the FID is decaying in time via T1 and T2* processes, the magnitude of the signal actually received will be less than the full nuclear magnetisation strength, by an amount determined by the time delay between excitation and measurement. When many slices are imaged in a single session, the constraints of time are such that the excitation of one slice affects its neighbours, effectively reducing the nuclear magnetisation that is available for excitation when their turn comes for imaging. In order to ensure that each slice is fully in equilibrium there has to be an interval TR, of at least 5T1 between excitations. In many types of imaging this is both undesirable and counterproductive to the image sought. Rather, faster imaging is achieved in a steady state condition with a much reduced effective nuclear magnetisation, and thus a reduced MR signal. Thus as rule of thumb "Fast Imaging " sequences give rise to noisier images than the more traditional MR methods that take many minutes to collect the data.

MR systems are designed so that the patient contribution to receiver electronic noise is the dominant factor. There are two separate contributions to patient noise. Biological tissue has an electrical conductivity close to seawater. Thus, the tissue absorbs RF energy and the applied RF field sets up circulating currents or eddy current in the patient body. The energy absorption by the patient acts like an extra fictitious resistor attached to the receiver coil. The eddy currents distort, and also attenuate, the RF waves as they progress into the body. These add to image noise through increased thermal noise and by introducing spatial variations into the spin tipping angle and the slice thickness that vary with anatomy. The second contribution from the patient arises from temporal variations in position, due to gross external movement and, more importantly, internal motions due to heart action, blood flow and peristaltic movements of the gut. Organised motions particularly of blood, are rhythmical and can easily introduce gross artefacts into the image, this is discussed below. Smaller random mechanical fluctuations lead to a more generalised degradation of the signal and thus appear as additional non-specific noise that smears the image.

Slice Selection

The first step in MRI imaging is slice selection. This step uses B and a gradient field to select a plane in the subject. Once selected, all spins within the plane are nominally precessing at the same Larmor frequency. Frequently, the selected plane is perpendicular to the Z axis and the slice selection step uses the Z gradient, G_{ZZ}. This is however is not the only choice. In a modern instrument, all three gradients can be combined to create a slice selective gradient along almost any direction in space. In several applications this is an important advantage for MRI over standard x-ray CT, in which the slicing has to be at an angle of less than 30° to the horizontal Z axis.

For simplicity we will assume that slice selection is carried out along Z using just the Z gradient. With the gradient applied, the Larmor frequency becomes linearly dependent on Z, so that at the position Z_s we have

$$f_s = \gamma(B_o + G_{zz} Z_s) \qquad \ldots 6.6$$

At positions a little in front of and behind Z_s, the Larmor frequencies are slightly lower and higher than this. The axial gradient field has effectively carved the subject into a series of parallel slices, each labelled with a slightly different Larmor frequency. Each slice can now be selected individually for imaging by using a pulse of RF that has just the right frequency to excite just one plane but leave the rest of the subject undisturbed. In section 6.2 we saw that a 90° spin flip could be achieved in NMR by making the correct choice of amplitude, B_{RF}, and time duration of the RF pulse. In principle slice selection is no different. There are however some practical issues that make slice selection a more complex issue. A pure sinusoidal wave turned on abruptly, held on for a time T seconds, then switched off abruptly, forms a square pulse of duration T seconds. This pulse necessarily contains a spread of frequencies with a range of $\Delta f \approx 2/T$ Hz. As the duration of the pulse becomes shorter, so the frequency range about the centre frequency of the RF wave increases. This is an inherent property of any wave packet or pulse, localised in time, see figure 6.14 and appendix A. In the presence of a magnetic field gradient, the spread in frequency content of the RF pulse leads to a spread in the range of positions along the Z axis at which proton spins are excited. In short, the width of the selected slice depends on the duration of the RF pulse. For example if we impose a gradient of strength G_{zz} =1mT/m , then a square pulse lasting 10 ms will have frequency spread of 200Hz and this will correspond to a distance along the z axis of

$$\Delta Z \sim \Delta f / \gamma G_{ZZ} \qquad \ldots\ldots 6.7$$

In our example, $\Delta Z = 200 / 42 \approx 5$ mm. In commercial MRI instruments, the radiographer can choose a slice thickness between 1 and 20 mm, essentially by altering the length of the selection pulse.

The frequency spectrum of a square pulse of waves has an irksome property for MRI slice selection. If the pulse is square in time then the frequency spectrum has the sinc shape (Sin(f)/f) in frequency. This is shown in figure 6.14. The undesirable features are the finite slope on the central section and the ripples at higher frequencies. The finite slope means that if the centre of the slice is flipped by 90°, then the edges will be flipped by slightly less, since there the RF amplitude will be smaller. The high frequency ripples can lead to serious image artefacts if left uncorrected since any excited tissue will give rise to a signal that will be picked up by the receiver coil. With the RF centre frequency set to excite a slice at say 24 mm from the centre, then in addition, other remote parts of the patient will be excited. If the entire signal were left unfiltered then these remote parts would appear as ghost artefacts, superimposed on the chosen cross section. Figure 6.14 also shows a standard partial remedy to these effects. If the profile of the RF pulse is itself altered into a sinc -like shape in time then the central slope in the frequency spectrum is increased and the relative amplitudes of the

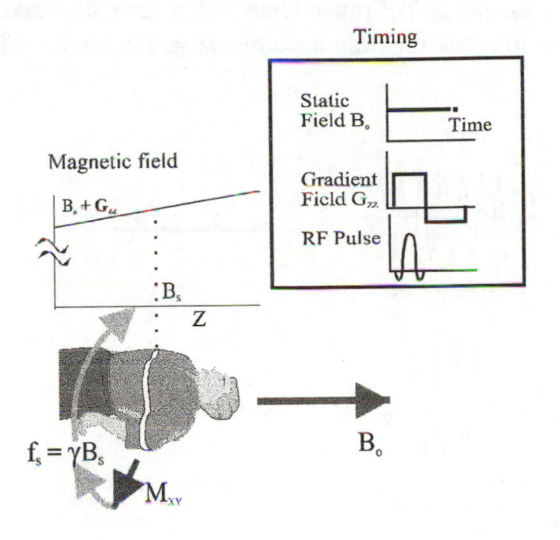

Figure 6.13: Slice selection using G_{ZZ} and an RF pulse. The chosen slice at $Z=Z_s$ corresponds to a Larmor frequency of $f_Z = \gamma (B_o + G_{ZZ}Z)$. The envelope of the RF pulse is approximately a sinc function in time so that its frequency spectrum is approximately square sided. The resulting volume of excited spins is a reasonably good approximation to a thin rectangular box with little overlap in space with adjacent boxes. The reversed section on the gradient pulse reduces phase differences in the slice along Z by making use of a gradient echo. After the gradient pulse has finished the transverse slice magnetisation precesses at the centre frequency, $f_o = \gamma B$.

high frequency ripples are decreased. In practice low pass analogue filtering is also employed to remove the inevitable residual spatially remote, high frequency contamination in slice selection.

Slice selection in a real MRI instrument is a good deal more complicated than the approximate description that we have just given. In order to achieve the best possible slice profile, the RF pulse would have to include at least 10 of the time ripples but this would make the duration of the RF pulse incompatible with rapid imaging. In practice there is an engineering trade off between imaging speed and quality of slice profile. Typically the RF pulse will mimic just the central three sections of the ideal envelope, shown in figure 6.14. In addition our description of the flipping process is actually only exactly true for small flip angles. At 90° and beyond the flipping process involves significant amounts of non-linearity and this requires the use of special RF pulse shapes that are calculated numerically and then synthesised in order to obtain accurate large flip angles. The non-linear

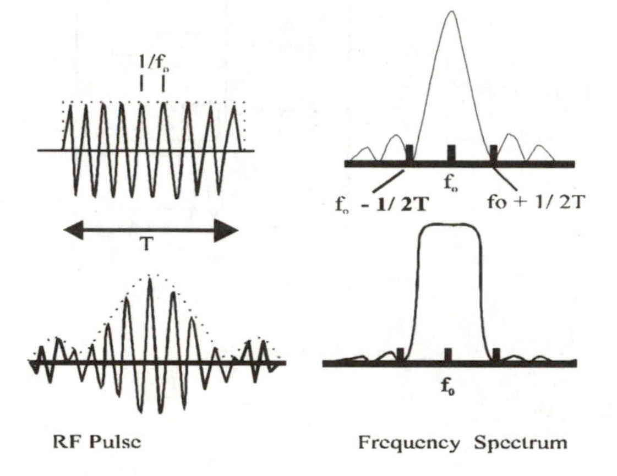

Figure 6.14: A pulse of pure sine waves lasting T seconds gives rise to a sinc function (sin (f)/f) frequency spectrum. The main central lobe of this spectrum occupies a range of frequencies, $\Delta f \sim 2/T$. This provides an approximate definition of the bandwidth and, in MRI, the slice width. There are however lower amplitude ripples extending to high frequencies. A pulse whose shape or envelope is "tailored" in time to approximate the central section of a sinc function yields a frequency spectrum with a steeper sided central section and lower amplitude ripples.

effects due to the finite RF skin depth of the patient increase in severity as the field , and hence the Larmor frequency increases.

The use of a gradient field in slice selection brings with it a small but undesirable spread in signal RF phase along the Z axis. The remedy illustrates the general MR method called gradient echo. The selected slice is necessarily of finite width. During excitation in the gradient there is a small spread in Larmor frequencies through the width of the slice. When the gradient is switched off, all the Larmor frequencies in the selected slice volume return to the centre frequency γB, the first term in equation 6.6. Although there is a single effective frequency for the rotating M_{XY} nuclear magnetisation following excitation, the different spatial contributions to M are somewhat spread in phase in the XY plane because the time histories of the spins differ across the width of the slice. The spread in RF phase, $\Delta\Phi = 2\pi\,\gamma\, G_{ZZ}\Delta Z\,\tau$ for a slice width of Δ Z obtained using a RF pulse lasting τ seconds. The remedy is shown in figure 6.14. After the RF pulse is finished the gradient field is reversed in direction and held on for a time-amplitude product calculated to be sufficient to just rephase the excited spins. In the reversed gradient the Larmor frequencies change (the sign of the second term in equation 6.6 is now negative) so that those components which had a slightly higher than average frequency during excitation now evolve with a correspondingly lower Larmor frequency. At the end of the reversed gradient period most of the spread in phase will be reduced to zero. This is the same principle as the gradient echo method used in fast imaging methods described below.

Frequency and Phase Encoding within a Slice

Once a plane of spins has been selected, the frequencies and phases of the Larmor precessions within the slice can be manipulated, by the application of further field gradients, so that each voxel within the slice is given a unique pair of co-ordinates, Larmor frequency and Larmor phase. The gradient encoding process depends on the fact that the precession frequency of the rotating coherent spin state can be altered reversibly, without significantly changing the strength of the rotating magnetisation. When the small extra gradient fields are turned on, the precession frequency increases from the central value f_o, and then slows down to the central frequency again, when the gradient field is turned off. Immediately after selection, all spins in the slice have the same frequency, f_o. When a gradient is applied each spin precesses at a rate determined by its local field. If the gradient alters the field just in one direction, then the slice is effectively carved into strips of different frequencies. Within each strip the precession frequency is constant with position.

Frequency Encoding

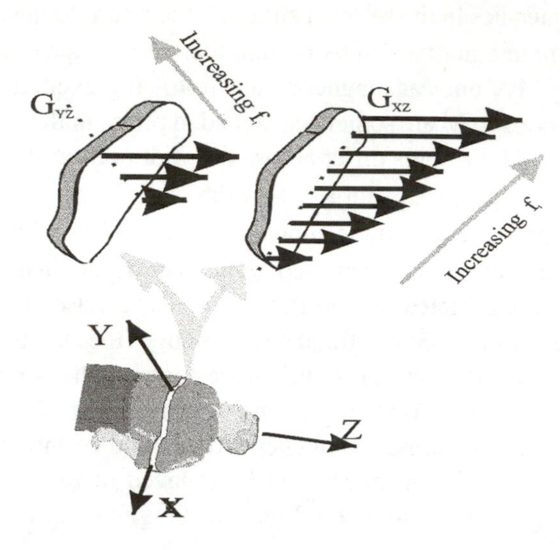

Figure 6.15: The use of gradient magnetic fields, G_{xz}, G_{yz} to encode the positions of nuclear spins within the selected slice. All gradients used in MRI are magnetic fields directed along the Z axis. G_{xz} is a field whose strength increases along the X axis. G_{yz} increases along the Y axis. In conventional MRI G_{xz} is held constant in time while the signal FID is being acquired. It is called the frequency- encoding gradient. G_{yz}, on the other hand, is pulsed on and off just prior to the signal acquisition in order to change the phase of the Larmor frequencies along the Y axis. It is called the phase encoding gradient.

More precisely, if a gradient field $G_{XZ} = dB_{XZ} / dX$ is applied, then the selected slice is carved into strips parallel to Y, with the frequency varying along X. Each strip contains spins precessing at $f_X = \gamma (B_0 + G_{XZ}X)$. If this gradient is maintained during data acquisition then the measured FID signal will contain a range of frequencies corresponding to the width of the subject in the X direction, multiplied by the strength of the gradient Gxz. The FID signal resulting from the entire slice, at this stage turns out to be one line through the Fourier transform of the T1, T2 weighted free water density over the whole slice. It is thus the direct analogue of the 1D Fourier transform of a single x-ray projection that we discussed in section 1.3 . We suppose that at each location within the slice at the point, (X,Y), there is an effective free water density which we will call ρ(X,Y) For simplicity we have assumed for the moment that variations in T1 and T2 are included in ρ. The contribution, ds, of a voxel with area dxdy to the RF signal, obtained in the presence of the x gradient, G_{xz} is given by

$$ds = A\,\rho(x,y)\cos(\gamma B_o t + \gamma G_{xz} X t)dxdy \qquad ......6.8$$

Where A describes the geometrical coupling efficiency of the RF receiver coil to the signal source at the co-ordinates X,Y. In reality A itself varies from place to place in the subject but this is generally taken into account in the computer and so we will drop A from all subsequent discussion. The gradient field varies only along the X direction and so all voxels in the same strip at X, have the same frequency. The contribution of the strip to the total slice FID is

$$s_{strip}(t) = \int \rho(x,y)dy \cos(\gamma B_o t + \gamma G_{xz} X t)dx \qquad .....6.9$$

If we add up the contributions from all the strips on the slice then we obtain

$$s_{slice}(t) = \int\int \rho(x,y)\,dy\,\cos(\gamma B_o t + \gamma G_{xz} X t)dx \qquad .....6.10$$

With just one more algebraic change we obtain the single slice theorem, described in chapter 1 and appendix B, once again. We notice that the frequencies have two parts, a constant high frequency part, γB and a variable low frequency part , $\gamma G_{xz}X$. The constant part is the centre frequency and this is removed at the demodulation stage of signal detection leaving the low frequency envelope of the signal.

$$Signal(t) = \int\int \rho(x,y)\,dy\,\cos(\gamma G_{xz} X t)\,dx \qquad ...\ 6.11$$

This is the Fourier transform of the projection of ρ(x,y) onto the x axis. The only difference between this and the x-ray CT case is the presence of time in the expression. As we step along the envelope of the signal in time, we are also stepping along a line in Fourier space, since we can write $K_x = \gamma G_{XZ} t$. This is only one line through the Fourier transform of ρ(x,y), but we need many projections, taken at different values of the orthogonal K space vector K_Y, in order to get a sufficient sampling of K space to enable us to reconstruct ρ.

The earliest MRI images were obtained in precisely this manner with each projection at a different angle being obtained by electronically altering the direction of the gradient field. Standard practice in modern MRI uses a slightly more convenient engineering approach, which amounts to the same thing mathematically. It simply alters the order in which points in Fourier space are sampled. This modification is called phase encoding . Instead of using just a single gradient field, which is held on during the time that the FID is measured we use two gradient fields, one maintained and one switched. These gradient fields are chosen to vary along the X, Y axes. We will arbitrarily assign the switched gradient, G_{YZ} to the y-axis. This is called the phase encoding gradient. The maintained gradient, G_{XZ} is assigned to the x axis and it is called the frequency encoding gradient. G_{XZ} does the job that we have described above, it produces a Fourier transform of a projection onto the x-axis, that is to say one line in Fourier space parallel to X. But we need many such parallel lines taken at different values of K_Y. This is the job of G_{YZ}.

Phase Encoding

When the gradient, G_{YZ} is turned on, the frequencies vary along the Y axis according to $f_Y = \gamma\,(B_\circ + \gamma\, G_{YZ})$. If the gradient is now turned off after a time T_Y, then all the frequencies return to the constant value $f_o = \gamma B$ but each row of spins at the same value of Y retains a phase shift,

$$\phi(Y) = 2\pi\gamma\; G_{YZ} Y T_Y \qquad \ldots 6.12$$

When the X gradient is turned on, the FID envelope signal (again after demodulation to remove the centre frequency) is given by

$$signal\,(t) = \iint \rho(x,y)\ \cos\,(\gamma\, G_{xz}\, X t + \gamma\, G_{yz}\, Y\, T_y)\, dxdy \qquad \ldots 6.13$$

This is another single line in K space , that is to say a set of points, again parallel to X, but taken at a value of $K_Y = \gamma\, G_{YZ}\, T_Y$. A complete sampling of K space is obtained by making repeated visits to the same slice, but at each visit, changing the strength of the Y gradient in increments. Each FID signal gives a single line along X but with a different offset, $K_y = \gamma G_{yz}\, T_y$, along Y ,(see equation B.6)

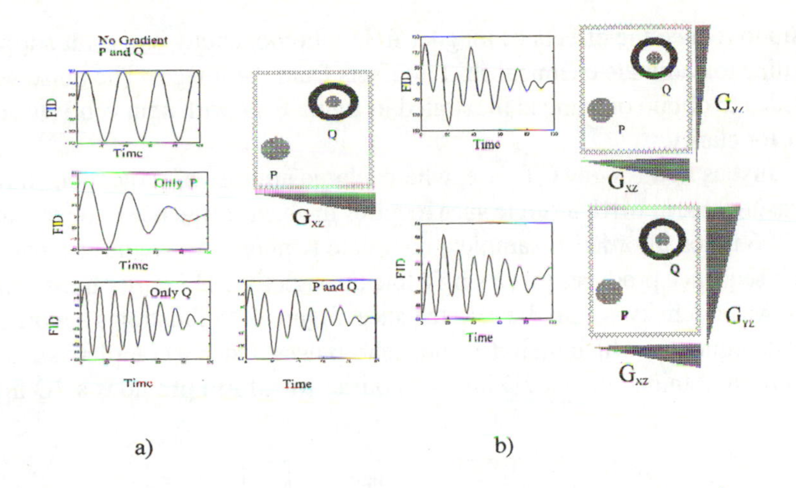

Figure 6.16a: With no gradient applied all objects in the slice have the same Larmor frequency. With a frequency encoding gradient along X, Gxz. The FID contributed by object P has a lower Larmor frequency than that contributed by object Q.

Figure 6.16b: Phase encoding using a finite duration pulse of Gyz held on for a time Ty. The effects of two different gradient strengths, G_{YZ} on the total FID contributed by both P and Q in the presence of the frequency encoding gradient , G_{XZ} are illustrated.

The sequence of events required to obtain a whole slice of data proceeds as follows

1) RF slice selection
1') Other Manipulations eg Spin echo , Fat suppression Inversion , Gradient echo
2) Phase gradient Gyz is applied for Ty seconds then switched off
3) Frequency, or readout gradient, Gxz switched on and held while data from the current FID is recorded.
4) Steps 1 to 3 repeated many times. Each repeat occurs after a time, T_R with a different value of Gyz , to obtain a complete set of lines covering the field of view in Fourier space.

The simplest scheme would leave out the step 1a, involving other spin manipulations, however many commercial imaging sequences make use of spin echo to reduce the effects of magnet field inhomogeneity and a fat suppression routine to eliminate chemical shift artefacts from the image. The pulse sequence needed to obtain one line is illustrated in figure 6.17 with spin echo etc steps left out for clarity.

Just as in the x-ray CT case, where a large number of projections have to be measured so, in MRI, a single slice requires the basic sequence to be repeated 128 or 256 times in order to sample a complete K plane. As we have already seen, each sequence produces just a single line through the 2D Fourier transform of the free water density within the selected slice. Figure 6.18 illustrates a more realistic pulse sequence, which includes spin echo pulses. Once a complete slice of data has been obtained then a 2D inverse Fourier transform provides a 2D image of

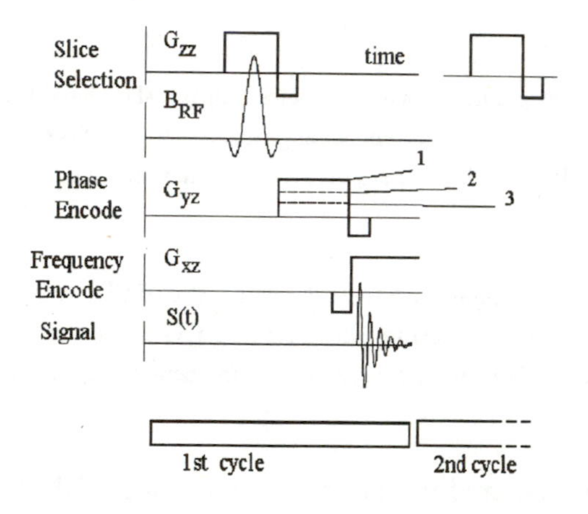

Figure 6.17: The simplest pulse sequence required for a single line through the 2D transform of the free water density. The phase encoding pulse G_{yz} is shown with different pulse heights labelled 1,2,3. These represent the different values used in successive repeats, in order to obtain a complete slice of data. The "tails" on the gradient pulse envelopes are used to narrow the initial precessional phase profile of the rotating magnetisation, using gradient echoes. The weighting of the image with respect to T1 can be achieved by altering the repeat time T_R

the variations in free water density across the slice. If the slice is to be allowed to return to equilibrium by waiting for about 5T1 between each reselection then it is clear that with T1 of order of 100- 500 ms, an entire slice will require several minutes in order to complete the sampling of K space in 128 or 256 lines. During this time the patient has to be immobilised in order to obtain fine spatial resolution. In practice, data from many slices have to be collected for a given scan. In order to save time, partial data is obtained from several slices interleaved

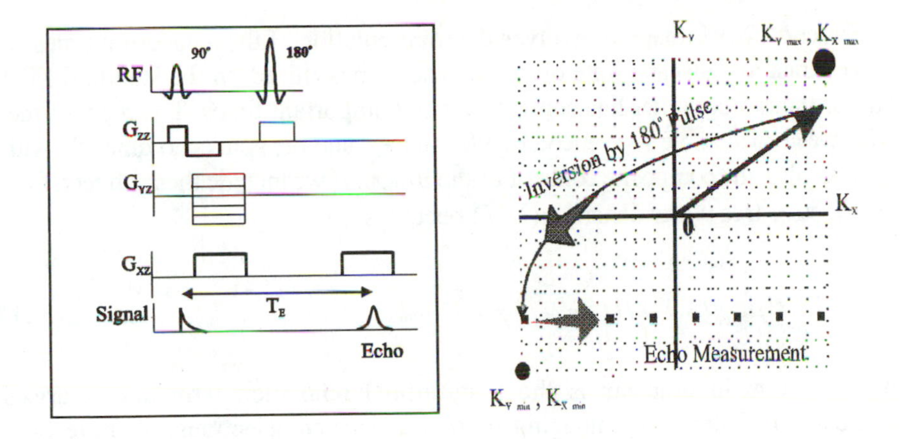

Figure 6.18a: One cycle of a simplified standard spin echo, SE imaging sequence. The key difference between this and figure 6.17 is the use of an additional 180° RF pulse to create the spin echo. Notice that the transverse magnetisation is allowed to evolve in both the X and Y gradients before inversion. The time series in each echo is one line through the Fourier transform of ρ(X,Y) taken with K_Y = constant. In this simple scheme the slice is revisited after a repeat time T_R to collect further lines in Fourier space.

Figure 6.18b : The trajectory in K space taken by the nuclear spin phases in Spin Echo MRI. During the initial evolution period the spins dephase in the X and Y gradients so that the K vectors are spread out along the line 0- A, marked "Evolution". The 180° RF pulse inverts the directions of all the spins $K_{x,y} \rightarrow -K_{x,y}$. The x gradient is then switched on and the FID is read out in time (along K_X).

in time so that while one region is returning to equilibrium another remote slice can be selected and measured. More modern schemes, which allow very much faster rates of data acquisition, are discussed in section 6.5. Although the pulse sequences used in these methods are often more complex than the standard SE scheme, the underlying principle remains that of sampling the Fourier transform of the free water density at a sufficient number of points to allow a tomographic reconstruction to be made.

T1, T2 Weighting

Practical MRI imaging involves the manipulation of the measuring times so that contrast in a disease state or tissue type is maximised. In the standard MRI scanner the radiographer has control over two important times. The repeat time, T_R between successive slice encoding 90 ° pulses, and the spin echo time, T_e, with which to alter the composite contrast of the image. If we include these effects, then our effective free water density ρ(X,Y) becomes

$$S \propto \rho_{free}(X,Y,Z)\,(1 - e^{-\frac{T_R}{T_1}})\,e^{-\frac{2T_E}{T2^*}} \qquad \ldots 6.14$$

The first term in brackets is the longitudinal relaxation term and it arises practically as follows. All imaging involves repeated selections, at finite time intervals, T_R, of a particular slice. After a few repeats at a steady rate (T_R<~ 5 T1), an equilibrium state is established in which the amount of nuclear magnetisation available for each re-excitation is reduced according to this term. Those spins that are still in the XY plane when the next selection pulse occurs do not form a coherent rotating magnetisation vector and thus do not contribute to the FID. The second term is the T2 relaxation factor. The FID is measured via the spin echo at the time 2Te after the selection pulse.

We illustrate how relaxation time weighting can change contrast in figure 6.19. Here we imagine that we have a region of tissue divided into two parts, A, B that have precisely the same free water density but differ either in T1, or in T2. Using a saturation recovery pulse sequence, we can differentiate between A and B using differences in T1. If $T_R = t_2 >> 5T1_{max}$, then both regions have time to return completely to equilibrium between selection pulses. Now there is no contrast between A and B since the free water densities in both compartments were assumed to be the same. If on the other hand $T_R = t_1 < T1_{max}$ then successive re-excitations establish a dynamic equilibrium in which the signal intensity from B is larger than A and thus there is a contrast difference between A and B.

Similar arguments can be used to differentiate between A and B using differences in T2 and Spin Echo sequences. If the echo time is set so that $2T_e = t_1$ and T_R is set to be at least 5 $T1_{max,}$ then there will be no T1 contrast but a maximum amount of T2 contrast. In general echo times, T_E, are in the range 5- 100ms and repeat times, T_R are in the range 50 -2000 ms.

An example of the effects of image weighting is given in figure 6.20.This shows a brain slice imaged using T_R =3000ms and a T_e of 80ms. This enhances the contrast for cerebrospinal fluid ,(T2 =300ms) compared with white, grey brain tissue,(T2 = 70 and 90 ms respectively); the latter contributions have largelydephased well before the measurement is made and thus make less

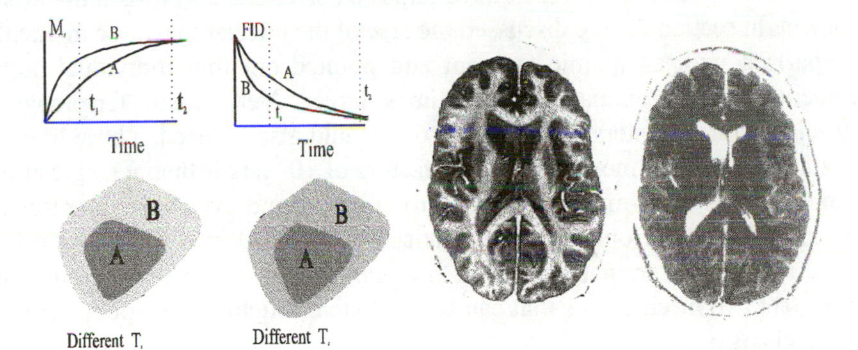

Figure 6.19: Two regions of tissue A,B are assumed to have the same free water density, but they can differ in T1, or T2. Choosing a slice selection repeat time T_R to be between the relaxation times of A and B produces a contrast between the regions determined by differences in T1. Similarly spin echo can be used to enhance T2 differences by the choosing the spin echo time $2T_e$ to lie between the transverse rela xation times of the two regions.

Figure 6.20: T1,T2 weighting in MRI Images. In each panel the brain slice is the same but the amounts of T1,T2 weighting differ. a) Heavy T2 weighting with TR=3000msec, Te=8 ms. b) Heavy T1 weighting with TR=600 ms, Te= 10 ms.

contribution to the overall signal. If the repeat time, T_R is reduced to a value that is comparable with the T1 values for the different tissues under consideration then the image is said to be T1 weighted. Now the equilibrium largely dephased well before the measurement is made and thus make less state is characterised by slightly different steady state amounts of magnetisation for each tissue type having different T1 times.Different disease processes and different soft tissue types each have their optimum ranges of the controllable parameters found from accumulated empirical experience rather than calculation .

MR Contrast Agents

Contrast enhancing solutions are used in MR, just as in x-ray radiography and CT, and they are used for the same reasons; to increase contrast in and around blood vessels and to enhance the visibility of tumours. The primary aim of the agent is to radically alter T1 or T2 * or both. This is normally achieved by using paramagnetic ions, which have unpaired electrons and thus a net atomic moment. In section 2.5 we discussed the size of the nuclear magnetic moment in comparison with an atomic moment and pointed out that individual atomic moments are approximately 1000 times larger. Very small ferromagnetic particles, with diameters of a few microns, could also be used. These have net permanent magnetic moments that are factors of 10^{8} larger than a single atomic moment. Thus local magnetic fields produced around even tiny quantities of paramagnetic or ferromagnetic substances, introduced into the body, can be expected to have a dramatic effect on the local relaxation of water protons There are in fact two mechanisms that can be exploited, roughly corresponding to T1 or T2 changes.

In the immediate vicinity of a strongly paramagnetic molecule, the local static field will be altered significantly over a distance of a few molecular diameters. This will radically reduce the local value of T2*. In addition, if the paramagnetic ion is tumbling at a rate close to the nuclear Larmor frequency then T1 will be radically reduced. To an extent the target process for contrast enhancement can be selected by an appropriate choice of carrier substance for the paramagnetic ion. Introduced carrier molecules that bind to bio-molecules can lead to relatively static field changes and hence reductions in the proton T2*. More inert substances with small molecular weights will be more mobile and with a little chemical engineering can be made to exhibit electron spin relaxation rates in the vicinity of the proton Larmor frequency and hence produce strong reductions in T1.

The paramagnetic ions are chosen from the transition metal (Cr, Mn , Fe) and rare earth metals (Gd) in the middle of the periodic table, see section 2.2. All of these have permanent atomic moments but not all are suitable as contrast agents. In fact all these metal ions are rather toxic, on their own, and thus have to be incorporated into chemical structures that render the metal ion harmless at the expense of somewhat reducing the effectiveness as a contrast agent. Even then, just as in x-ray radiography, a small minority of patients produces very strong and unpredictable allergic reactions to these substances. Gd-DPTA is a commonly used agent. Its use is discussed in chapter 8

MRI has the potential for intrinsic contrast enhancement using simply elapsed time, in addition to the standard T1,T2 weighting described above. Thus in flow imaging a bolus of flowing spins can be excited and then imaged at a later time after it has flowed into the imaged region of interest. This is discussed in more detail in section 6.6 below.

6.5 Fast Imaging Methods

We saw in section 6.4 that the data acquisition for a conventional SE image will take several minutes, especially if long TR intervals between repeated slice selections are used to allow T2 weighting to flavour the image contrast. There are two related reasons for significantly reducing the imaging time. First over a period of several minutes, the patient will inevitably move to some extent, even if the body surface is effectively clamped. In addition the internal motions associated with heart action, blood flow, breathing and peristaltic motion in the gut will all serve to limit the spatial resolution of any image that involves moving tissue. In the worst case, uncompensated movement can actually introduce completely spurious features or motion artefacts into the final image. Since both heart action and breathing are regular periodic motions, their effects can effectively be eliminated by monitoring the mechanical motions and then ensuring that each successive projection is obtained at exactly the same phase in the body cycle. Bowel action, on the other hand, is essentially random and thus compensation by timing will not be successful. If the imaging time can be radically reduced, then all unwanted motion artefacts can be reduced since a short snapshot in time is used to capture the image.

Very fast imaging also opens up the possibilities for imaging very ill patients, watching in real time heart valve motion and, in principle, ionic activity within the brain. Roughly speaking there are two different engineering approaches to fast imaging. One family uses repeated echoes to collect many lines in Fourier space at each selection, the other uses small tip angles and very short gradient echoes to create a complex steady state. Both families of methods are capable of bringing the total imaging time down to very much less than one second per slice. Each family has its own set of engineering problems and strengths and weaknesses in terms of the diagnostic information that it can yield. Fast imaging methods in general are relatively new and still under development so that we cannot describe a single most commonly used method comparable to SE in conventional MR imaging

Turbo Spin Echo

Our description of the standard SE method for conventional imaging in section 6.4 showed that a single line in Fourier space was collected at each slice selection. This is unnecessarily wasteful in time since, even after the first echo, there remains a rotating nuclear magnetisation that can be refocussed, using another 180° pulse. In fact for times up to about T2 many useful echoes can be

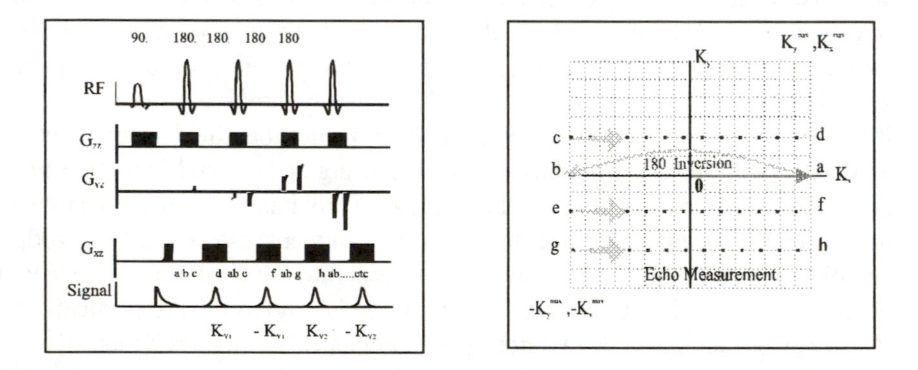

Figure 6.21a: Turbo spin echo pulse sequence

Figure 6.21b: The K space trajectory used in Turbo Spin echo. Evolution in the first period of X gradient produces a projection corresponding to half the field of view with $K_{Y=0}$, point, a, in figure 6.18. An inverting RF pulse together with a Y gradient increment together take the spin system to the beginning of the first reading along K_{Y1} of the first echo. After the reading is complete, a decrement in the Y gradient returns the spin system to its starting point, a, . The cycle repeats with larger Y increments. Generally only a partial coverage, eg 16 or 32 lines of K space, is attempted at one excitation. The slice is then reselected to complete the sampling.

formed in a train. If the phase encoding gradient G_{YZ} is altered between echoes then each successive echo will actually correspond to a new line in Fourier space. A typical pulse sequence and the corresponding K space trajectories are shown in figure 6.21a and 6.22b In figure 6.21a we have labelled the 180° RF pulses with X and Y, corresponding to the direction of the applied RF field. This compensates for the inevitable slight inaccuracy of any 180° pulse. Perfectly accurate large angle, (>=90°) pulses, delivered in the presence of a field gradient, are impossible. Small deviations in tipping angle create small-unwanted transverse magnetisations of imprecise phase. These partially incoherent components build up, with successive inversions about the same axis, causing the transverse signal to die out too rapidly. If however the initial selection pulse is delivered along the X axis but the subsequent inversion pulses are delivered along Y, then a rephasing takes place which restores the coherence of transverse magnetisation almost independently of the accuracy of the inversion pulse. This scheme was invented for NMR long before the advent of MRI and is called the Carr,Purcel,Meibohm, Gill (CPMG) sequence.

Just as in SE imaging, increments along K_Y are obtained by allowing an evolution in the Y gradient, G_{YZ}. In turbo spin echo, TSE these increments are created by pulses of G_{YZ} delivered after each inversion pulse. Notice that the sampling sequence alternates between $\pm K_y$ and that after each line in Fourier space is obtained, the state of spin evolution is returned to the starting point (point a in figure 6.22b). This arises because a pair of G_{YZ} pulses of equal magnitude but opposite sign encloses the signal readout part of sequence. This is essentially the same as a gradient echo since all extra Y dephasing created by the positive pulse is exactly undone after the negative pulse.

In principle this process could be carried on until all of Fourier space had been sampled. There is however a problem arising from T2 decay; the later echoes, being further in time from the original excitation, will be systematically lower in amplitude, as result of T2 decay, and this will introduce bands into the Fourier transform, leading to artefacts in the image. In practice 16 or 32 lines, forming a segment of data, are collected after one excitation, then the slice is revisited to collect second, third and so on segments until the complete plane is sampled. Reference is sometimes made to the turbo factor, which is simply the number of lines of data collected at each excitation. This provides a measure of the speed increase over conventional SE imaging.

Gradient Echoes

Echoes, or spin rephasing are produced by the application of short, paired, positive and negative pulses of a gradient magnetic field to the rotating nuclear magnetisation. While a gradient field, G_{YZ} is applied, nuclear spins acquire slightly different Larmor frequencies according to their position along the Y axis. If the gradient is turned on, held constant for a time, T_Y, then switched off, the nuclear spins acquire an altered phase, $\Delta\phi = 2\pi\,\gamma G_{YZ} Y T_Y$, in exactly the same way that we described for phase encoding in conventional MRI in the previous section. In general the extra phase acquired, over a time interval, while a varying gradient is applied is given by

$$\Delta\Phi = 2\pi\gamma\, Y \int_0^{T_Y} G_{YZ}(t)\, dt \qquad \ldots 6.15$$

After an interval during which the net or integrated field-time product is positive, the nuclear spin phase will have advanced with respect to the centre Larmor frequency. After an interval of net negative field-time product, the phase will have lagged. If a bipolar pair of positive and then negative gradient fields is applied in succession, any phase advance in the first interval will be reversed in the second interval. At any time that the net field time product sums to zero, the original nuclear spin phase distribution will be restored and an echo signal formed in a very similar manner to the spin echo described above. Although gradient echos are similar to RF echoes, there is a key difference. The RF spin echo flips the directions of all spins in the XY plane by 180°. This has the important consequence that spin dephasing from any static field operating between the times of selection and inversion is re-focussed at $2T_e$. This includes dephasing due to magnet inhomogeneity and any patient magnetic susceptibility effects. In short the RF spin echo removes most static contributions to $T2^*$. The gradient echo on the other hand, is formed only from the spins that dephased in the previous time intervals of gradient field application. All other sources of spin dephasing are left undisturbed. Gradient echoes are therefore inherently sensitive to $T2^*$ effects, whereas spin echoes are sensitive to T2 effects. It turns out that gradient echo signals are also sensitive to motion, an extra contrast mechanism in some applications but the cause of artefacts in others. This is discussed below in section 6.6.

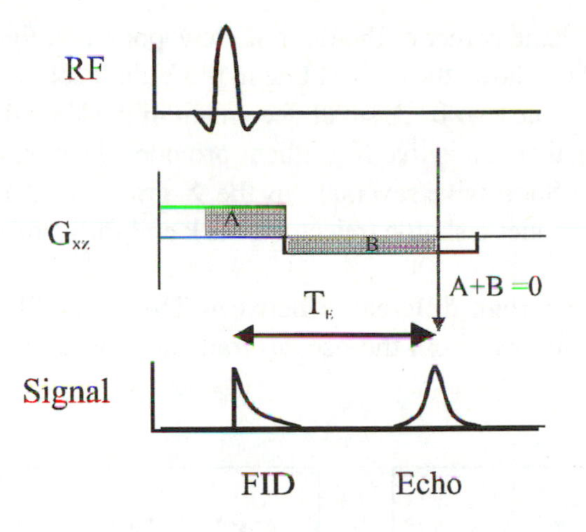

Figure 6.22: Gradient echo production. After the excitation using the RF pulse the total transverse magnetisation dephases in a magnetic field gradient field G_{YZ}. . If this gradient is abruptly reversed in direction then all the dephasing which took place when the field was positive is undone during the period of reversed gradient. After a time T_{ge} the FID will reform, apart from $T2^*$ decay to produce an echo. It should be noted that the use of gradient echos does not remove the effects of magnetic field inhomogeneities that arise either from the magnet or tissue susceptibility effects. The advantage is that T_{ge} can be considerably shorter than in conventional spin echo methods.

Echo Planar Imaging

Echo planar imaging holds the record for speed of data acquisition, <100ms for a single slice of data. It is very similar to turbo spin echo imaging, the key difference being the use of repeated gradient magnetic field echoes rather than RF spin echoes. Each line in K space is acquired from one gradient echo. Short pulses of an orthogonal gradient field are used to step between the K space lines. Figure 6.23a illustrates the sequence of pulses used and the K space trajectory during data acquisition. The repeated gradient echoes of the X projection (K_X) are produced by the alternating, G_{XZ} gradient field. The interleaved pulses of Y gradient, shift the phase of each echo and hence the value of K_Y for each data acquisition. It is clear that in order to complete the whole sampling in less than 60 msec the X gradient must be switched with a frequency of about 10kHz.

After slice selection, phase evolution in both X and Y gradients takes the K vector to the top right hand corner of the field of view, point a in figure 6.23b. A reversed X gradient then allows the highest line in the Y direction to be measured backwards. This is the line a $\rightarrow$ b . A small decrement in the Y gradient takes us to the Fourier point c, then a positive X gradient provides the next line down in Fourier space, c$\rightarrow$d. Successive reversals in the X gradient interleaved with decrements along Y produce a shuttle trajectory, back and forth and down across the Fourier plane.

There are two important differences between TSE and EPI. As we have already mentioned it follows from the use of gradient echoes in EPI and spin echoes in

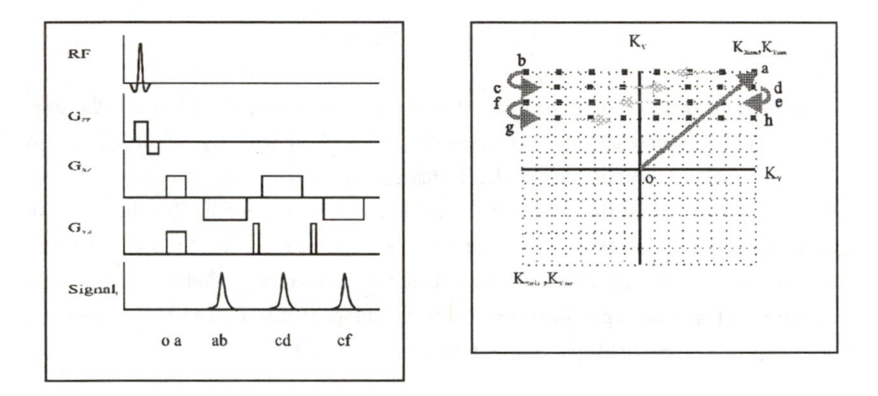

Figure 6.23a: The Pulse Sequence used in echo planar imaging, EPI.

Figure 6.23b: The K space trajectory in EPI. The EPI sequence uses gradient echoes to complete a slice using just one FID. The gradient Gxz is switched at K to create a train of gradient echoes. Between each echo a gradient pulse Gyz increments the phase so that each echo corresponds to one line through the Fourier transform of the free water density. Specifically immediately after selection the spin phases evolve in the combined X, Y gradients to reach point a. The first line is read out along a to b then a Y gradient pulse shifts K_Y by one step. The next gradient echo provides the readout line c to d. The whole process repeats until all the data in the FOV have been acquired.

TSE that EPI is sensitive to $T2^*$ and TSE is sensitive to T2. In addition the standard EPI trajectory in Fourier space starts at the highest K magnitudes, passes through zero and proceeds to the highest K magnitudes of opposite sign. Normal TSE does the opposite, lowest K values are collected first. Roughly speaking it is low K values (long wavelengths) that contain most global contrast information but the high values (short wavelengths) convey fine detail and determine spatial resolution. Given that the Fourier samples that are measured earliest in time following excitation are the most accurate, it can be seen that EPI and TSE serve two slightly different masters. EPI has been used extensively in recent years in the search for functional contrast in the brain arising from blood susceptibility differences, precisely because this is thought to produce $T2^*$ effects. Both conventional spin echo MRI and TSE would abolish any contrast arising from susceptibility effects.

Steady State Gradient Echo Methods

Fast MRI imaging can also be accomplished by dramatically shortening the repeat time, T_R , between successive slice selections and using excitation flip angles that are much less than 90^0. If T_R is chosen to be much less than the average T1 for the tissue in the field of view, then complete thermal equilibrium is never re-established during the data acquisition. Instead, after a few selection pulses, a dynamic equilibrium is established in which only a fraction, depending on T1, of the total nuclear magnetisation is available for imaging. The remainder stays, precessing in the XY plane but partially dephased. Small flip angles are required in order to ensure that the steady state level adopted by longitudinal magnetisation is actually high enough for the successive RF pulses to produce some coherent signal. There is an optimum flip angle, α, for such situations, called the Ernst angle. This is determined by the expression

$$\cos(\alpha) = \exp(-TR/T1) \quad \ldots 6.16$$

If T_R is shortened even more so that it is less than T2 then a more complex equilibrium results. The generic feature of steady state methods is the presence, immediately before the next excitation pulse, of magnetisation components in the Z direction, whose magnitudes vary with position according roughly to the spatial variation of T1, together with components in the XY plane, left over from several previous excitations. The leftover components remain and thus accumulate over several cycles and thus the instantaneous signal is a complex sum of several previous excitations. If no other steps are taken a steady state image will be degraded by spurious interference bands running across the image that result from the mixtures of phases and T1 weighted magnetisations. A plethora of methods, with acronyms such as FISP, FAST, GRASS, PSIF, FADE have evolved from the first of these schemes called FLASH , fast low angle shot .The different schemes deal in different ways with the problem of remaining transverse spin coherences and thus remove the spurious banding from the resulting images. In some schemes the periodicity of the banding is reduced below one voxel size so that each voxel represents an average over several bands and hence several T2 weightings. This is achieved by using large amplitude gradients to create very large phase dispersions across the field of view. In other methods deliberate attempts are made to "spoil" that is to say dephase the remaining XY component. Finally there is a group of methods that aim for complete refocussing of the XY components just prior to each excitation pulse. A more detailed treatment of the various idiosyncrasies of these methods is mathematically well beyond the remit of this book and thus we leave this subject in this cryptic form and refer readers to dedicated texts on MR.

3D or Volume Image Acquisition

Throughout this book we have repeatedly described tomography as a slicing process leading to 2D cross sections as the basic imaging elements . Both x-ray CT and MRI have developed what are effectively 3D acquisition schemes that have certain advantages. In section 4.5 we gave a brief description of spiral CT imaging , the most modern development and described how this allows 2D cross sectional images to be reconstructed at virtually any angle to the scanner axis. There the spiral trajectory through K space is achieved by a simultaneous translation of the patient, while the x-ray generator and detector are revolving. 3D MR imaging is also employed on modern scanners and is achieved rather more simply by the inclusion of a third variable encoding gradient, G_{ZZ} in addition to G_{ZX} and G_{ZY} . Clearly the pulse sequence is rather more complicated since one of the gradients (most simply the axial gradient, G_{ZZ}) is first used to select the slice and then to encode along the Z axis. Volume acquisition is particularly useful in MR time of flight angiography, which is described in the next section .

6.6 Imaging Movement and Flow

Movement of tissue, fluid flow and whole organ or limb movement is the bane of all medical imaging, leading to blurring and the formation of ghost artefacts. At the same time the imaging of particular types of motion within the body is actually the purpose of some examinations. Most types of MRI are particularly sensitive to motion and thus MRI images can be both degraded by unwanted motion and intentionally used to display motion. We can treat both cases by considering the effects of movement in producing a contrast in the MRI signal. Two distinct categories of effect can be identified, modulus contrast or "time of flight" and phase contrast. Both arise because the coherent rotating nuclear magnetisation is relatively long lived and any imaging protocol lasts a finite amount of time, during which any movement can lead to significant shifts in coordinate position so that excited spins travel through the gradient fields acquiring changes in phase that differ from truly static tissue.

Time of Flight Methods

The time of flight effects are best understood and most frequently employed with respect to flowing blood. When a particular segment of a blood vessel is excited, the water protons in that segment are set precessing. At the same time that segment of blood is moving along the vessel with an average velocity v. At slightly later times, other vessel segments, downstream receive the precessing protons, so that the total nuclear magnetisation there is a mixture of yet unexcited and in-flowing, already excited, protons. Similarly the initial excitation region will lose some or all of its precessing spins through flow. Depending on the time delay chosen from the initial excitation, the downstream in-flowing component will be some way along its T1 and T2 relaxation curves. A number of different time of flight schemes can be used to obtain flow velocity dependent contrast. We will illustrate the principle with a simple idealised example of blood flowing at constant velocity, v, along a vessel that lies along the Z axis. Using conventional SE, an upstream volume of blood can be excited with a 90° RF pulse. An 180° RF pulse corresponding to a downstream location is applied after a time T_e . This involves a frequency shift between the two RF pulses of $\gamma G_{ZZ} \Delta Z$ where ΔZ is the distance along the vessel between the excitation region and the imaging region. At

$2T_e$ an echo will form only from that volume of blood which has flowed the distance ΔZ in the time T_e. This is called an in-flow method. It will of course be noted that at the echo time, $2T_e$ the blood volume giving rise to the echo signal, will not be physically located at the imaging region, but a further distance $\Delta Z\ T_e$ downstream. A complimentary method applies both RF pulses to the same volume. Now the excited out-flowing blood will be lost, creating a signal reduction. This is called the "black blood method".

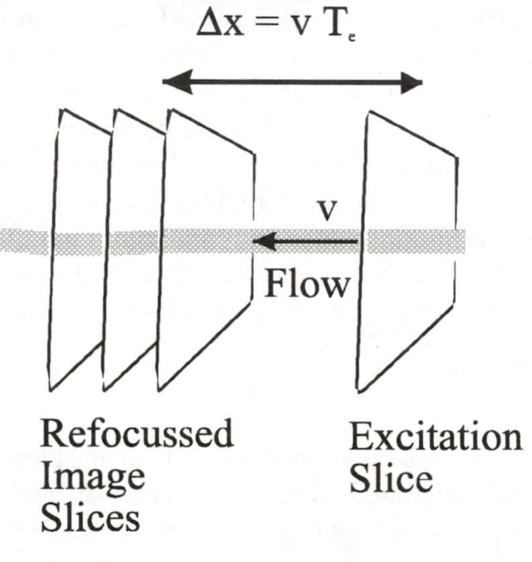

Figure 6.24: Time of flight Flow imaging. An inflow contrast is created by selecting a slice upstream of the desired imaging region. Blood excited in this region flows along the vessel with velocity v. If downstream regions are interrogated with each refocussing after a delay of vT_e then at a time $2T_e$ an image of the blood flowing into the imaging region will be formed.

Time of flight, ToF, MR angiography , MRA is generally achieved using 3D volume acquisition. We will illustrate the method using the common example of MRA imaging of the carotid arteries leading to the brain. An excitation region is chosen at about the level of the shoulders. The image volume then comprises the cervical neck and lower parts of the skull. Generally an additional spin saturation pulse is applied , covering the upper part of the skull and brain. This pulse effectively saturates venous blood moving downwards into the imaging volume so that venous structures do not appear in the final images. After the 3D acquisition, a stack of 20-50 very thin contiguous parallel slices are constructed , each one showing a horizontal cross section . In each slice the in-flowing blood appears very bright and thus the cross sections of the vessels containing fast moving blood are by far the strongest signals. These most intense voxels in each slice are then selected to produce a MIP – Maximum Intensity Projection.. The ensemble of these voxels from all the slices can then be reconstructed into 2D projection images at virtually any angle chosen to show the course of the arteries under investigation. Figure 6.I.6 was constructed in this way. The general procedure that we have just outlined can be applied to virtually any part of the vasculature .

Phase Contrast Methods

All MRI imaging methods make use of gradient magnetic fields in order to localise the component signals. Stationary objects, within the field of view, ideally experience gradient fields whose magnitudes vary with time only to the extent dictated by the field controller. In these circumstances a precise knowledge of the field at every point in the field of view translates into a precise knowledge of Larmor frequency and phase and finally precise signal source localisation. If the gradient fields seen by a particular volume of tissue alter with time in a manner not prescribed by the field controllers then localisation becomes ambiguous, leading either to unintentional misregistration, that is to say image artefacts, or an additional velocity and acceleration dependent contrast mechanism.

Moving structures necessarily experience a time variation in a static applied magnetic field gradient since their motion takes them from one field strength to another. Again we will use a simple example to illustrate the general principles of phase contrast flow imaging. We consider a vessel running parallel to the Y axis containing blood flowing at velocity v. It will be apparent that blood flowing along the Y direction will experience a modulation in RF phase. The size of the effect depends on the switched field gradient, G_{YZ} applied along the Y axis. Stationary objects experience no net change in MR phase after a balanced positive - negative gradient switching sequence but objects, moving along the X axis will experience a net field time product. Thus the moving objects acquire a different MR phase from the stationary objects after the sequence a,b,c shown in figure 6.25. Typically $v = 0.2$m/s in arteries but much slower in veins. Thus for a given gradient strength the arterial will change much more than the venous signals. In a conventional GE sequence, a G_{YZ} gradient is applied for a time T_Y

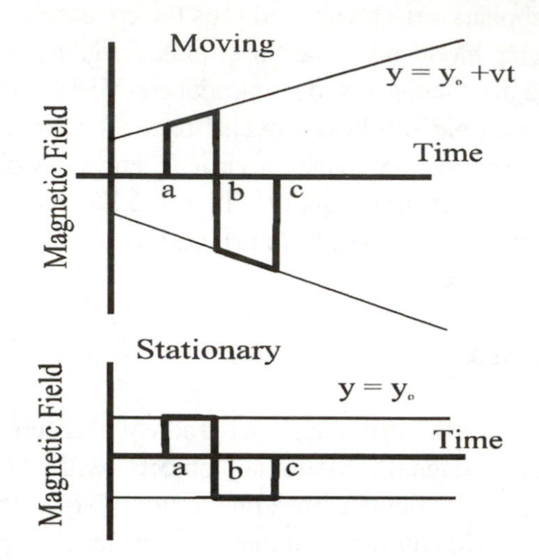

Figure 6.25: The magnetic field experienced by moving and stationary tissue. Stationary tissue exposed to a bipolar gradient pulse experiences no net phase shift. Moving structures, on the other hand, move through the gradient field and so exact compensation of the positive and negative field time products does not take place. The net phase shift after such a bipolar gradient pulse is proportional to flow velocity.

in order to perform phase encoding along the Y axis, see section 6.4. During this interval, the blood flows a distance $\Delta Z = vT_Y$. If we consider the small elements within this time then at each one the precession frequency is $\gamma G_{YZ}\, Y(t)$. In a short interval dt between t and t+dt these spins acquire an MR phase shift of $d\phi$. Adding all such elements over the time T_Y, the total relative phase shift of the moving volume with respect to the stationary ones is given by

$$\Delta\,\phi = 2\pi\gamma \int_0^{T_Y} v\ \ G_{YZ}\ (t)\ dt \qquad \ldots 6.17$$

In general phase modulation angiography involves the use of three or more images of the same field of view. First a velocity insensitive image is obtained which encodes the positions of the vessels. Then two or more images are obtained with velocity sensitive gradients oriented in two orthogonal directions. The small phase differences between velocity sensitive and velocity insensitive images are then calculated and these differences, being proportional to velocity modulate the vessel signals according to their local flow velocity

6.7 Image Artefacts in MRI

We have repeatedly emphasised in this chapter that localisation in MRI depends entirely on the linear relationship between Larmor resonance frequency and applied magnetic field. As long as the fields are known precisely at all positions inside the patient then localisation proceeds smoothly to the production of excellent high resolution images. In practice the situation is not so clear cut. Even the designed fields deviate slightly from their specification both in space and time. In addition patients can themselves be the source of weak localised field perturbations as a result of surgical implants, magnetic dust in the lungs, even small quantities of ferromagnetic particles in clothes and cosmetics. As we saw in the last section blood flow, respiration and peristaltic motion create continuous small internal body motions that can introduce false detail into the MRI image. The complexity and increasing sophistication of MRI in its many forms, together with the inherent sensitivity of MRI to even tiny field changes makes the subject of MRI image artefacts very important. We will not attempt here to give a full description of the many different artefacts and their common remedies. Instead we will provide an overview of the generic types, most of which in one way or another, boil down to the effect of a field or fields deviating from a known specification. There are other sources of artefact arising from small chemical shifts in the Larmor frequency between water and fat, real non-ideal RF pulses changes and the effects of $T2^*$ signal decay in fast imaging.

Static Field Distortions

The large static polarizing field of about 1T is generated by a number of coils of superconducting wire. We have already emphasised that it is impossible to create an entirely uniform magnetic field over an arbitrary volume using windings of finite size. The design compromises, used to create the region of uniformity, inevitably lead to very rapid field deviations outside the specified region. The field profile will look rather like a table cloth draped over a circular table. Within the circle the field (the height of the cloth above the floor) is relatively constant or flat but starts to drop very rapidly at the edge. If a particular field of view includes this fringe region then severe distortions in the image will result. Distortion due to main field inhomogeneity manifests itself most strongly in the frequency encoding direction if small amplitude gradients are used. Throughout our discussion of MRI we have always assumed that the applied gradient fields, our G_{IZ}, are precisely linear. Again it is impossible to create a perfectly linear gradient over an arbitrary volume and so that a compromise is sought with a specified linearity over the central volume. Gradient non-linearity manifests itself as an image distortion in both the phase encode and frequency encode directions.

The combination of a severe main field spatial variation at the periphery of the central volume together with an applied gradient can create an ambiguous image. The effect is one source of ghosting or aliasing. Returning to our table cloth analogy, a combined main field and gradient would correspond to the table top being tilted very slightly along one direction. This inevitably means that there are pairs of places, diametrically opposite each other across the table, that have exactly the same height above the floor. These places in the MRI magnet correspond to the same Larmor frequency. Thus regions outside the intended field of view can be superimposed on the intended portion of the image. The image will then take on the appearance of a dense clear image with a ghostly, more transparent and slightly fuzzy impression of another part of the anatomy superimposed. In some circumstances signal coherence between the intended central region and the artefact, arising from the periphery leads to a strongly banded interference pattern. Such ghosting effects can clearly hamper image interpretation but sometimes known common effects can be put to use in diagnosis.

In addition to spatial variations in the applied fields, patients themselves are the weak sources of magnetic field. Just like the effects of motion, this is a source of artefact in some images but a source of contrast in others. Apart from obvious gross field disturbances produced around imbedded steel objects, the tissues themselves are very slightly magnetically polarised in the main field. Thus there is a very slight variation in the total local magnetic field adjacent to nasal sinuses where there is contrast between slightly polarised tissue and the essentially zero polarisation of air. The extra polarisation shifts the Larmor frequency very slightly leading to both frequency and phase mis-registration. Recently the effect of tissue magnetic susceptibility has been utilised in functional MRI, fMRI see chapter 8.

Time Varying Fields

In section 6.6 we discussed the effects of motion in giving rise to both artefacts and contrast in MRI. Motion artefacts can be seen as part of a general category of time varying fields. Apart from saving electrical energy, the use of superconducting magnets in MRI also leads to a very stable main field. In fact the current in the magnet, after its initial charging, flows in a continuous closed loop of superconducting wire without appreciable dissipation. The superconducting properties of zero resistance and flux constancy ensure a very high degree of temporal stability in the main field. Fields generated by permanent magnets and conventional resistive magnets for specialised MRI applications vary in time to a very much greater extent than the now conventional superconducting variety. Any uncompensated drift in the main field shifts the Larmor frequency away from the assumed RF chain central frequency. This leads to phase and resonance offset induced artefacts in the image.

All tissue, stationary and moving is actually subjected to small unintentional temporal field variations in an MRI system. These arise from eddy currents induced by switched gradient fields in the metallic parts of MRI supporting structures. These induced fields tend to persist for many ms after any abrupt applied field change and thus upset the precise relationship between position and frequency. The currents themselves are rather hard to predict in detail and since they are changing in time their effects are rather complex and differ with imaging sequence.

Finally the human body is itself a weak electrical conductor and so the applied RF fields themselves induce eddy currents in the tissue which both heat the subject and distort the applied RF field. This is referred to as the skin depth effect because at high frequencies, electrical currents tend to flow in a very thin surface layer of a conductor. At 1T (43MHz) the skin depth in soft tissue is about 20 cm. As the frequency rises, the skin depth decreases inversely as the square root of the frequency. The main consequence of eddy currents is the tissue heating but in addition the applied RF field magnitude and phase are altered significantly with increasing depth into the subject. This upsets the precise definition of a slice since the degree of nuclear spin tipping will alter into the subject. Additionally in principle, since tissue conductivities vary anatomically, there are phase changes in both the applied RF fields and the resulting MR signals that give rise to misregistration and signal magnitude changes and thus artefacts. It appears that at present these anatomical effects are small in comparison with other sources of error.

Water Fat Resonance Offset

All protons in the human body can resonate at frequencies close to 43MHz/T. However in general as we have seen in section 6.3, it is mainly free water protons that contributes to the MR signal on conventional imaging time scales of tens of ms after excitation. At the increased field strengths of modern MRI 1.5-4T another effect called water fat resonance offset becomes important. Protons in fatty tissue have a gamma value that is about 3ppm different, as a result of the chemical shift from aqueous free water protons. The fat protons are excited and their T2 values are comparable with aqueous molecules and thus they contribute to the signal. The 3ppm chemical shift produces a difference in frequency of about 145 Hz which leads to a misregistration in the frequency encode direction. When a gradient strength of 1mT/m is used, the fat shift corresponds to a position shift of about 3 mm. Fatty tissue in images can be differentially displaced with respect to its surroundings. Fat in bone marrow is sometimes seen on the wrong side of thin bone sections, as a result resonance offset errors.

Fat signal suppression can be accomplished by a number of techniques. A common method uses the fact that the fat and water signals, differing by 145 HZ are periodically (T= 1/145 = 4.4ms) exactly in and exactly out of phase. By choosing the echo time to be a multiple of this period, it is possible to enhance or reduce the effect of the fat signal on the image. Another method exploits the difference in T1 between water and fat. Using an inversion pulse and a suitably chosen interrogation time, the imaging signal can be taken just at the point in time when the fat signal is crossing zero. See figure 6.6 for an illustration of the inversion recovery method.

6.7 Images from MRI

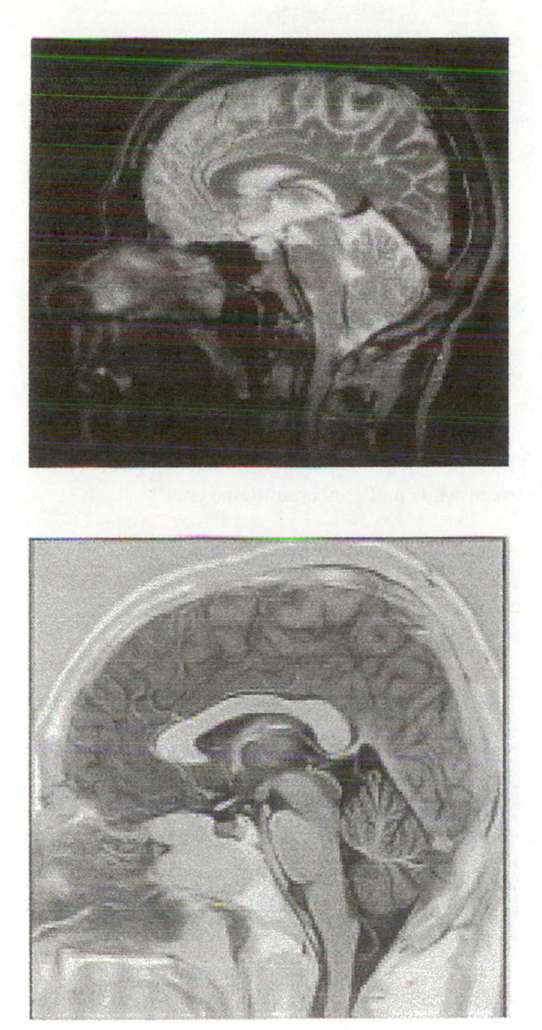

Figure 6I.1: High resolution saggital MRI brain images obtained using different sequences (Philips).

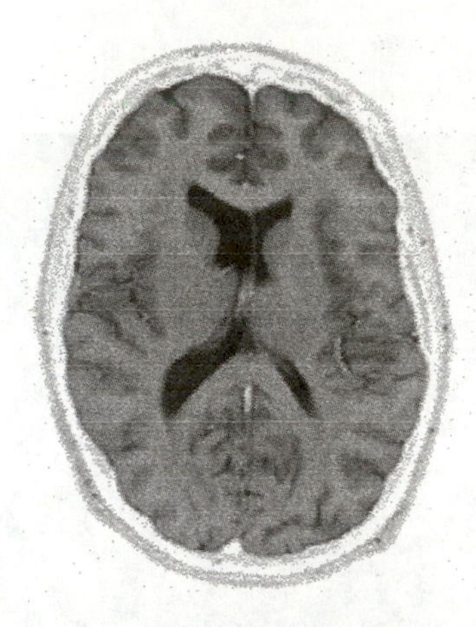

Figure 6I.2: A typical T2 weighted MRI Brain slice image showing the strong contrast between white and grey brain tissue (IOP)

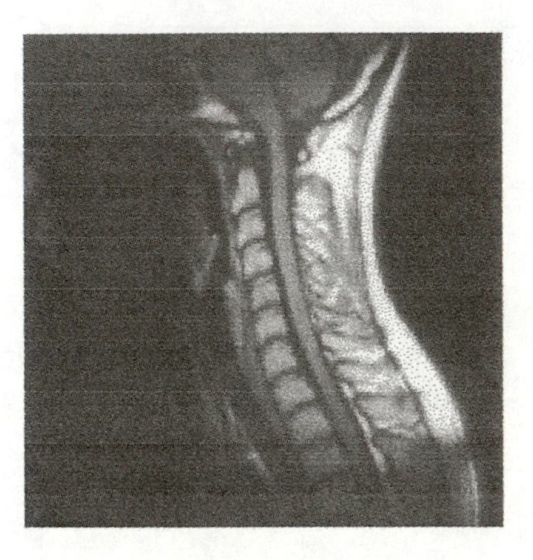

Figure 6I.3: High resolution MRI image of the spine obtained using a spine coil (IOP)

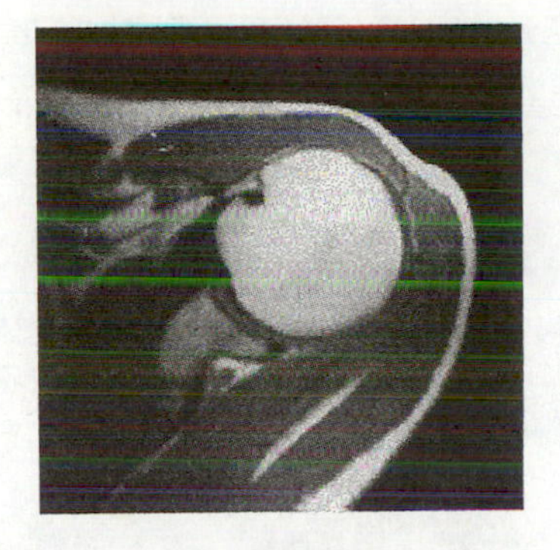

Shoulder

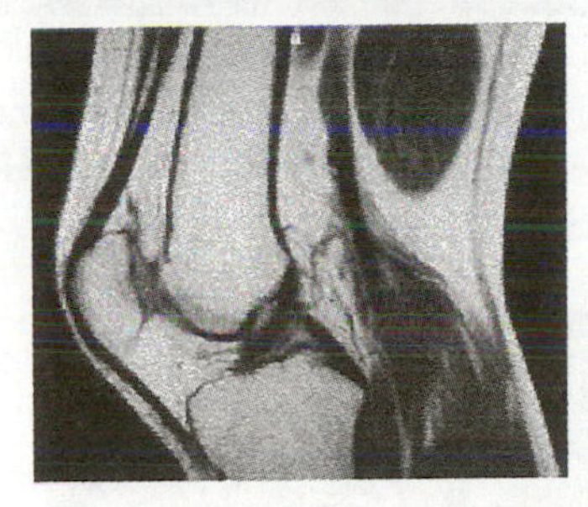

Knee

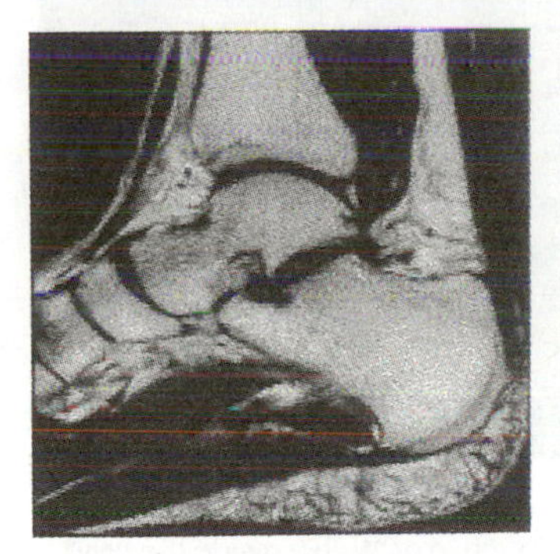

Ankle

Figure 6I.4: High resolution MRI images of the tissue around major joints (Seimens)

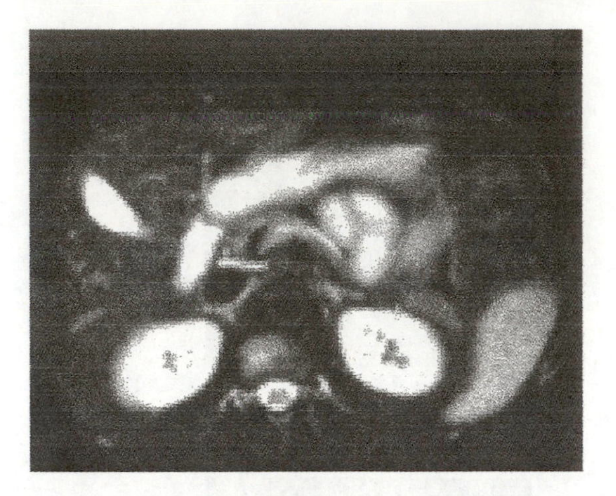

Figure 6I.5 MR image of lower abdomen showing cross sections of liver, kidneys and small intestine. This can be compared directly with the equivalent x-ray CT image shown in figure 4I. 5 (Seimens)

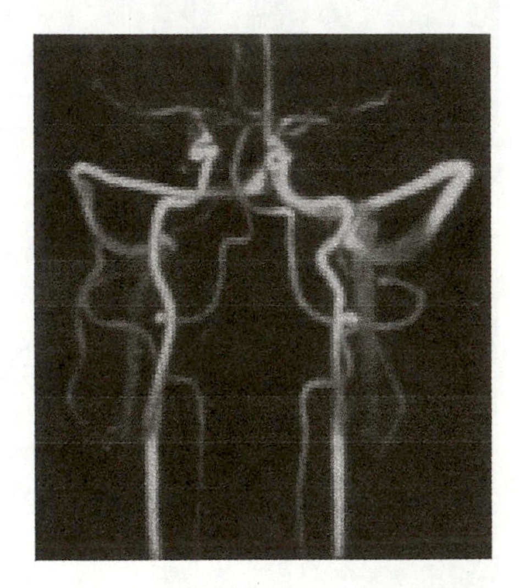

Figure 6I.6 : MR angiogram of circle of Willis, inside the head , where the ascending carotid arteries join and then branch off into the middle , posterior and anterior cerebral arteries to feed separate sections of the cortex and brain stem (Seimens)

Image Artefacts

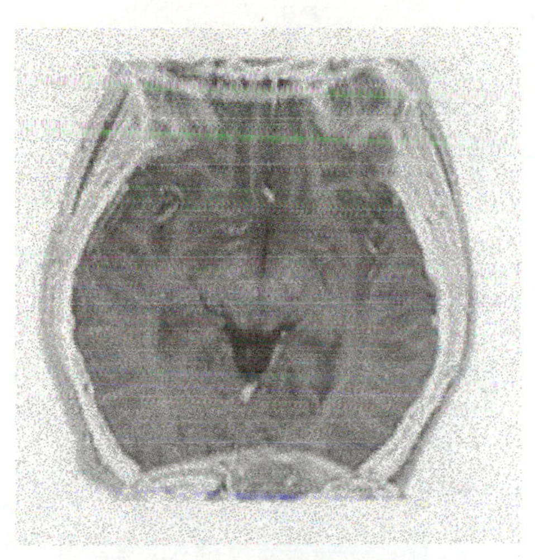

Figure 6I..7 : MR Wraparound artefact in brain imaging (IOP)

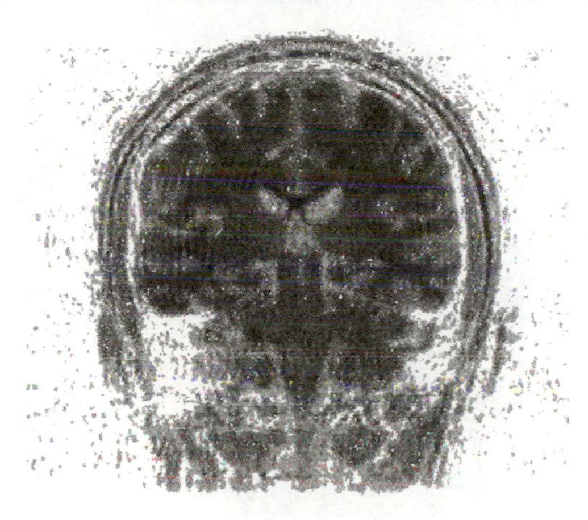

Figure 6I.8: MR artefacts caused by gross subject movement during the data acquisition (IOP)

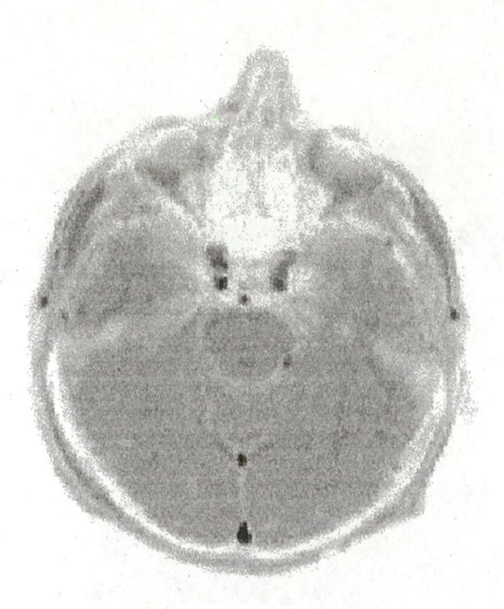

Figure 6I. 9: The effect of flow in MRI. The discrete very dark patches arise from flow in the sinuses and arteries. This slice is transverse to the image in 6I.6, just below the circle of Willis . (FLASH MR sequence) (Charing Cross)

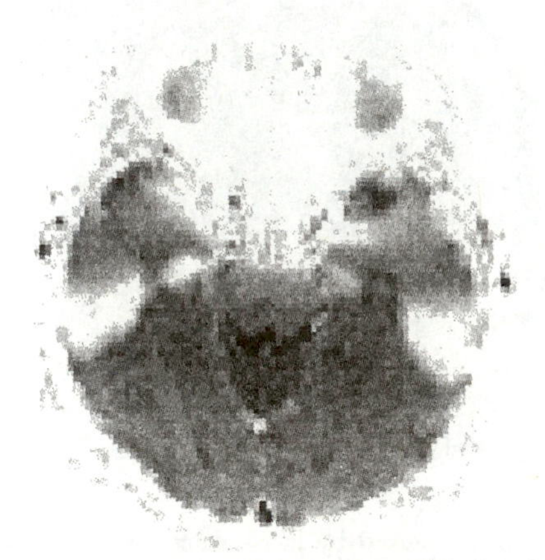

Figure6I.10: T2* EPI image of same slice as figure 6I.9 showing the effects of small field distortions produced at the front and temporal regions by the air cavities . Flow effects in the posterior sinus are again visible in this image (Charing Cross)

Questions and Problems

1. Explain briefly the meanings of the terms 1) free water density 2) longitudinal relaxation time. Explain why the skull and other solid parts of the human skeleton do not give rise to any observable signal in conventional MRI.
2. What is the Larmor resonance frequency of water protons in magnetic fields of 1.0T, 1.001T and 0.999T.
3. Calculate the range of water proton resonance frequencies produced by a human head placed in a magnetic field gradient 5mT/m.
4. Estimate the pulse width in seconds required to produce a 90 tipping angle if the peak RF field strength, $B_1 = 10^{-5}$ T this pulse were used for slice selection in MRI, what steps could be taken to obtain an approximately square slice profile?
5. A continuous time varying signal containing all frequencies between 1 and 20 kHz is to be digitised. What is the minimum sampling rate to avoid aliasing? At this minimum sampling rate how many samples must be collected in order to achieve a frequency resolution of 100 Hz? In practice why would a higher sampling rate be desirable?.
6. Calculate the factor by which the bulk nuclear magnetisation changes when there is a 5K decrease in the temperature of the living human tissue.
7. Explain what is meant by the term free induction decay in MRI, and how it can be produced using a combination of static and pulsed magnetic fields.
8. Consider two adjacent regions of soft tissue ,A, B that have the same free water density, the same transverse relaxation time but different longitudinal relaxation times , Ta= 100 ms, Tb= 400 ms.

   a) Write down expressions for the envelopes of the time varying signals arising from these two regions of tissue following a 90°RF excitation pulse.

   b)Explain how a particular value of repetition time, TR can be chosen to maximise the contrast between these regions. Estimate a suitable value for TR that will produce this maximum contrast.
9. a) Explain how the modern MRI scanner achieve slice selection and spatial encoding of spin positions within a given slice

   b) Discuss the main changes in instrumentation required in order to perform spatially MRI on a bee with a typical overall dimension of 10 mm, at the same fractional resolution as human MRI.

10. In conventional clinical imaging the Z gradient coil is used for slice selection ,the X gradient for frequency encoding and the Y gradient for phase encoding. Sketch the trajectory in K-space taken for 2D images obtained in this manner. How could the X and Y gradient fields be combined to obtain a radial K-space trajectory, the standard in CT and one originally used in MRI.

7 ULTRASOUND IMAGING

"*The application of echo-ranging techniques to the determination of structure of biological tissues*"

The title of Wild and Reid's 1952 paper

7.1 Introduction

The several forms of diagnostic ultrasound imaging come second only to x-ray radiography in their frequency of use and range of application in clinical medicine. Nearly every child born in Europe and North America after about 1970 has probably had at least one ultrasonic examination whilst still in the womb. Ultrasound imaging, like standard x-ray radiography is found in even the smallest hospital since general practitioners are taught to interpret the results of these standard investigations. Ultrasonic imaging does not produce any ionising radiation and, at diagnostic power levels, is essentially free from hazard to the patient. The main contrast mechanism in ultrasound arises from variations of sound velocity and this produces sufficient contrast to image soft tissue boundaries. Before the advent of x-ray CT, diagnostic ultrasound was an important non-invasive technique for imaging soft tissue differences outside the lung. The technology, after forty years of continuous development, is well understood and the apparatus is very cheap to produce. It is in fact the cheapest of the all methods discussed in this book by a factor of at least ten.

With this pedigree, the reader would be forgiven for thinking that diagnostic ultrasound makes the very expensive developments in tomographic methods, discussed elsewhere in this book, rather unnecessary. This is not the case; ultrasound is not a tomographic technique and the very best of ultrasonic imaging can only obtain a spatial resolution of a few millimetres and, it turns out, is very limited in its application to some parts of the body such as the brain and the lungs. However in many ways this chapter will not do justice to many of the modern methods which Ultrasound supports. We will not actually venture much beyond a simple description of the Sonar principles that are used in echo ultrasound. At this level, ultrasonic imaging amounts to measuring the elapsed time between the emission of a sound pulse and the reception of an echo from an internal body boundary. This is illustrated in figure 7.1. Inside the body the various soft organs and bone have different mechanical tissue properties, giving rise to different sound velocities and these give rise to wave reflections from the boundaries between different tissue types. Sound emitted or reflected from a moving object suffers a change in frequency. This is called the Doppler effect. Doppler ultrasound imaging allows measurements to be made of bulk mechanical motions within the body, such as blood flow within veins and arteries and the mechanical motions of valves and muscles within the heart.

The chapter begins with a general introduction to the physics of sound, as a mechanical wave disturbance. Here we describe how the velocity of sound propagation varies with tissue type and how boundaries reflect and scatter the incident waves. Using ultrasound to investigate human anatomy requires a particular combination of engineering techniques. These are described in general terms in section 7.2. The final section 7.3 we describe in turn the standard B-mode imaging scheme and the methods used in Doppler ultrasound.

7.2 Physical Principles of Ultrasound

The Velocity of Sound

The phenomenon of sound is familiar to all of us with intact hearing since it is the vehicle for the most important method of everyday communication between human beings and indeed many other animal species. Children unknowingly play games with sound echoes. They shout at distant walls to create echoes and they count the elapsed seconds between a lightning flash and the roll of thunder to estimate the distance to the storm. During WW2 ultrasound, SONAR was developed as an underwater compliment to RADAR for submarine detection and location. Electromagnetic, radio waves are attenuated rapidly with depth as they travel through electrically conducting salt water, but sound waves can travel significant distances with relatively little attenuation. All species of whale have evolved a means of underwater sound communication that allows them to communicate over large distances. Similarly bats, having poor sight, use sound echoes to locate and catch flying insects, and avoid obstacles in their path.

An ultrasound wave, like all sound waves, is a propagating mechanical disturbance of the matter through which it passes. Sound of any description, unlike electromagnetic waves cannot propagate through a vacuum. The sound

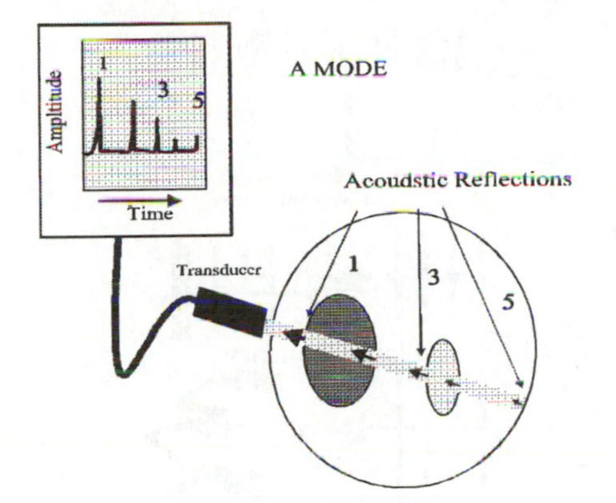

Figure 7.1: The principle of Ultrasound imaging. A brief pulse of ultrasound is emitted from a piezoelectric transducer. This pulse enters the body and is partially reflected from internal surfaces marking boundaries between mechanically different tissues. Those reflected waves that travel exactly back along the incident wave direction are detected by the receiver. The round trip time for an echo, created in uniform soft tissue, increases with depth from the transducer at a rate 0.67μs mm^{-1}. The simplest medical ultrasound scan provides a single line through the patient. This is called amplitude or A-mode scan.

wave creates pressure disturbances that accelerate and displace the atoms in its path, figure 7.2. Generally there is no net permanent displacement of particles, rather a local oscillatory disturbance is passed along from one group of atoms to the next along the direction of travel of the wave. This is just like the nearly circular motion of a surfer, who bobs up and down in the water with each passing small wave, waiting behind the surf, for the next good breaker. Sound is a wave of mechanical vibrations like those of a weight hanging on the end of a spring. The frequency of vibration of the weight depends on both the stiffness of the spring and on the mass. Similarly the frequency and velocity of travel of sound waves depend on the bulk elastic properties of the material and its density. Solids, on the one hand, and liquids and gases on the other can support two different types of sound wave. In general a definite stress is required to change the shape of a solid without changing its volume; this is called shear. All solids have a finite shear modulus. The shape of a volume of liquid on the other hand can be changed without the creation of an internal stress and the shear modulus of a

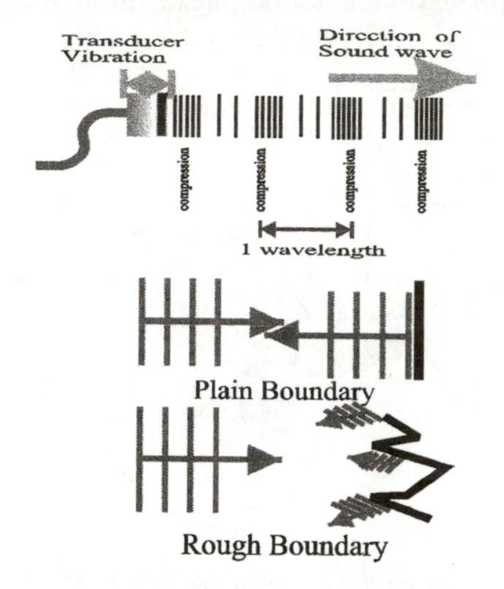

Figure 7.2: A longitudinal or compression sound wave produced by the oscillatory excitation of a piezoelectric transmitter. The wave consists of an alternating pattern of compression and rarefaction in the medium. Sound waves are reflected from boundaries to produce echoes. In tissue boundaries are rough and thus the "reflection" is diffuse. An Ultrasound system only uses the back-scattered components to form an echo image

liquid is zero. All substances have a finite bulk compressibility and so longitudinal, compression waves will propagate through solids, liquids and gases but transverse, shear waves can only propagate in solids. All soft tissue and body fluids behave like liquids (really gels of varying viscosity) and ultrasound is propagated in these as a longitudinal wave. Bone, being a solid, can support both longitudinal and transverse matter waves. In uniform or homogeneous materials such as sea water or air it is possible to write down a very simple wave equation that describes the propagation of sound through matter in terms of the bulk properties, density and elasticity, of the material. More precisely the velocity of propagation for compression waves is determined by the density and the bulk modulus, K, so that the speed of sound ,c is given by

$$c = \sqrt{\frac{K}{\rho}} \qquad \text{... 7.1}$$

The bulk modulus of a material is the reciprocal of its compressibility. The bulk modulus of air is 2×10^4 times smaller than water and its density is factor of 1000 smaller than water. Together these give rise to a factor of five between the sound velocity in air and water; c_{air}= 332 ms^{-1}, c_{water}= 1500 ms^{-1} . The speed of sound in soft biological tissue is about the same as water, c= 1500 ms^{-1} and this makes it is quite possible to measure the round trip time taken for a sound wave to cross,

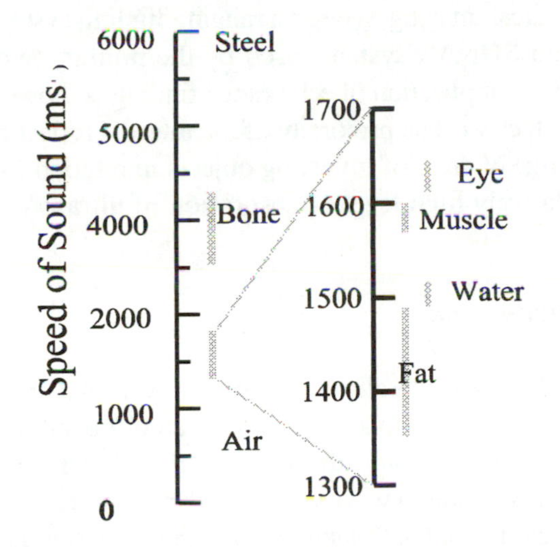

Figure 7.3: The velocity of sound in a range of biological substances. Notice that the sound velocity for most soft tissues only varies by a few percent about that of pure water 1498 ms^{-1}. This is thought to arise from variations in compressibility rather than density of the tissue. By contrast bone and air have very different sound velocities as a result of the large density and compressibility differences.

say the thorax, and return after reflection to the receiver. Taking a typical thorax dimension to be 30 cm, then the echo will take about 0.4 milliseconds, well within the capabilities of electronic timing devices. Ultrasound is generally taken to be any sound disturbance above the bandwidth of human hearing. In normal adults this is 10 to 30,000Hz; standard middle C has frequency of 264 Hz. The general relationship between velocity, frequency and wavelength of any wave is

$$c = f \times \lambda \qquad \dots \quad 7.2$$

A sound wave at middle C has a wavelength of 5682 mm but an ultrasonic wave at 1 MHz has a wavelength of 1.5 mm.. Both sound waves travel through air at the same speed. Spatial resolution in ultrasound imaging systems, just like their optical analogues, is determined ultimately by the wavelength and thus ultrasound frequencies in the radio frequency range, 1-10 MHz are used in imaging to obtain fine spatial resolution. Theoretically spatial resolutions on the order of 1mm are possible with ultrasound.

Ultrasonic medical imaging, seen as a ranging finding system, is very similar to the RADAR and SONAR systems used by the military and civil transport agencies. The medical application of echo range finding is, however, complicated by three important factors. The proximity of the internal reflecting surface to the transmitter, the range of sizes of reflecting objects in relation to the wavelength and finally the relatively high level of absorption of ultrasound in all types of tissue.

Ultrasound Radiation Field

Aircraft and ships are literally miles from a RADAR transmitter but the internal organs of the body are relatively close to an ultrasound transmitter. The radiation field in front of any transmitter of coherent waves, be it of electromagnetic waves or sound waves, is divided into two regions; the near field, within a few wavelengths of the transmitter, and the more remote, far field. Close to the transmitter the radiation field is very complex, changing relatively rapidly both in amplitude and phase with position. By analogy with light optics and microwaves this is called the Fresnel Zone. At any field point in front of an extended transmitter, the total amplitude of the radiation results from a summation of wave contributions originating from different points on the transmitter's surface, each one arriving at the field point with a different phase, as a result of the different distances between the point of emission and the field point, (see appendix A for description of wave concepts such as phase). Close to the transmitter, these phase differences are relatively large, since the differences in path length are comparable to the shortest path between transmitter surface and field point. Small changes in the position of the field point can then make large differences to the overall sum. In the far field, differences between path lengths are small in comparison with the shortest distance from transmitter to

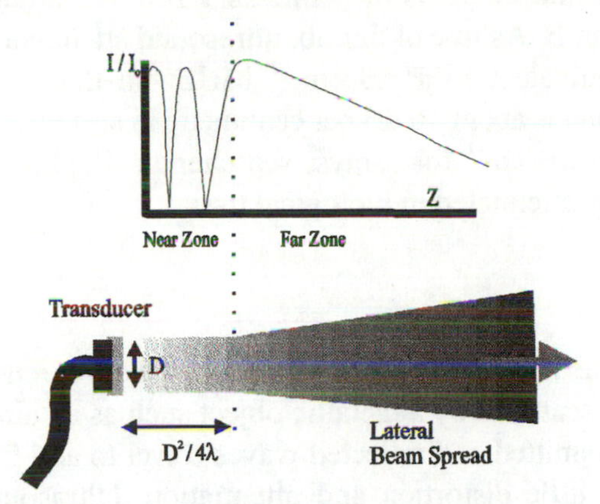

Figure 7.4: The variation of ultrasound intensity in front of a transmitter. In the near field region the intensity (on the axis of the transducer) varies rapidly with distance, Z. The width of the beam remains fixed and is determined by the lateral size of the transmitter element, D. In the far field region, beyond a distance $D^2/4\lambda$, the intensity becomes more smoothly varying and the width of the beam increases with distance.

field point, thus all contributions arrive with nearly the same phase. Small changes in the field point here make little difference to the overall sum and amplitude stays relatively constant with position along or across the sound beam.

Attenuation

Ultrasound matter waves, just like electromagnetic radio or light waves are absorbed, scattered and reflected by the media through which they pass. The total reduction in intensity of an ultrasound beam, on passing through a thickness of material, is determined by the amount of energy absorbed and the amount scattered away from the direction of the beam. Each type of material has an empirical attenuation coefficient, just like x-rays. The amount of transmitted intensity follows an exponential law ,$I = I_0 e^{-\mu X}$, where $\mu = \mu_{scatter} + \mu_{absorption}$, and X is the distance travelled in the material, see figure 2.11 and section 4.2. Conventionally the degree of attenuation in ultrasound is measured, in a relative fashion, in terms of decibels, dB. Generally the ratio of two intensities, A, B is defined in decibels as 10 Log_{10} (A/B). An attenuation of 10 dB is the same as an intensity ratio of 1:10 and 20 dB is the same as 1:100. Negative dB numbers mean that A is less than B. As rule of thumb, ultrasound attenuation coefficients are approximately equivalent to 1 dB cm^{-1} MHz^{-1} so that at 10 MHz the attenuation of ultrasound is about 10 dB per centimetre in soft tissue. This should be compared with 0.8 dB cm^{-1} for x-rays, see chapter 4. Thus ultrasound is comparatively strongly attenuated in biological tissue.

Absorption

A RADAR beam is relatively weakly absorbed and scattered by the air but strongly reflected and scattered by a metallic object such as an aircraft. Thus in RADAR both the transmitted and reflected waves travel to and from the target object with relatively little distortion and attenuation. Ultrasound is heavily absorbed by the biological tissue through which it passes en route to and from a reflecting boundary. The organised, imposed motion of the sound wave induces a variety of motions of the very small cellular and sub-cellular structures that make up human tissue. These motions are heavily damped by viscous friction and hence transform the organised ultrasound wave energy into random thermal energy or heat.

Reflection of Ultrasound and Impedance
An Electrical Analogy

Electrical waves travel along conducting cables, producing a pattern of electrical currents that vary with both time and distance. The ease with which this propagation takes place is characterised best by the impedance of the cable. For direct electric currents the impedance is the same as the electrical resistance, Resistance = Voltage/Current, giving a measure of how much applied electrical force or voltage is required to produce a given electrical displacement or current. Although electrical waves involve alternating voltages and currents, the concept of impedance still holds good but now it involves not only the resistance but the reactive properties of the cable; the capacitance and inductance per unit length. Anywhere on a long cable, the ratio of instantaneous voltage to current, V/I, is the impedance of the cable and this is conventionally given the symbol, Z. If the cable has the same properties, all along its length, then an electrical wave, launched into one end, will travel to the other end more or less intact but simply reducing in amplitude along the way, in accordance with how much absorption of energy or loss takes place. What happens at the far end of the cable depends on the impedance of the terminating component. If this exactly matches the characteristic impedance of the cable then the entire wave will enter the termination. If however the termination differs in impedance, then a reflection will take place and a component of the wave will start travelling back down the cable. Nearly all travelling waves are treated in physics in this way, using the concept of impedance to characterise both the transmission through a uniform medium and the degree of reflection from a boundary. For ultrasound matter waves, the characteristic acoustic impedance is given by

$$Z = \rho \times c = \sqrt{K\rho} \qquad \text{since} \quad c = \sqrt{\frac{K}{\rho}} \qquad \ldots\ 7.3$$

The density of soft tissue is approximately constant and so the impedance and therefore the amount of reflection by an idealised internal boundary is determined by the sound velocities on either side of that boundary.

Scattering

Scattering of ultrasound waves comprises a wider class of phenomena than the reflection from an ideal smooth plane boundary. A variety of broadly similar mechanisms, classified roughly according to whether the size of the scattering object is larger or smaller than the sound wavelength, are at work to attenuate and redirect the passage of a ultrasound wave through the human body. Ultrasound imaging makes direct use only of that scattered wave component which preserves the organised wave pattern and returns along the incident beam direction to be received as an echo. In fact the pattern of reflected and scattered waves produced by the body is extremely complex and so it is only the back-scattered component that is recovered. It represents tiny fraction of the total information contained in this acoustic field.

On the scale of a fraction of a millimetre, internal boundaries between organs are far from smooth and so the back reflected ultrasound wave is very much less intense than the incident wave. A good analogy is the scattering of light from the surface of the sea. Under conditions of complete calm, the surface becomes a smooth mirror and the reflected light will form a sharp inverted image of objects such as boats sitting in the water. As the surface becomes ruffled by the action of the wind, the image becomes progressively more fragmented, as photons scattered from different patches of sea, making different angles with the sun, dart off in random directions, away from the observer. The total amount of light reflected remains about the same, but it is now diffusely scattered. Eventually the surface becomes randomly corrugated on a small scale and the sea takes on a dull sheen. There is still a small amount of light scattered into any angle and this is sufficient to detect that there is a diffuse boundary between air and seawater. Similarly, in ultrasound, the small amount of diffusely scattered sound wave energy that happens to return along the incident wave direction is used to detect the presence of a boundary in tissue. The signal obtained from blood in Doppler ultrasound can hardly be described as a reflection from a surface since it arises entirely from the combined signals scattered from a collection of separate red blood cells in motion in the flowing plasma.

Reflection

If we focus attention on the useful, organised scattered component and idealise the boundary to a mirror then we can estimate the relative amounts of reflection, and transmission in terms of the mechanical material properties on either side of the boundary. We have to consider an idealised situation in which a plane wave with an amplitude, A, is directed normally at a flat boundary between two media 1 and 2. At the boundary, a fraction R of this wave motion will be reflected and the remaining fraction, T, will carry on into the next region. At the boundary itself, the three waves, A,R,T must together satisfy two important conditions. Just on the left side, the total wave amplitude is A+R and on the right side it is T. Right at the boundary we must insist that the amplitudes of left hand and right hand waves match. If this were not the case then the amplitude at the boundary would be discontinuous. This is simply unphysical since it would require matter to be in two places at once. Similarly we have to insist that the local pressures produced by waves on the left and right match exactly at the boundary. After a little algebra it is found that the reflection coefficient, given by the ratio, R/A, at the boundary between two media whose characteristic impedances are Z_1, Z_2 is given by

$$\frac{R}{A} = \frac{Z_1 - Z_2}{Z_1 + Z_2} = \frac{c_1 - c_2}{c_1 + c_2} \quad \text{... 7.4}$$

here, in the second expression, we have assumed that the density in the two regions is approximately the same. These expressions provide an upper limit on the reflection coefficients between different media ie blood-muscle, fat-water. In general the real reflected intensities are less than this expression would predict, as a result of the roughness of boundaries. Nevertheless the diaphragm, large round cysts and the eye give results close to this prediction. It is evident that the largest reflected intensities arise from the biggest differences in velocity (impedance). Thus, bone-soft tissue and air-soft tissue boundaries give the largest reflections with dB values of about –5 to -10 dB (relatively large), fat-water has a coefficient of -28 dB (relatively small) and blood - muscle a value of -32 dB(even smaller). The simple expression for the reflection coefficient, R also provides a justification for some of the limitations of ultrasound imaging. At the air-skin boundary there is a large mismatch in velocity and thus a large reflection coefficient, R. Unless the transmitter is in good, airless, contact with the skin, very little ultrasound energy will actually be transmitted into the body. Similarly the mismatch between soft tissue and bone is very large so that most ultrasound, incident on the skull, is reflected at the inner scalp boundary, leaving little wave amplitude to use for imaging the internal structure of the brain. In spite of this, before x-ray CT, ultrasound was used to demonstrate asymmetries of the brain midline caused by tumours. A modern technique called Trans Cranial Doppler provides valuable information on blood velocity and disease processes in cerebral arteries.

The Ultrasound Doppler Effect

The Doppler effect is a general phenomenon that can appear in association with any type of wave disturbance. The effect concerns changes in wave frequency and wavelength caused by transmitter or receiver motion. The effect of motion on the pitch of the sound of a siren on a passing train is a familiar experience. The siren emits a sound with a definite pitch or frequency and a definite corresponding wavelength. When the transmitter, emitting a frequency, f Hz, is at rest with respect to the observer, the regions of maximum sound compression, separated in space by one wavelength, enter the observer's ear at the rate f per second, figure 7.2. When the transmitter is receding at a velocity v, the maximum sound compressions are still emitted at f per second but in one period, 1/f, between compressions, the transmitter has travelled an extra distance $v \times 1/f$. Thus the wavelength of the sound in the still air is not $\lambda = c/f$ but $\lambda' = c/f + v/f$. In the still air between train and observer the wave still travels at $c = f' \times \lambda'$ and so the frequency, f' heard by the observer is given, after a little algebra, by

$$f' = \frac{f}{(1+\frac{v}{c})}$$

$$\text{since } \frac{v}{c} << 1 \text{ then : -} \qquad \ldots \quad 7.5$$

$$f' - f = \Delta f = \frac{-fv}{c}$$

Receding transmitters give rise to a decrease in sound frequency, $\Delta f = -f v/c$ since $v > 0$, but approaching transmitters produce an increase in frequency, $\Delta f = +f v/c$.

The Doppler effect in ultrasound imaging is slightly different because we have to consider reflections from a moving boundary and the fact that in general the sound beam makes a finite angle with the velocity vector of for example the blood flowing in a vessel. We can however use the same arguments. We consider a boundary moving away from the transmitter; it receives a wave with a decrease in frequency, which it reflects back towards the receiver. The receiver records a Doppler shifted version of the reflected wave, which is already a Doppler shifted version of the wave emitted by the transmitter. Thus overall, the frequency shift is $-2fv/c$ for receding reflecting boundaries and $2fv/c$ for advancing boundaries. This reflection Doppler shift effect is used extensively by the police for monitoring and the accurate recording of vehicle speeds on the roads. Notice that in this case $v/c < 1$ since a car travelling at 120 mph $= 54\ ms^{-1}$ has a speed which is very much less than the speed of sound in air, $c = 300\ ms^{-1}$. In medical

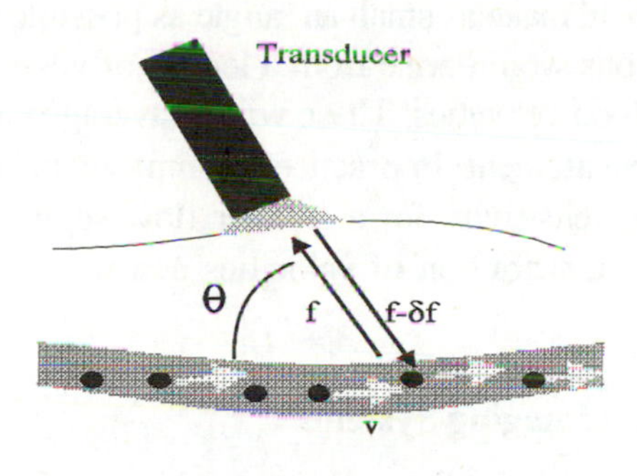

Figure 7.5: The principle of Doppler ultrasound. Some of the ultrasound wave emitted at a frequency f_0 and wave speed c is reflected back to the transducer at a lower frequency from a boundary moving to the right, at a speed v. The change in frequency is proportional to the speed of the moving reflector, v, resolved along the direction of the ultrasound beam. The frequency shifted, scattered signal from blood arises mainly from scattering from red blood cells rather than plasma.

ultrasound imaging the moving objects of interest are blood flowing in the veins and arteries and the motion of muscle tissue in around the heart. In all these cases $q/c \ll 1$ and so the approximation used above is a very good one and there is a simple linear relationship between the speed of travel of the reflecting object and the consequent change in frequency of the reflected sound wave.

The effect of the finite angle between the ultrasound beam and the direction of object motion can be taken into account simply by noting that it is only the component of object motion along the sound beam that produces an acoustic Doppler effect. Thus our expression for the change in frequency becomes

$$\delta f = \pm \frac{2 f v \cos(\theta)}{c} \qquad \ldots 7.6$$

Where θ is the angle, shown in figure 7.5, between the ultrasound beam and the direction of the travel of the object. This angular effect has an impact on the use of Doppler to measure flow velocity. Ideally the ultrasound beam would cross the vessel at right angles, $\theta = 90^o$ but then there would be no Doppler shift, (cos $(90^o) = 0$) and hence no frequency shift. Thus for maximum sensitivity the ultrasound beam should make as small an angle as possible with the vessel. Now however the reflections would arise from a length of vessel that will in general contain a range of blood velocities. These will be averaged together and degrade the flow velocity measurement. In practice a compromise is sought, with $\theta \sim 60^o$, which trades flow sensitivity for a smaller flow segment and thus a better representation of the distribution of velocities across a particular vessel cross section.

7.3 Ultrasound Echo Imaging Systems

Ultrasound Generation and Detection

Without exception all ultrasound waves are generated and detected by piezoelectric materials. In general the same element, called a transducer, is used to both transmit a series of ultrasound pulses and detect the returning echoes. The success of an ultrasound system depends above all on the quality of this component. The active material is generally cut from a thin crystal of the ferroelectric ceramic, lead zirconate titanate, PZT or a thin piece of the plastic, polyvinylidene difluoride, PVDF. PZT is polarised in a strong electric field during its manufacture, so that applied electric voltages change the crystal dimensions and conversely, applied mechanical strains induce in it an electrical voltage. The former process is used for ultrasound generation and the latter for detection. A typical crystal will shrink (along the direction of the applied electric field) at a rate 0.3 $\mu m\ kV^{-1}$. The same crystal will produce an electric voltage, under the influence of applied stress at a rate 0.0026 $V\ m\ N^{-1}$. The active part of the transducer is manufactured by depositing silver electrodes on the front and back surfaces of the crystal so that it can be electrically polarised, uniformly across its thickness, figure 7.6. Just like a drum, this element will have a number of mechanical resonances at frequencies determined by the size and shape of the transducer. For PZT a thickness of 0.4 mm gives a resonance at 5 MHz. The element is generally mounted on a backing material both for mechanical support and resonant damping and is faced with a matching layer, like the blooming of a camera lens, to reduce reflections between the transducer and the skin tissue at the body surface. Theoretically this is achieved best using a matching layer thickness of a quarter acoustic wavelength and a material with impedance $Z_{match} = (Z_{transducer} \times Z_{skin})^{1/2}$. A matching layer made from aluminium powder cast in epoxy resin is often used with PZT. Since the acoustic impedance of plastic is quite close to human tissue, matching layers are unnecessary with PVDF.

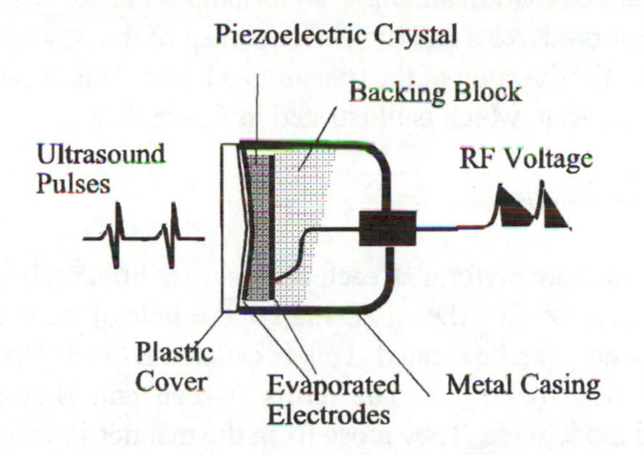

Figure 7.6: A single element ultrasonic transducer construction. As a transmitter, the piezoelectric crystal is excited by electrical voltage pulses applied to the surface electrodes at radio frequencies (1-10 MHz).The alternating electrical polarisation of the material (PZT for example) causes the crystal to shrink and expand, producing alternating compressions (longitudinal sound waves) in the surrounding media. As a receiver, the element converts returning echoes into electrical signals at RF frequencies that are conducted, via the electrodes, to the amplification chain of the instrument. For maximum efficiency, the transducer is operated at a mechanical resonance frequency of the element.

A-Mode Measurements

We have already discussed the complexity of ultrasound scattering and reflection processes and alluded to the fact that practical ultrasound imaging systems necessarily have to reject a great deal of the information in the received signal in order to produce an unambiguous image. An ultrasound beam, launched into the body at a particular position and angle, will encounter a large number of reflecting and scattering boundaries making oblique angles with the incident beam. In general these will send reflected waves off in a wide range of directions.

In principle one could imagine a system that recovered a 3D image from the totality of this scattered wave field. This turns out to be a very complicated task and so in practice only those waves that are reflected, by boundaries at right angles to the incident beam, back along the incident direction of travel, are collected and analysed to form an image. In its simplest form, a single ultrasound echo measurement produces a one-dimensional map of the positions of reflecting boundaries, along the direction of the transmitted beam. This is called an A-scan, or amplitude mode scan, which is illustrated in figure 7.1.

Standard B-Mode Imaging

If many A scans are performed, each one using a different beam direction or different transmitter location then a 2D map of the field of view, covered by the collection of A scans, can be created. This is called a B-scan (Brightness mode) and is illustrated in figure 7.7. The terms A-scan and B-scan are slightly confusing in the modern era. They arose from the manner in which display was

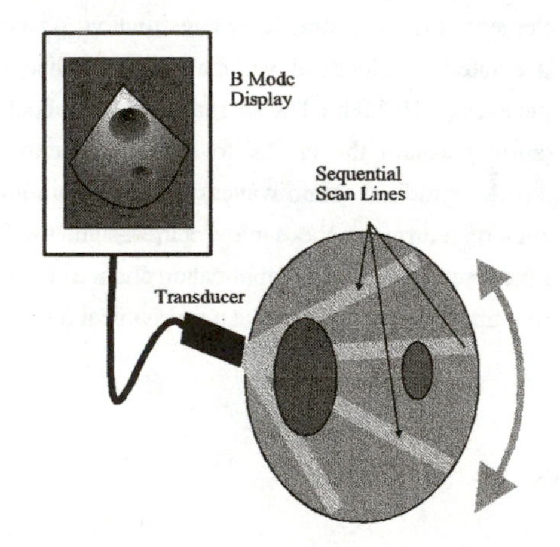

Figure 7.7: The standard B-Mode ultrasound imaging geometry. The 2D scan is composed of many A-Mode scans each taken with a different beam position or angle with respect to the field of view within the patient. In the earliest scanning systems a single transducer was translated or rotated between each A mode line. In modern instruments, a multi-element transducer together with variable delay pulse sequences is used to steer (or focus) the beam electronically, see figure 7.9

originally presented on an oscilloscope. In the A-scan mode the horizontal axis of the trace corresponds to depth (round trip echo delay time) and the vertical axis records the relative amplitude of the echoes. In the B-Scan mode both the X and the Y directions on the screen correspond to anatomical distances. Each component scan line is plotted with trace brightness modulated by the variation in signal amplitude along the scan direction. Digital imaging removes the need for this clever analogue display system but the terms have remained. A-scan now refers to single line investigations and B-scan to 2D composite images constructed from sequential A-scans, obtained over a range of directions, called sector scans or positions, called rectilinear scans. The generic 2D scheme, which is referred to as the B scan, is by far the most commonly used imaging scheme and so the term, ultrasound imaging, is nearly synonymous with B scan imaging. In its earliest forms the B scan instrument used a single transmitter/receiver element whose position or orientation was altered mechanically to create the array of beam directions. More recent instruments use a stationary multi-element transmitter/receiver and electronic delay system to steer the composite beam over a range of angles. Similar methods are widely used in RADAR to steer beams over a wide angular range or in Radio Astronomy to receive incoming radio signals from a larger effective synthetic aperture. Figure 7.9 illustrates this technique.

M Scan or Time Position Plot

Although the simplest of A-scan measurement is now seldom used, a modification called M scan or TP, time position scanning is still used, where rapid motion monitoring is required. The M scan consists of repeated A scans, along a single direction intersecting a moving boundary such as a heart valve. Boundary movement causes the echo time to fluctuate and this fluctuation is displayed as a displacement along the vertical axis of a monitor, whose horizontal axis is a time base. Thus, as the cursor scrolls across the screen, at a constant rate in time, the fluctuations of the moving boundary are laid out as a position –time plot. This facility is often incorporated into general purpose B-scan instruments. The 2D image is used to select an appropriate line and then this line can be displayed as a time–position plot. The repetition rate of the scan is the sampling rate of the image and this has to be high enough to avoid aliasing effects, (see appendix A).

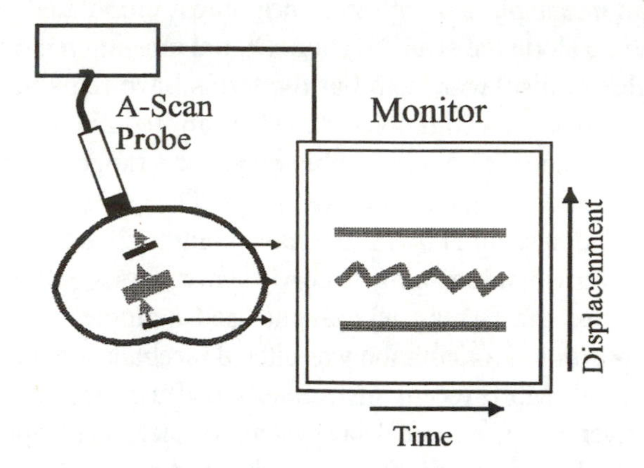

Figure 7.8: The M-mode or time position scan. The single selected A-scan shown here produces three prominent reflections, the middle one is from a moving boundary such as a heart valve. The time variation of the echo times is displayed on the vertical axis while the cursor scrolls across the screen at constant rate in time.

B-Scan Designs
Scanning Mechanisms

B mode ultrasound imaging necessarily requires that there should be a sequence of A scan lines, each taking a slightly different path through the subject. Originally this was achieved using a single moveable transducer attached to a gantry that encoded the transducer position and orientation. The image was then assembled from the series of individual lines arranged according to the trajectory and angle of the transducer. This cumbersome scheme was replaced by smaller but still mechanical designs in which either a single or a small group of transducers were rotated or wobbled so that over time a sector of the subject was scanned in a sequence of A scan lines. Since about 1974, electronic transducer arrays have largely replaced these electromechanical schemes. The electronic designs are cheaper to produce and, containing no moving parts, they are more robust. They also have the advantage of flexibility since any one A scan beam is produced by the synthesis of many smaller elements, which can be combined electronically to synthesise beams with different properties such as direction and depth of focus. The principle is illustrated in figure 7.9. The acoustic radiation field is produced electronically by an array of very small transducers. This is fabricated from a large piezoelectric crystal that is cut into nearly identical sub elements, each of which is supplied with its own RF excitation electrodes. Each element, excited on its own, produces a nearly hemispherical radiation field. When these elements are excited in combination, using slightly different pulse times, constructive and destructive interference between the individual waves results in a synthesised beam travelling in a definite direction with a definite depth of focus

The excitation pulses, supplied to each element, are derived from the same oscillator and thus have the same frequency and RF phase but these are supplied

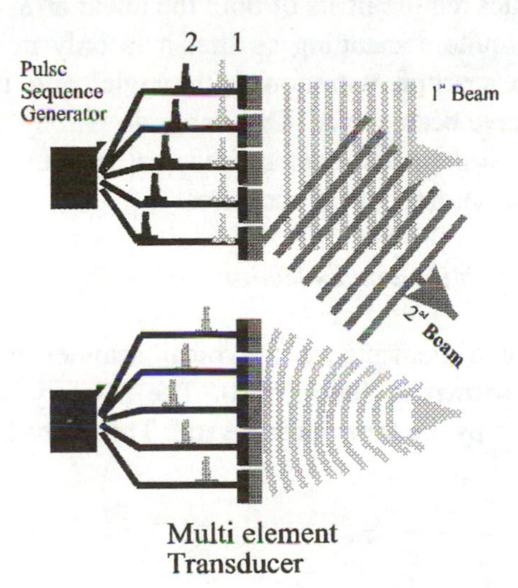

Figure 7.9: An illustration of ultrasound beam steering and focussing using an array of transducers each with its own phase delay. In practice the pattern is a combination these two so that a partially focussed directed beam is produced.

to the elements via an array of electronic delay lines that introduced calculated patterns of phase (delay) differences between the elements. If there are no relative delays between pulses, the whole array acts a single large transducer, creating a beam which propagates at right angles to the transducer. If on the other hand a progressive time delay between pulses is introduced across the elements, then interference between the component beams creates a single composite beam propagating at an angle to the transducer. Changes in the delay increment between elements causes the composite beam to be steered through a range of angles, sweeping out the field of view. Beam focusing is carried out using the same principle but a different pattern of delays, as shown in the lower part of figure 7.9. The same principles are applied to a variety of array geometries, each with its own advantages and disadvantages. A flat linear array can produce a sequence of beams, by successively exciting just a small subset of the array elements. Since there is no beam steering here, the array has to be quite long in order to obtain a reasonable field of view. A curved array is also employed, using the same linear excitation principle. This produces a fan of beams and thus wider fields of view. In general the electronics requirements of both the linear array and the synthetic aperture systems are quite demanding so that it is only in recent years that problems of effective aperture variation with angle, and interference from inevitable side lobes have been solved. Only now do wholly electronic devices achieve better resolution than the older mechanical systems which had large, accurately crafted, constant transducer apertures.

The Electronic Amplification and Detection

The main electronic elements of a typical scanner transmission and reception system are shown in figure 7.10. The primary reception of the ultrasound echo is made by the transducer crystal. This turns the mechanical

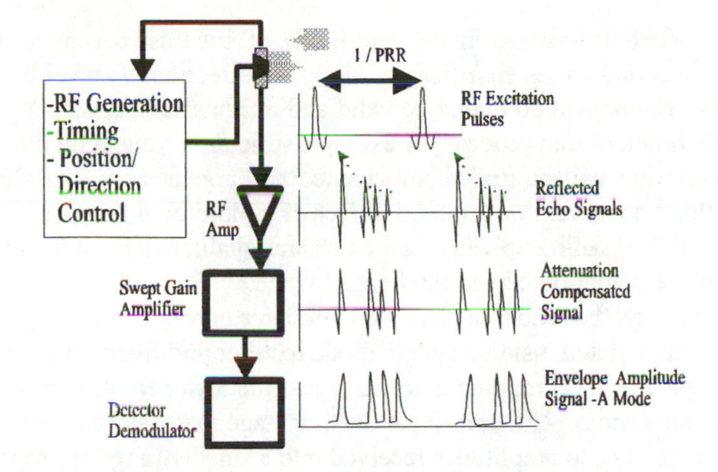

Figure 7.10: A schematic diagram of the main components of a typical ultrasound imaging system. RF pulse generation, pulse repetition rate, PRR and beam position or angle are accurately controlled by a master oscillator. The echo signals received in response to each ultrasound pulse are amplified before signal processing. The amplified signals are first compensated for attenuation using the swept gain amplifier. The envelope of the echo signal train is recovered by RF detection (rectification + low pass filtering) to create a single line scan or A-mode trace. 2D scans are composed of many A mode lines, each one taken at a different beam angle or position.

compressions, produced by the returning sound wave into radio frequency pulses. In general each emitted pulse is analysed on its own so that stray late reflections/scattering from preceding pulses do not corrupt the image. Thus the pulse emission, echo reception and processing or at least data storage of each scan line is completed before the next pulse is emitted. The same principle is used both in single element and array scanners. In the latter however the reception is carried out by the whole array or many elements with introduced phase delays that ensure primary sensitivity to the back-reflected echo from the most recent emitted pulse. Each echo is first amplified by a conventional RF circuit and then passed to the swept gain amplifier to compensate for the increased ultrasound attenuation with increasing depth of reflection. To a first approximation the returning train of echoes decrease exponentially in amplitude with delay time. The swept gain amplifier makes an approximate compensation for this by increasing the gain logarithmically in time. This compensates for the average exponential decrease in the amplitude of successive echoes. If the attenuation of sound varies significantly along the direction of travel then this approximate scheme will no longer be valid and image artefacts will appear as anomalously bright or dim echoes. For example structures lying behind a region of high absorption will be undercompensated and appear as a dim shadow. Curiously this known and understood defect of simple gain compensation is exploited by skilled radiographers to gain a clearer qualitative mental picture of the nature of particular objects in the field of view.

Before display, the echo train is passed through a non-linear, scaling amplifier and finally demodulated, using a simple diode rectifier and filter, which removes the high frequency RF components to leave just the lower frequency amplitude envelope of the echoes. The non-linear scaling stage matches (compresses) the very wide range of echo amplitudes received into a smaller range appropriate for 64 or 256 grey scale display.

The Pulse Repetition Rate

As we have seen both the 1D M mode and the 2D B mode ultrasound images are built up from a series of line scans, each line comprising an echo train with a definite duration corresponding to the maximum distance of a significant reflector from the transducer. In the A and M scan methods the measurement cycle is repeated many times each second either to allow averaging, and hence reduce noise, or to monitor the temporal changes position of objects in the field of view. In 2D imaging a series of pulses, each with a different direction of travel is required in order to build up the 2D image. In all schemes any ambiguity, arising from a late echo from a previous pulse getting muddled with the current echo train, would severely hamper image interpretation and thus there is an upper limit on the pulse repetition rates that are used. The maximum pulse rate or PRR_{max} is related to the maximum depth of the image by PPR_{max} (kHz) = $1500/2D_{max}$, where D is in millimetres. Generally a margin of error is employed and repetition rate of about 1kHz is used. At this repetition rate in the B scan mode, a single frame comprising 512 lines will take 0.5 s to acquire, making it rather slow for some cardiac applications seeking to follow heart valve motion. M-scan monitoring is often used for this purpose since, although only a single line with the width of the beam is displayed, the repetition rate of 1 kHz is sufficiently rapid to follow all mechanical motions of the heart.

Spatial Resolution

Echo imaging depends on timing the round trip delays of reflected pulses. Spatial resolution, along the direction of travel, depends ultimately on the wavelength used. In addition high spatial resolution requires short pulses so that separate echoes do not overlap in time and the temporal centre of an individual echo is precisely defined. Since the average sound velocity in soft tissue amounts to a distance of 1.5 mm of travel in a microsecond, the pulse length needs to be about one microsecond in duration. In turn this means that an ultrasound pulse at 10 MHz would only contain about ten complete electromechanical oscillations. A good deal of engineering has to go into both the mechanical and electrical designs to ensure that such pulses can be produced. Pulse shaping is needed, just as in MRI, to limit the frequency spread of short duration pulses. The mechanical design has also to trade off high sensitivity resonant operation with its inevitable long ringing time against the need for short pulsed operation. Typically the RF circuit has to produce a voltage pulse of 100-200 V with a rise time of about 30 ns.

7.4 Doppler Imaging

In the earliest and simplest Doppler systems, continuous wave ultrasound, rather than pulsed ultrasound, was used to monitor blood flow in superficial vessels. The problem for continuous wave Doppler ultrasound is that it cannot resolve the depth of the moving reflector and thus cannot distinguish between two overlaid vessels. This problem has been resolved in recent years by using a hybrid pulsed Doppler system that combines the echo range finding property of pulse echo imaging with the motion sensitivity provided by the shifted frequency components in the reflected pulses. Extraction and localisation of the Doppler signal is far from straightforward. The main elements of a single transducer pulsed Doppler system are shown in figure 7.11. The main effort goes into extracting the small amplitude Doppler frequency shifted signals from the larger amplitude, non-specific junk in the returning signal at the imposed carrier frequency. From our simple Doppler equation we see that at 10 MHz, a blood flow velocity of 10 cm s^{-1} produces a maximum frequency shift of about 670 Hz. If a continuous wave were used, then signal extraction could be done simply by digital Fourier analysis but the need for spatial location means that pulses of ultrasound have to be used and the signal is necessarily more complicated. The scheme illustrated in figure 7.11 separates location from frequency shift analysis in the following manner. A continuous wave ultrasound signal is chopped, using an electronic gate, with a repeat rate of 1/T to produce a series of pulses, each one containing a few cycles of the original pure continuous sine wave. A particular echo, corresponding to a selected moving boundary, is selected using an electronic gate. This step produces the spatial localisation of the Doppler signal. The selected echo contains a normal f_o signal together with a motion shifted component, f_o+2vf_o/c. Synchronous demodulation separates out the wanted Doppler shifted signal from the unwanted normal reflection. Each pulse provides one sample in time of the Doppler signal that is now an instantaneous measure of the boundary velocity, v. The repetition at 1/PRR seconds then gives a digitally sampled record of the boundary velocity (a series of spikes at intervals of 1/PRR). This time series is passed through a sample and hold amplifier and low pass filtered to give a continuous measure of variation of boundary velocity with time.

Duplex Scanners

Many modern clinical systems combine a sector imaging system with a Doppler system so that the vessel under investigation for flow can be imaged and the best orientation for the Doppler measurement ascertained. In fact the two components complement each other in flow studies. The identity of putative pulsatile vascular structures can be confirmed by the Doppler measurement The 2D image supplies angular information and vessel size so that the Doppler shift signal can be quantitatively interpreted in terms of total blood flow and local flow velocity. Typically the duplex scanner presents the two types of information in single convenient image by superimposing a colour coded Doppler signal on a grey anatomical background. The anatomical and Doppler information are obtained using the same transducer, but at different times. The anatomical image clearly requires a swept beam, but the Doppler measurement must use a single line of sight. Thus a line is selected from the anatomical image for further study using Doppler. After both sets of data have been collected and processed, the final composite image can be produced.

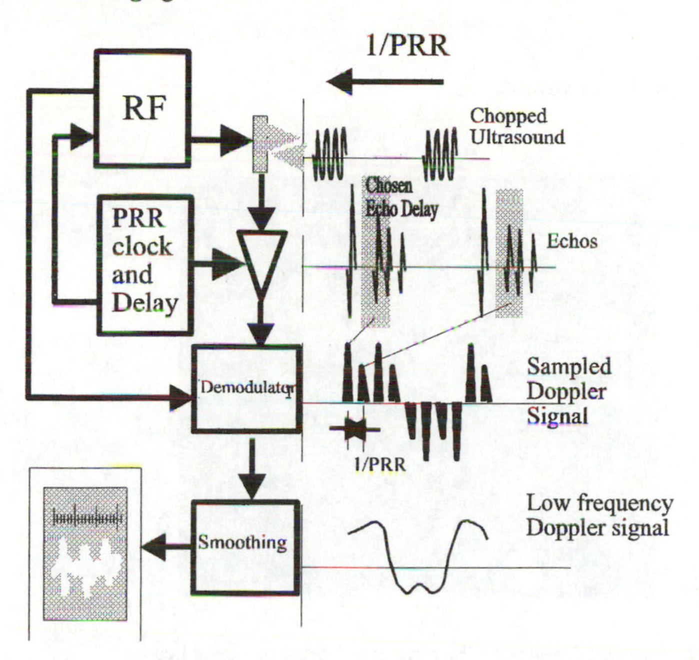

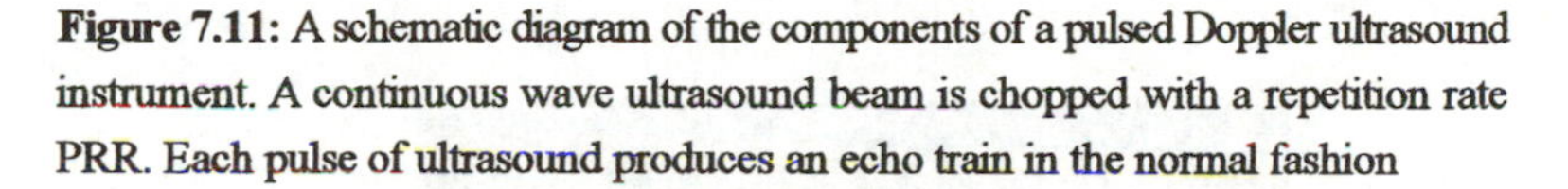

Figure 7.11: A schematic diagram of the components of a pulsed Doppler ultrasound instrument. A continuous wave ultrasound beam is chopped with a repetition rate PRR. Each pulse of ultrasound produces an echo train in the normal fashion

Miniature Invasive Ultrasound Probes

Many important small organs of the body, particularly in the reproductive and digestive tracts lie relatively deep inside the body. The acoustic radiation from an external, body surface transducer has to traverse several centimeters of absorbing soft tissue and bone before reaching these structures. As a result these structures are often poorly imaged from outside the body. However many structures such as the ovaries and embryos early in gestation can be more closely approached using the natural body orifices. Recent advances in transducer miniaturisation have allowed B scan ultrasound imaging to be carried out from inside the body. This development results in greatly improved image quality both in terms of spatial resolution and contrast. Commercial systems are in clinical use in gynecology and similar systems are under development for insertion into the major arteries.

7.5 Images From Ultrasound

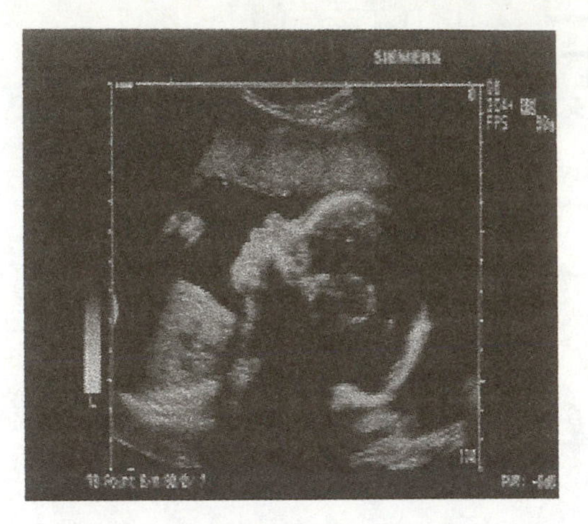

Figure 7L1: A typical Foetal scan showing good detail of foetal head

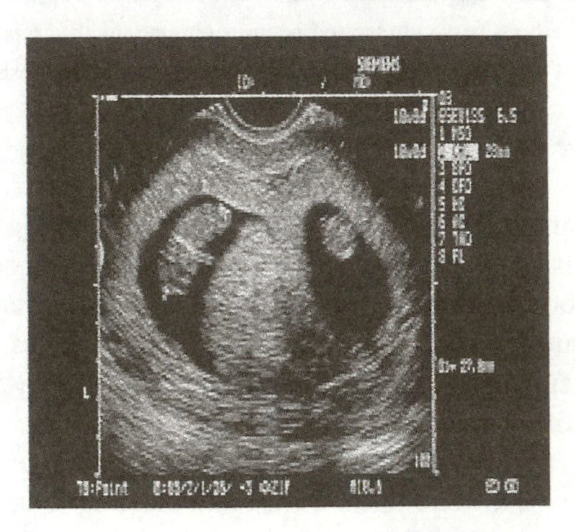

Figure 7L2: Twin gestation in bicornuate uterus

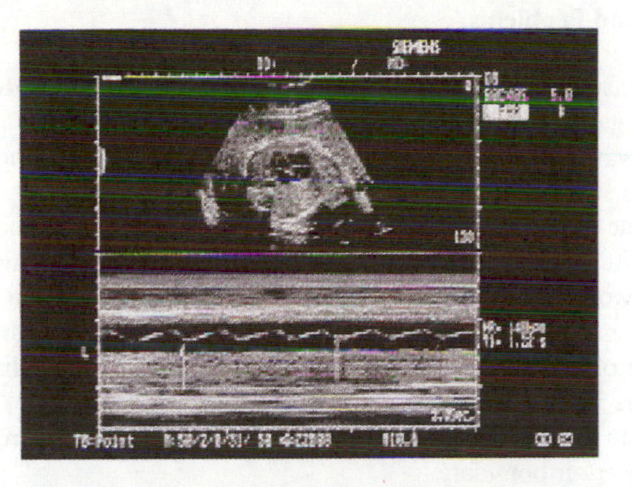

Figure 7I.3: Simultaneous B and M mode display

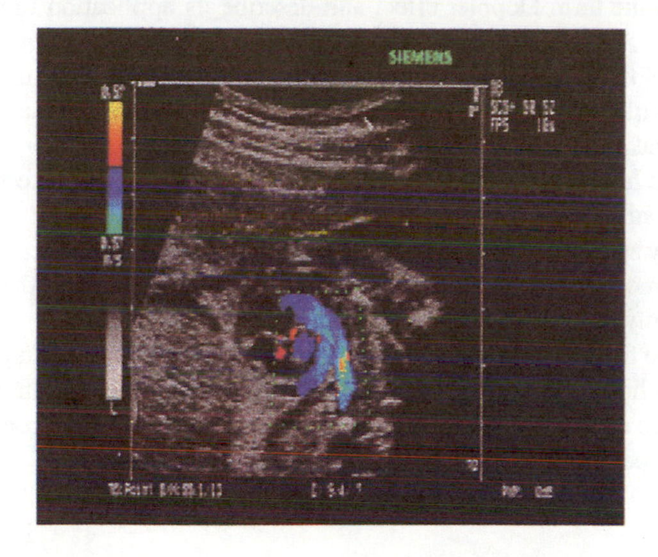

Figure 7I.4 : Colour coded Doppler scan of foetal heart

Questions and Problems

1. Electromagnetic waves (photons) and sound waves can travel through solids liquids, gases. Electromagnetic waves can also travel through a vacuum but sound waves cannot. Explain this difference and describe the differences between sound waves in a solid and in a liquid.
2. Estimate the maximum round trip time for a pulse of ultrasound to pass across the human thigh and return to the transmitter. How many principal echoes would be produced and what would be their spacing in time ?.Make a rough estimate of their relative amplitudes. [Assume the thigh to be a right cylinder of soft tissue with radius 10 cm enclosing a right cylinder of bone of radius 3 cm]
3. Estimate the ratio of reflected to incident ultrasound energy reflected by
   a) a water-air boundary
   b) a soft tissue- bone boundary.

   Explain why ultrasound imaging of the brain is relatively very difficult
4. Explain the term Doppler effect and describe its application to ultrasonic imaging.

   Estimate the frequency shift, produced by blood flowing at 1m s^{-1}, on an incident ultrasound beam whose wavelength is 10mm when the ultrasound beam makes an angle of 60° with the blood vessel.
5. Explain how ultrasound pulses can be both generated and detected by the same transducer
6. Broadly speaking Ultrasound methods fall into the two categories, A-Mode and B-mode. Explain the differences between these. Which of these is most commonly used in clinical practice?
7. Describe the main components of a typical B mode ultrasound system.
8. Explain how a simple Doppler imaging system can include the virtues of pulsed as opposed to continuous wave operation. What vital role does the pulsed operation perform?

8 IMAGING IN CLINICAL PRACTICE

"Baked, boiled, grilled or fried.. Show us what's in Mog's inside".
Meg made an X-ray spell to see what was in Mog's tummy
"Mog's Mumps", Helen Nicoll and Jan Pienkowski 1972

8.1 Introduction

The preceding chapters have described the scientific principles that underlie a range of imaging techniques. We have followed the imaging wave, be it x-ray, γ–ray, or radio-frequency, from its genesis to final destination on a photographic film or computer monitor. But the scientific description is deficient in one important respect: it has made little mention of the patient or doctor. Indeed, the non-clinical reader may be puzzled as to why one imaging modality would be chosen above another or how the decision has been made to image a patient in the first place. This chapter will attempt to redress the balance and place the techniques within a wider clinical perspective.

8.2 Dramatis Personae

The Patient and the Doctor

From the patient's perspective, the whole process of imaging starts long before an x-ray or γ–ray is generated and ends long after the photographic film is developed. It starts when he or she is seen by a doctor. Depending on the nature of the problem, the doctor may be part of a specialist department within a hospital, or be the patient's own General Practitioner. Whether the consultation takes place in the casualty department, in the G.P.'s surgery or in the patient's own home, its key elements can be caricatured in the following way: (i) the patient describes a particular problem and is examined; (ii) a decision is made as to whether the problem warrants further investigation - imaging being one of many investigative options; (iii) the results of the investigations inform the decision as to what to do next. Any abnormalities revealed by the tests will themselves initiate the next cycle of investigation. In some cases the medical image will identify the nature of the problem (an x-ray revealing a bone fracture after a fall, for example). However, more commonly, the image merely confirms that the patient has a problem rather than identify its nature (a soft tissue mass on an x-ray image has many causes and is not in itself a diagnosis).

In general, the imaging is not carried out by the doctor that saw the patient but takes place in one of the several specialist imaging departments found within most hospitals. All doctors can interpret very simple x-ray, MRI, gamma or ultra-sound pictures; however, the subtlety of the changes associated with many pathological states and the range of imaging techniques available has resulted in the development of a number of imaging-related medical specialities. The details of how the image is actually obtained and the interpretation of the results are left to the specialist who, for most routine imaging investigations, does not have any direct contact with the patient.

The Imaging Specialities

The first medical images were produced using x-rays and thus the first doctors who specialised in imaging were appropriately named radiologists. Today, x-rays form a much smaller part of the radiologist's workload, which now includes ultrasound and MR imaging techniques; however, the name has stuck. Radiology is the largest and most important imaging speciality and the breadth of the field has led to further sub-specialisation within radiology itself (for example, some radiologists specialise in brain and spinal cord imaging - neuroradiology). As pointed out above, in most cases the radiologist will not have direct contact with the patient being investigated; however, in some interventional radiological procedures, the radiologist is more akin to a surgeon, performing operations with special instruments under x-ray or ultrasound guidance. Gamma imaging is associated with a different medical speciality and hospital department - nuclear medicine. Specialists in nuclear medicine use radionuclides to produce a variety of medical images and to treat certain cancers. Ultrasound investigations are performed by doctors from many specialities. For example, echocardiography, imaging the heart with ultrasound, is normally performed by cardiologists (heart specialists) rather than radiologists.

The Radiographer

Medical images, be they ultrasound, MR, x-ray or gamma-images, are usually acquired by specialised technical staff called radiographers - the name, like that of radiology, reflecting the part played by x-rays in the development of medical imaging. Radiographers can themselves sub-specialise in the acquisition of MRI, CT ultrasound or gamma images (particularly PET and SPECT). It is the radiographer who explains the details of the procedure, prepares and positions the patient, adjusts the imaging parameters, takes the image and develops the final picture.

The Medical Physicist and Radio-pharmacist

Many MR units will have a medical physicist to help develop imaging protocols and monitor the performance of the scanner. Most gamma imaging facilities will have an associated radio-pharmacist to produce radio-nuclides.

Imaging Facilities

The distribution of imaging facilities is defined largely by the expense of the equipment and the local need for a particular investigation. As x-ray and ultrasound equipment and gamma cameras are relatively cheap and the need for such investigations so great, most hospitals have these facilities. The more specialised and expensive imaging tools such as CT and MRI are only available in larger hospitals. PET scanning is only found in a handful of units with on-site cyclotrons.

8.3 Clinical Investigations

We have described how the request for an imaging investigation is initiated, and the different medical personnel responsible for its acquisition and reporting, but so far we have not addressed the question of why a particular image has been chosen above all others. How does the referring doctor decide to request an MRI rather than an x-ray, CT or ultra-sound image? In fact, there is no single answer to this question; the investigation of any given condition changes as new technologies become available and a consensus view develops as to the optimal strategy. Cost, measured both in economic terms and in terms of potentially harmful side-effects, has an important influence on the decision. In general, the investigation chosen is the simplest, cheapest and safest able to answer the specific question posed. In the following sections we describe a range of clinical questions and the way in which imaging is used to answer them.

X-rays

For many clinical problems, the first image obtained will be a single projection x-ray. Because of the high contrast between bone and soft tissue, the x-ray is particularly useful in the investigation of the skeletal system. Two views of an area are often taken in the same session to give the clinician an idea of the problem in three dimensions. For example, an image of the lower back taken from the front of the patient will often be combined with a side-on view. X-rays can also provide clinically useful information about soft tissues. An x-ray image of the chest, for example, reveals a remarkable amount of information about the state of health of the lungs, heart and the soft tissues in the mediastinum (the area behind the breast bone). In fact, for some conditions, the early stages of breast cancer, for example, an x-ray is the most sensitive method of detecting soft tissue disease. In contrast, soft tissue organs such as the spinal cord, kidneys, bladder, gut and blood vessels are very poorly resolved by x-ray. Imaging of these areas necessitates the administration of an artificial contrast medium to help delineate the organ in question. For example, the outline of the gut is usually enhanced by administering a solution of barium, either orally or rectally, depending on the region under investigation. The barium is usually combined with compounds that effervesce to improve the contrast between the gut and surrounding soft tissue by producing a gas, which distends the gut lumen. Iodine-based contrast media can be injected into a blood vessel to image specific arteries and veins, such as those supplying the heart (the coronary arteries), those supplying the brain or those passing from the heart to the lungs (pulmonary arteries). The urinary system is also commonly investigated with iodine-based contrast which, when excreted, delineates the kidneys, the ureters (the tubes passing from kidney to bladder) and the bladder, each region being visible at a different time after the injection. Other uses of artificial contrast are the myelogram (injection of contrast into the fluid filled space surrounding the spinal cord) and the arthrogram (injection of contrast into joints, particularly the shoulder and wrist).

The dose-related effects of the ionising radiation limit X-ray investigations. The clinician and radiologist have to weigh up the possible risks to the patient, resulting from the procedures themselves and the benefits of answering the clinical questions posed. For example, the risk associated with the small x-ray dose required to image the chest is outweighed by the potential benefits of revealing an unexpected chest or heart problem. On the other hand, the potential risk of a small dose of ionising radiation to a foetus is not outweighed by the benefit of imaging the abdomen; great care is taken not to inadvertently expose pregnant mothers. It would be impractical to debate the necessity of every single x-ray image, thus the decision as to what constitutes a justifiable risk is largely pre-determined by legislation. In fact, the risks associated with x-ray images are not all radiation related. Another factor to consider is the possibility of an allergic reaction to injected contrast. Despite careful screening, a small number of patients have unpredictable and life threatening reactions that pose a far more serious and immediate threat to the patient's well-being than the condition being investigated. Radiology departments are always prepared for the emergency resuscitation of a patient with an unexpected allergic reaction

X-ray CT

In general, CT images are only obtained after a problem has been identified with a single projection x-ray or ultrasound image; however, there are clinical situations (a head injury, for example) in which the clinician will request a CT image as the first investigation. CT is particularly useful when imaging soft tissue organs such as the brain, lungs, mediastinum, abdomen and, with newer ultra-fast acquisitions, the heart. As with conventional x-ray images, tomographic images are obtained both with and without artificial contrast to aid the identification of blood vessels. CT has an added advantage over the single projection x-ray in that it contains information about tissue type. Since the images are based on estimates of the linear x-ray attenuation coefficient, the radiologist can identify different tissues (bone, fat and muscle, for example) by comparing the coefficients at specific locations to known values of different tissue types.

Gamma Imaging

Like x-ray images, gamma investigations are limited by the dose-related effects of ionising radiation and their spatial resolution, even with tomographic enhancement, means that they are poorly suited for the imaging of anatomical structure. However, the technique has found an important niche in the imaging of function that is to say how well a particular organ is working. In practice, function equates to the amount of labelled tracer taken up by a particular organ or the amount of labelled blood-flow to a particular region. The radionuclide is usually injected into a vein and activity measured after a variable delay depending on the investigation being performed. A quantitative difference in 'function' provides the contrast between neighbouring tissues, allowing a crude image to be obtained. Gamma images are requested far less frequently that x-ray or ultrasound images as they are reserved for specific clinical situations; however, they provide quantifiable information difficult to obtain using any other method, often avoiding the necessity for a far more invasive procedure. Table 8.1 shows some of the more common radionuclide investigations. In practice, many of these investigations are obtained tomographically by rotating the gamma camera around the patient to produce a SPECT image.

The uptake of ^{201}Tl and ^{99m}Tc is compromised in poorly perfused or infarcted heart muscle causing an area of decreased uptake in the gamma image (an infarct is the damage caused by a block in one of the coronary vessels). Typically patients undergoing a heart scan will have one image taken at rest and one after exercise. In the investigation of pulmonary emboli,

ERRATA

An Introduction to the Principles of Medical Imaging

Chris Guy and Dominic ffytche

Dr Guy has informed us that Figure B.2 on page 341 is incorrect. Please find below the correct version of the figure. We regret the inconvenience.

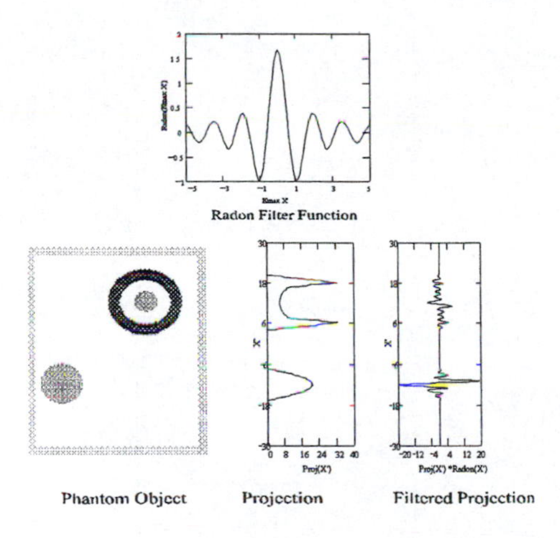

Figure B.2: a) The Radon or Ramp Filter: R (x') is the 1D Fourier transform of K forK<K_{max} . The function R (x) is convoluted with P (x') to yield the filtered backprojection. b) The Filtered Backprojection process. A single projection from the test slice on the left of the diagram is shown in the middle panel. The filtered projection obtained by convoluting this projection with the Ramp filter, R(x') is shown in the right panel. Notice that the filtered projection has both positive negative parts.

a^{133} Xe aerosol is inhaled to image the distribution of air in the lungs and small micro-spheres labelled with ^{99m}Tc are injected intravenously to assess the

Table 8.1 Common radionuclide investigations

Target	Radionuclide	Indication
Heart	^{201}Tl	Diagnosis of coronary artery disease and assessment of heart function
Lung	^{133}Xe ^{99m}Tc-DTPA	Assessment of pulmonary emboli - clots Compromising blood supply to the lungs
Bone	^{99m}Tc-MDP	Investigation of tumour seedlings
Kidney	^{99m}Tc-DTPA	Assessment of kidney function
Gut	^{99m}Tc-RBC	Assessment of gut blood loss
Thyroid	^{99m}Tc-pertechnate	Investigation of nodules in the thyroid and thyroid over-function
Brain	^{99m}Tc –HMPOA	Localisation of epileptic focus
Infection	^{99m}Tc-HMPOA WBC	Searching for a focus of infection

pulmonary distribution of blood. The clinician is looking for areas that are ventilated but not perfused to indicate the presence of a significant blood clot. In the bone scan, ^{99m}Tc labelled methylene diphosphonate (MDP) is injected and the whole body scanned 3-4 hours later to assess the spread of a tumour from its primary site. In kidney scans, an intravenous injection of ^{99m}Tc labelled diethylenetriaminepentaacetic acid (DTPA) helps quantify the ability of each kidney to extract and excrete the tracer. In a gut scan, the source of a slow blood leak can be identified for up to 24 hours after the injection of ^{99m}Tc labelled red blood cells. Similarly, in patients where an infection is suspected but no source is found, an injection of ^{99m}Tc -hexamethylpropyleneamine oxime (HMPAO) labelled white blood cells can help localise a hidden abscess. In the investigation of a thyroid nodule, a tumour appears as a low uptake defect after the injection of a ^{99m}Tc -pertechnate tracer. Finally, the pre-surgical work-up of patients with epilepsy often includes a ^{99m}Tc -

HMPOA or a ^{18}F fluorodeoxyglucose PET scan to localise the epileptic focus.

PET

When CT and, a decade later, MRI first became available, their place in the investigation of a wide range of clinical conditions was immediately obvious. The relatively high spatial resolution of these techniques together with their ability to image in 3 D allowed the clinician to ask new clinical questions. The demand for the investigations far exceeded the number of scanners and their importance is reflected in the fact that most, if not all, of the larger hospitals in the United Kingdom have CT and MRI facilities. Similarly, the simple and relatively cheap modification of existing gamma cameras to allow the acquisition of tomographic images has meant that most hospitals have a SPECT facility. In contrast, PET, first proposed in the 1950's, has taken much longer to be accepted as a clinical tool. The problem is related in part to the cost of the scanner and its ancillary services - the cyclotron and radiopharmacy - and in part to the absence of a defined clinical niche. Thus, while PET has a number of theoretical advantages over SPECT such as its higher spatial resolution and its use of a number of biologically interesting radionuclides, in practice, it remains a research tool, found in a handful of national specialist centres, used in the investigation of tumours, heart and brain function.

MRI

MRI is still developing as a clinical tool and its place in the investigation of many disorders has yet to be established but, for several reasons, it has already found a particular place in the imaging of the brain and spinal cord. One reason is its ability to detect subtle changes in cerebral and spinal cord anatomy that were not resolvable with CT (a slipped disc pressing on a spinal nerve or a small brain tumour, for example). This advantage of MRI over CT is due in part the superior spatial resolution of the technique and in part to the fact that MR images are insensitive to bone - in CT, the proximity of bony vertebrae to the spinal cord make this region difficult to image as a result of partial volume effects, see section 4.5. Secondly, somewhat fortuitously, a range of common pathological brain lesions which are invisible to x-rays, in particular the demyelinating plaques found in multiple sclerosis, have T2 characteristics that differ from normal brain tissue. Thirdly, unlike standard CT, MR images can be acquired in a variety of different planes - parallel to the length of the spinal cord for example. With the increasing availability of MRI scanners, MR imaging has also found a place outside neurological and neurosurgical practice. Although high resolution MR images of the chest and abdomen are limited by the unavoidable smearing effects of breathing movements and heart beats, the loss of resolution can be reduced by triggering the scanner at a particular point in the respiratory or cardiac cycle or by obtaining a rapid sequence of images using gradient echoes. Cardiac MRI is a less invasive alternative to the traditional x-ray investigation of the heart which requires high concentrations of iodine-based contrast to be injected into the heart chambers. It is used to image congenital heart diseases, heart tumours and tears of the aortic wall - the main artery from the heart to the rest of the body. As in CT, MR images are often acquired both with and without an artificial contrast to help identify blood vessels; although, in this case, the contrast medium contains the paramagnetic element, gadolinium, see section 6.4.

MRI offers an improved spatial resolution without the risk of ionising radiation; however, there are several reasons why it is unlikely to make other imaging techniques obsolete. Not all subjects can be investigated with MR, some find it difficult to tolerate the tight, enclosed environment of the scanner while others are too large to fit in the scanner bore. Furthermore, patients with pacemakers, artificial joints or surgical clips cannot be scanned and there are technical problems in scanning unconscious patients that require monitoring or artificial ventilation. While the insensitivity of MRI to bone is an advantage when imaging the brain, in certain situations it can also be a disadvantage. Small precipitates of the bone mineral, calcium are found in a variety of long-term inflammatory or cancerous lesions. Radiologists have long been able to use the calcification to help identify particular pathologies; however, in MR imaging, this inflammatory signature is not detectable. MRI is also at a disadvantage when imaging the lungs because of the low density of the tissue and the consequent low concentration of free water.

Ultrasound

Ultrasound is an effective and safe investigative tool. It offers only limited spatial resolution, but can answer a number of clinical questions without the use of ionising radiation and, unlike MRI, the equipment required is portable, compact and relatively inexpensive. It has found a particular place in the imaging of pregnancy, but it is also used to image the liver, spleen, kidneys, pancreas, thyroid and prostate glands as well as a screening tool in interventional radiology (see below). The ease with which an ultrasound image can be obtained makes it ideally suited for use in clinical situations where an immediate diagnosis is required. For example, kidney failure in a particular patient may be the result of an outflow blockage - a problem in the bladder or ureters - disease of the kidney itself or a problem with the blood supply to the kidneys. Each of these possible causes will lead to different treatment strategies and the clinician needs to know, with some degree of urgency, which treatment path to follow. Ultrasound images provide an instant answer to the question by revealing a fluid filled distended ureter or bladder. In contrast, x-ray or gamma images will, at the very least, take several minutes to provide the same answer. Ultrasound plays an important role in the investigation of the heart and blood vessels - the echocardiogram can provide a surprising amount of functional as well as structural information. However, there are a number of specific clinical situations in which ultrasound cannot be used. Structures surrounded by bone, such as the brain and spinal cord, do not give clinically useful images, and the attenuation of the ultrasound signal air/tissue boundaries means that the technique is not suitable for imaging structures in the lung or abdominal organs obscured by gas in the overlying bowel.

Interventional Radiology

Many of the diagnostic procedures traditionally undertaken by the general surgeon, biopsies of inaccessible lesions for example, have moved from the operating theatre into the radiology department. The trend is due in part to improvements in imaging techniques and radiological expertise and in part to the fact that, compared to an operation, the radiological procedure results in a shorter stay in hospital, a smaller scar and obviates the necessity for a general anaesthetic. In the past, fluoroscopy was used to guide the radiologist, but increasingly ultrasound images, CT or combinations of all three techniques are being used, the exact combination depending on the particular procedure being performed. Whatever the technique employed, the resulting images help the radiologist to guide specialised instruments to the desired location and to perform a variety of investigative and therapeutic procedures. For example, by guiding a thin tube (a catheter) from a small incision in the leg to the heart, the radiologist is able to widen the bore of a narrowed the coronary artery. While sparing the patient a major operation, such procedures are not without their own risks. If complications do arise, interventional radiological procedures often lead to emergency operations.

8.4 Clinical Examples

The fictional stories of three patients illustrate a typical sequence of events that lead from the initial presentation of a symptom to the generation of a medical image and its consequences.

Case 1 - X-ray, Fluoroscopy, CT, ^{99m}Tc -MDP Bone Scan

A 65 year old man with a long-standing 'smokers cough' noted that his phlegm was, on occasion, bloodstained. His wife thought that he was looking rather thin and encouraged him to visit his GP. The GP examined him and found no obvious abnormalities but made an appointment for him to see the consultant chest physician at the local hospital and organised a chest x-ray, some blood tests and an examination of his sputum.

A week later the patient was seen by the chest physician with the results of his investigations. The blood tests and sputum sample had not revealed any abnormalities but the chest x-ray showed a small, soft-tissue mass in the periphery of the right lung. The consultant explained that the mass may be a tumour but that this diagnosis could only be confirmed by examining a small sample of the tissue - a biopsy. One way of obtaining the sample would be to pass a thin, flexible fibre-optic tube (a bronchoscope) along the airways. However, this particular soft tissue lesion was located in a part of the lung that was difficult to access with this method. After discussing the problem with a radiologist, it was decided that a sample would best be obtained by passing a needle through the chest wall into the lesion using a fluoroscope to guide the procedure.

The following Monday the patient arrived in the x-ray department, having been asked not to eat or drink that morning. After discussing the procedure with the patient, the radiologist identified the soft tissue mass with a fluoroscope and the chest wall was numbed with a local anaesthetic. The patient was asked to hold his breath for a few seconds while a needle was passed through the chest and into the lesion. After the procedure, the patient had a chest x-ray to make sure that the lung had not collapsed and stayed in hospital for the remainder of the day.

One week later the patient returned to the chest physician. The biopsy had revealed the lesion to be a particular type of tumour and the chest physician discussed the possibility of an operation with a surgical colleague. The surgeon suggested that the patient had a CT of the chest to reveal any spread of the tumour to the lymph nodes that drain the lung (the hilar lymph nodes behind the breast-bone) and a bone scan to establish whether or not the tumour had spread to the bones (see figures 5I.1, 5I.2). The patient returned to the hospital on two occasions during the next week, on the first occasion he attended the nuclear medicine department and was given an injection of ^{99m}Tc bound to methylene diphosphonate. 3 hours later he was asked to empty his bladder before the start of the scan to prevent the high urinary concentrations of tracer from obscuring the view of the pelvis. His whole body was scanned with a gamma camera. The scan lasted about 45 minutes and he returned home on the same day. On his second visit he returned to the x-ray department where his chest was imaged in the CT scanner.

The following week the patient returned to the cardio-thoracic surgeon who was pleased to inform him that there was no evidence that the tumour had spread and discussed the benefits and risks of surgery. The patient decided to have an operation to remove the tumour and was admitted 3 days later. The operation was performed under a general anaesthetic. The chest wall was opened to expose the right lung and the lobe containing the lesion was removed. Lymph node samples were taken at the same time to look for microscopic spread. After the operation, a temporary tube inserted through the chest wall allowed the lung to re-inflate. The tube's position and the amount of air within the chest was assessed with a series of chest x-rays over the next four days. When the lung had re-inflated, the tube was removed.

The lymph node biopsies taken during the operation did not show any evidence of spread. The patient was reviewed by the chest physician and, five years after the operation, remained in good health.

Case 2 CT, Fluoroscopy, MR, DSA

A 25 year old man went to his GP complaining that he was developing a limp. The GP noted that the patient's right leg was stiff, with brisk reflexes, and had lost strength in specific muscle groups. He was referred for a neurological opinion at the local hospital and was seen by a neurologist the next week. The neurologist was concerned that limp may be the result of a brain lesion and arranged some blood tests and a CT scan of the head. The CT scan was performed later that week and the radiologist that supervised the scan noted a lesion on the left side of the brain which enhanced after a small volume of intravenous, iodine-based contrast was injected into the patient's forearm. The patient returned to the neurologist one week later to discuss the diagnosis and treatment options. He was informed that the lesion was a vascular malformation, an abnormal knot of blood vessels that had probably been present since birth. There was a significant risk that the abnormal vessels would bleed at some time in the future, resulting in a brain haemorrhage. Two alternative treatments were available. The first was for the patient to undergo an operation in which the skull would be opened and the abnormal vessels removed. The second was for a radiologist to attempt to cut-off the malformation's blood supply from within the arterial system. The aim would be to pass a catheter from an incision in the leg up to the lesion, using a combination of fluoroscopy and CT to monitor its position. Once in position, a blocking agent (either small spirals of metal or a type of glue) would be introduced, stopping the flow of blood in the malformation which, in time, would shrink to form a small, fibrous scar. If successful, the radiological intervention would obviate the necessity for an operation. The patient opted for the radiological procedure and was admitted for further tests: a high resolution MR image of the head, both with and without gadolinium contrast, and a digital subtraction angiogram (DSA, see figure 6I.6) to show the vascular anatomy of the malformation's feeding vessels. During the procedure itself the patient was lightly sedated and the catheter manipulated from the leg to the brain lesion. The glue-like blocking agent successfully cut off the blood supply and the patient was released from hospital 3 days after the procedure after undergoing a second DSA. He had a series of MR images over the next year to monitor the size of the malformation and, with the help of physiotherapy, the stiffness and weakness of the right leg gradually improved.

Case 3 - MR, EEG, ^{18}FDG-PET

A 20 year old women with a history of epileptic seizures dating back to early childhood, was referred to a neurosurgeon for evaluation and possible surgical treatment. The pattern of her epilepsy had been constant for the last 15 years and had not responded well to drug therapy. In a typical attack she would smell something burning for a few seconds and, if talking, experience difficulty in finding her words before loosing consciousness. Her family reported that after her speech slurred, she would go into a trance-like state for a minute before her body tensed and she started to shake. The neurosurgeon admitted the patient for a series of electrophysiological and imaging tests. An MRI of the head showed that a region in her left temporal lobe was slightly smaller than the equivalent region on the right, with some evidence of scarring while her electroencephalogram between attacks showed an occasional abnormal spike over the left hemisphere. During an attack the electroencephalogram revealed a localised electrical discharge over the left temporal lobe which spread to encompass the whole brain. Images of brain function were also obtained with a ^{18}*FDG-PET* scan which revealed an area of decreased metabolism in the left temporal lobe. Given the concordance of clinical, electrophysiological and imaging evidence, the neurosurgeon discussed the possibility of an operation to remove the focus. This involved opening the skull, under a general anaesthetic, localising the electrical focus directly on the cortex and removing a portion of the left temporal lobe. The patient agreed and, after surgery, had a series of post-operative MRI's to monitor progress and check for complications. Her seizure frequency improved from two - three per day to one every few months.

8.5 Research Progress in Neuroimaging

Throughout this book we have emphasised the progress in medical imaging made in the past twenty five years, following the introduction of CT. In the main this progress has been confined to anatomical imaging. It is only in recent years that attention has been focused on very rapid imaging methods which can provide movies of the functioning human body. An important limitation of modern imaging is revealed by a very simple question. Does an imaging technique tell a clinician whether a patient is alive or dead? The answer in most cases is actually probably not. Neither CT nor conventional MRI can easily distinguish a living patient from a cadaver, except by inspection of the more subtle consequences of blood flow or tissue movement that are actually considered to be artefacts in conventional investigations. The recent developments in very fast imaging technology in both CT and MRI now allow mechanical heart action to be imaged directly. Metabolism can now be studied on a timescale of minutes using nuclear techniques. However, an important area in which current imaging techniques fall short of the ideal of real-time movies is that of cerebral function. To date, no single technique has a spatial and temporal resolution sufficient to image the functioning brain. PET, SPECT and fMRI still provide relatively "static" pictures of which brain regions are correlated with particular cerebral processes. In musical terms, current imaging techniques establish the key in which a piece of music is played, identifying the different notes on which the piece is based, but cannot discern the rhythm or even which note follows another; they cannot discern the melody. It would be reasonable to assume that significant progress in understanding cerebral processing on a macroscopic scale requires a means of detecting electrical brain events, or their metabolic consequences, on a time scale of a few milliseconds with a spatial resolution of about 1mm. Figure 8.1 summarises the present state of play.

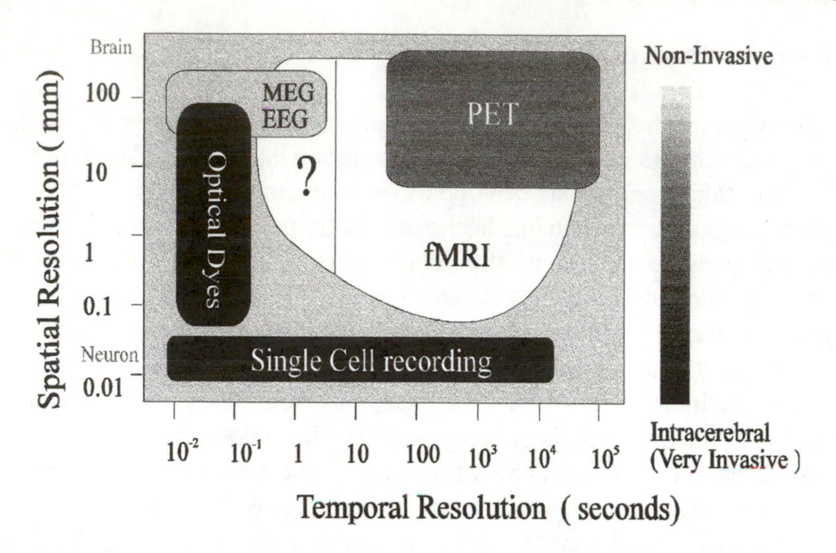

Figure 8.1: The domains of spatial and temporal resolution obtained from present techniques used in neuroimaging. The pivotal role to be played by fMRI, combining high spatial resolution (1-5 mm) with moderately good temporal resolution (2- 10 sec) is made clear in this graph. The box labelled "optical dyes" refers to a recent technique involving the measurement of optical reflectance from the surface of exposed cortex, both with and without voltage sensitive dyes to enhance the response Both "single cell recording" and optical reflectance require access to the cortex, PET involves injection of a radioactive tracer.

We will focus our discussion by considering a rather simplistic hypothetical analogy. Suppose that computer company A has clandestinely acquired a revolutionary new machine to be marketed by a rival company B. Company A already has a clear idea of the computer chip architecture because they know that they use the same chips. Simply switching on and operating the machine allows company A can find out, without leaving too many tell-tale signs of their investigation, how fast the machine executes instructions and what it does with standard supplied programmes. Company A will probably have at least a working hypothesis of what is different about this new machine and have constructed a plan for an investigation using completely noninvasive electronic listening devices. They give the computer a range of simple, carefully designed tasks and attempt to unscramble the resulting emitted signals. They must be able to localise the signals both in time and space in order to reveal which chips and which parts of chips, in what order, are dealing with the test inputs.

This analogy is essentially the long term programme for brain research. The present sticking points in neuroscience are the absence of a non-invasive technique capable of sufficient simultaneous temporal and spatial resolution and an adequate understanding of what we should be measuring. The brain's transistors are neurones and these are available for study from tissue taken from the brains of animals. Single cell and now even multiple cell recordings have provided a vast literature and a deep understanding of most of the important chemical and electrical processes that occur at the single cell level in all animal brains. Beyond this however the going gets progressively harder, as the level of integration increases. The transduction, encoding and transmission of signals related to the stimulation of peripheral sense organs, be it the eye, the ear, the nose or the skin for example, are understood and the main brain receiving stations of visual, tactile, auditory and olfactory signals are known reasonably well. One could envisage that, given a technique with adequate temporal and spatial resolution, one might be able to provide movies of these low level functions. However perhaps the most important aspects of brain function, memory, consciousness, emotion, problem solving, language and the binding of separate senses into a coherent percept of the world, are still shrouded in mystery. At this level it is not even clear what aspect of cerebral activity we should be measuring. For example should it be the level of activity within an area or specific neural codes or correlated activity between different areas? Thus, in spite of the remarkable progress that has been made throughout this century in our understanding of nature, the organ that is responsible for this intellectual enterprise is almost as mysterious now as it was one hundred years ago. We have no reason to be embarrassed by our ignorance since the brain is undoubtedly the most complex system known to mankind.

A great deal, some might say a disproportionate amount, of research activity in macroscopic brain studies is devoted to the study of vision and the associated brain area at the back of the brain called the visual cortex. This is reflected in our choice of research example. There are two main reasons for this emphasis. First, in the human brain much more neural machinery is devoted to the processing of sensory visual information than any other of the senses. Second, throughout this century, detailed information has accumulated from careful animal studies and human patients with discrete brain injuries has accumulated and been synthesised into a quite detailed coherent picture of how a complex visual scene is deconstructed into a collection of different attributes such as angles of lines, contrast changes and apparent motion across the field of view. Many such visual attributes can, to an extent, be isolated by the use of a particular visual stimulus and thus the associated strands of brain processing can be studied in relative isolation. It is hoped that, although a particular study is specific to the visual sense, in fact the researcher is dealing with a generic brain processing mechanism that eventually can be applied to other sensory areas and higher brain functions which cannot so easily be deconstructed. The study of human vision might then provide the means to reveal how the whole human brain works.

Figure 8.1 refers to five different methods, each of which can provide partial information restricted either in temporal or spatial resolution. We will limit our discussion to the most non-invasive trio, PET, fMRI and EEG/MEG. The principles of PET and MRI measurement have been covered in earlier chapters. Here we need only describe the modifications required to provide a contrast mechanism related to brain function. EEG and MEG are not imaging techniques in their own right; they have not been discussed elsewhere in this book and thus these require a small amount of preliminary description.

EEG and MEG

The technique of electroencephalography, EEG was invented in 1926 by Burger who was the first to record a variety of electrical voltage disturbances on the scalp, which seemed to be related to clinical behaviour and thus brain function. The technology is very simple, consisting of electrodes, held in conductive contact with the scalp, and connected to high gain electronic amplifiers. In general the human body acts as an antenna for all electrical disturbances in its vicinity and so a differential voltage measurement is made from pairs of electrodes. Spurious low frequency effects, mostly caused at the electrodes themselves, limit the useful low frequency range to frequencies above 1-2 Hz. There is no technical upper frequency bound but measurable brain activity contains very little signal power above 50 Hz. Typical normal voltage amplitudes range from 0.1 to 10 μV and these can easily be detected after electronic amplification and filtering.

All humans exhibit some rhythmical electrical activity, both when they are awake and during sleep. For example when most people close their eyes or relax, a large amplitude, nearly sinusoidal voltage disturbance with a frequency of about 8-10 Hz appears. This is the alpha rhythm. It is widely distributed over posterior parts of the head. The EEG of normal alert subjects consists of a complex low amplitude pattern of voltages containing frequencies between 2 and about 40 Hz. The pattern changes both in time and in spatial distribution across the scalp, with task and the degree of attention. At first sight it would appear that an analysis of these spontaneous signals could provide access to underlying brain function, since scalp voltage changes must be generated in some way by the synchronised groups of neurones involved in particular aspects of that function. So far however normal spontaneous electrical activity has proved too complex to relate to function in any detailed manner. Gross disturbances in the normal EEG, caused for example by epilepsy, were for several decades the only quantitative means of diagnosis of that condition. From about 1960 onwards the related technique of evoked potential, EP measurement has been used extensively by neuroscientists and psychologists to investigate the time course of scalp signals related to specific peripheral stimuli. If a given stimulus, for example a flashing light is repeated, there results a small electrical scalp signal which can be revealed by averaging the results of say 50 to 100 repeated stimuli. In general these signals contain frequencies up to about 40 Hz and are generally more restricted in their spatial extent than spontaneous activity. Extensive efforts were made in the 1960's and 1970's to use the spatial scalp patterns of these evoked signal to localise the brain centres involved in processing of the specific stimuli. It is now widely accepted that spatial localisation of a single small group of say 10^5 synchronised neurones using EP is possible with a resolution of about 10 mm. This scenario is however rarely met in practice since every stimulus evokes responses in either rather large patches of neurones spread across one region of the cortex or two or more remote regions acting in concert. It turns out that in this more realistic case it is effectively impossible to turn a map of measured surface fields into a map of underlying current generators. When this is attempted the result is often unphysical, with putative current generators positioned unphysically in white matter or the ventricles. Finally when a detailed comparison of scalp recordings is made with simultaneous intracerebral recordings, it becomes clear that the surface field variations are rather poor reflections of a very much more complex underlying electrical activity actually taking place in the brain. It is assumed that this arises from the spatial smoothing effect of the surrounding skull, which has an electrical conductivity 80 times less than that of soft tissue. Thus EEG and EP, although able to provide useful correlates of brain function, only reflect a small fraction of the true complexity of the underlying electrical brain activity.

Starting in 1970 the related area of magnetoencephalography, MEG was developed as a possible way of circumventing some of the shortcomings of EEG. Scalp magnetic fields are extremely small, ranging from 2 to 0.01 pT, some six orders of magnitude smaller that earth's magnetic field. MEG has to use special sensors, called **SQUIDS** that only operate in liquid helium, and special magnetically screened rooms to remove the very much larger interfering magnetic fields produced by electrical equipment and moving steel objects. Commercial MEG systems are comparable in price to the million or so pound cost of an MRI machine. Good EEG data, on the other hand, can be obtained with equipment costing a few thousand pounds. To justify its cost, MEG would have to demonstrate an ability to obtain significantly different information from that already provided more cheaply and simply by EEG. It is often stated, incorrectly, that MEG makes direct measurements of neuronal electrical activity via the magnetic field produced by the moving ions directly involved in brain function at a neuronal level. Furthermore the resistive scalp is said to be transparent to magnetic fields and thus MEG provides a better localisation than EEG. After twenty years of development and study it is now clear that MEG is just as susceptible to brain conductivity variations as EEG and suffers exactly the same mathematical problem of inversion when multiple sources are active. The two methods provide roughly the same information and suffer from exactly the same pitfalls. The main generators of the scalp magnetic field are thought to be confined to sulci (groups of neurones aligned tangentially to the scalp) whereas the main generators of the EEG are oriented radially, on the gyri. The two methods could provide complimentary information about very specific brain areas, involving just one brain fold, but this is not of paramount importance at present. Commercial interest in MEG has withered on the vine, especially following the apparent progress made in fMRI. MEG may return in the future, when more detailed electromagnetic investigations are indicated; for example its ability to measure absolute fields and low frequency changes. Both EEG and MEG can provide extremely good temporal resolution of an averaged version of the activity of large groups of neurones but any localisation is extremely suspect. Returning to the music analogy they are able to describe the rhythm of a piece but not the component notes. It is doubtful that the tomographic principle could ever be applied to either method and thus both these areas, used on their own, will remain in the wings.

Considerable progress has been made in brain function localisation, and thus imaging, by taking a step back from the direct physical consequences of ionic currents and turning to the consequent changes in blood flow and oxygen metabolism that accompany neuronal activity. Regional cerebral blood flow, rCBF, measurements using PET, carried out over the past fifteen years have provided a valuable body of information, some of it new, some merely confirming what was already known from previous careful clinical studies of the brain. fMRI operates in the same area but makes a slightly different connection to the metabolic changes associated with brain function.

PET Studies

As described in chapter 5, PET has a number of useful attributes such as sensitivity to metabolic activity and blood flow, but suffers from several important drawbacks, such as the administration of significant doses of charged particle ionising radiation, and a limited spatial resolution. It is probably not capable of temporal resolution on a scale of less than 100 seconds. It has been used over the last 15 years to image areas of the brain that are differentially activated by specific stimuli. In a typical experiment subjects are scanned 12 times over a period of three hours, each scan being acquired after an injection of water labelled with about 500 MBq (~11 mCi) of the positron emitter ^{15}O resulting in a total exposure of 5 mSv. Because the radioisotope has a half -life of 2 minutes, scans are separated by 10

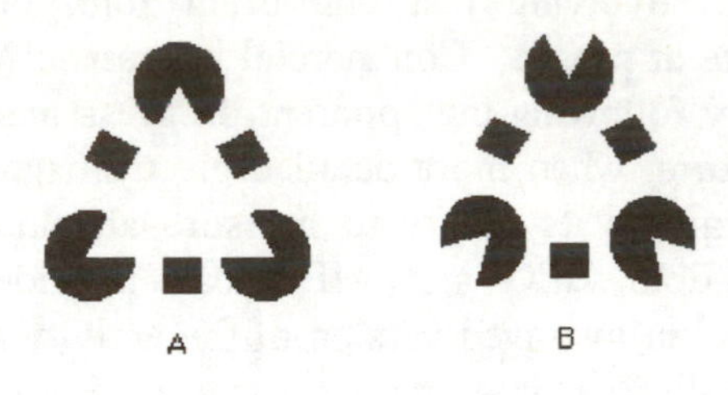

Figure 8.2: A visual stimulus used in an ^{15}O PET experiment. The arrangement. of shapes in A produces the percept of a white central triangle whereas the same shapes reoriented, as in B do not. Stimulus A is shown for 2 minutes during which time PET gamma counting takes place. After an interval of 10 minutes stimulus B is shown for 2 minutes, again with PET counting. The difference PET map computed from these two experiments provides a measure of the extra regional blood flow involved in processing stimulus A with respect to stimulus B.

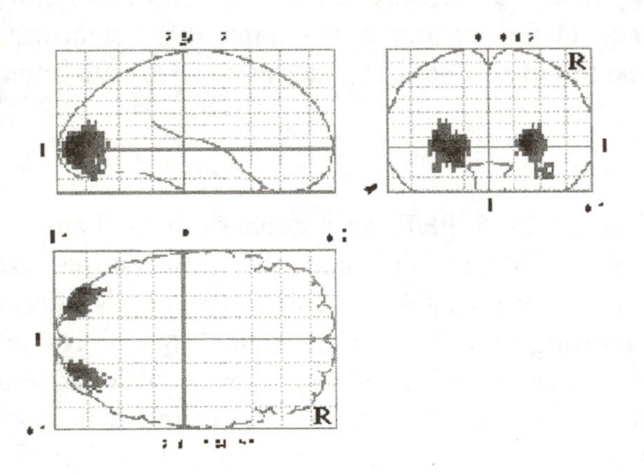

Figure 8.3: Three orthogonal views of the difference activation PET map produced by the A and B stimuli shown in 8.2.

minute intervals to allow background activity to decay. Subjects will be presented one type of stimulus for six of the twelve scans (for example stimulus A in figure 8.2, in which the black elements are arranged in such a way that a white triangle is perceived) and another type of stimulus for the remaining scans (for example stimulus B in figure8.2). Difference maps calculated from the two conditions provide a measure of extra blood flow to brain areas activated by the stimulus. Typically the counting time for each scan is 1-2 minutes and so nearly all temporal resolution is lost. In common with all gamma imaging, PET is limited by low photon count rates; the resulting statistical uncertainties are further amplified by the differencing between activation and rest conditions. The difference activation maps have a spatial resolution of about 15mm; adequate for many areas of study at this stage in our understanding. The dose

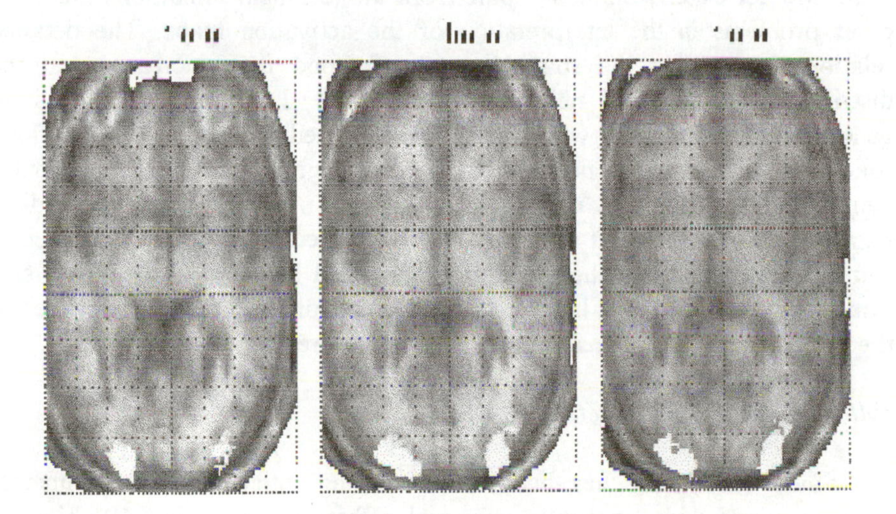

Figure 8.4: The results of the visual stimulation PET experiment superimposed on three slices of the "average" brain of the four subjects used in the study constructed from individual anatomical MRI images. After the normalisation, scaling and realignment of the individual brains the resulting average brain is smeared as a result of the normal anatomical variability between subjects.

to a PET subject is about one year's recommended limit dose and thus repeat studies of the same subject are not possible on a shorter time scale. Improved statistics are sometimes obtained by using several volunteers and then averaging their results. For example, Figure 8.3 shows the increase in rCBF for stimulus A over stimulus B averaged over four subjects. The averaging requires sophisticated re-mapping and scaling of the individual brains, using MRI images. The smoothed and averaged data refer to a smoothed averaged brain which may have little detailed resemblance to any particular brain since brain anatomy is notoriously variable from individual to individual. The extent of the inter-subject variability is apparent in Figure 8.4 where the increases in rCBF have been superimposed upon an average brain of four of the subjects taking part in the study. Despite the fact that the brains have been remapped and scaled into a standard space, the average anatomical brain image is rather blurred as equivalent features, such as the ventricles or skull, are not exactly aligned. Apart from the technical limitations there are other problems in the interpretation of the activation maps. The detailed relationship between the spatial extent of blood flow increases and the distribution of activated neurones is uncertain. Furthermore there is no general agreement about why there is any immediate increase blood flow, following an increase in neural activity. Some groups claim that the extra neural energy required by cerebral processing is provided, not by immediate extra oxygen delivery, but rather by an in-situ metabolism of stored glucose. Thus it is possible that the blood flow recorded by PET is actually a side issue with only a tenuous link to the neurophysiological activity. Echoes of the same debate are apparent in the fMRI literature.

fMRI and the BOLD Effect

Since 1990 there has been an increasing interest and experimental activity, world-wide, in subtle temporal effects observed in MRI signals. These signal changes are associated with the brain activity resulting from periodic stimulation of, for example, the visual cortex. The earliest work appeared to hold the promise of nearly real time functional imaging of the brain without the use of any ionising radiation. Although conventional MRI can take several minutes to acquire data for a single slice, the use of Echo Planar Imaging, EPI, seemed to offer the possibility of a time resolution of about 50 ms. Quite early on in this history the Blood Oxygen Level Dependent effect, BOLD was mooted as the likely contrast mechanism and this, with certain reservations, offered the possibility of a rather more direct measure of brain function than that provided by the regional blood flow changes measured in PET.

In a typical experiment a subject is exposed, for example, to a visual stimulus consisting of a repeating cycle of 30 seconds of visual noise and 30 seconds of darkness. In the earliest experiments, a single MRI slice was selected to intersect the likely cortical region and 2D EPI images were made at intervals of about 2.5 seconds, more recently, improvements in data acquisition and storage have allowed 14, 32 or even 64 brain slices to be acquired, the larger number of slices requiring longer sampling intervals. Figure 8.5 shows an example of the MRI response to visual noise in two subjects, scaled onto a standard brain and averaged. Examination of the time series of MRI signals, within the activated region, reveal amplitude fluctuations of between 1 and 2 %. Since the effect is relatively weak and EPI inherently noisy, within the canon of MRI techniques, correlation between the temporal periodicity of the imposed stimulus and the resulting MRI signal has to be used to achieve statistical significance. It is now accepted that the fMRI signal waveform differs significantly from that of the stimulus, and, there is an apparent delay of between 5 to 10 seconds between the onset of the stimulus and the peak of the resulting fMRI signal. These relatively long, and method dependent, time delays of the putative physiological signals somewhat reduces the potential of fMRI and brings into the question the relationship between fMRI signals and neuronal activity. Although the data acquisition time bottleneck has been removed by EPI, the physiological delay imposes its own temporal constraint on the method

The BOLD hypothesis explains the observed signal changes roughly as follows. When significant areas of the cortex are stimulated into quasi synchronous activity, energy is used and this elicits an increase in the local input of fresh, oxygenated blood. It appears that the oxygen conversion rate does not match the increased flow and thus there results a local increase in the concentration of oxyhemoglobin delivered to the activated area. The deoxy molecule is relatively strongly paramagnetic but the oxygenated molecule is only weakly diamagnetic. As the local blood pool becomes more

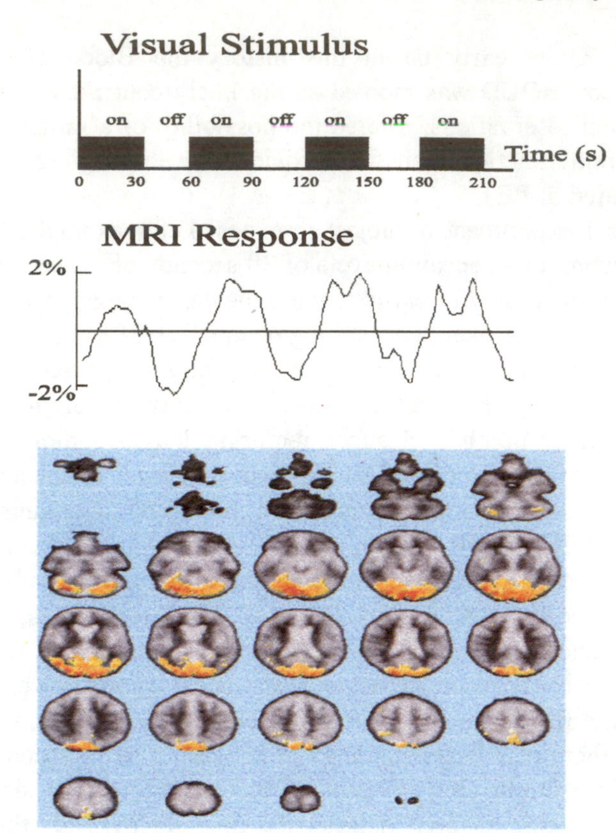

Figure 8.5: The MRI response to periodic visual stimulation. The visual stimulus is ON during the time intervals indicated by the shaded rectangles and OFF in the intervening times. The MR response can be seen to be filtered by the physiological process, to leave just the lowest frequency sine wave of the stimulus temporal variation. In addition there is a time delay between stimulus and response. The areas of the subject's brain found to be responding to the visual stimulus are superimposed in colour on an anatomical MRI image of the brain depicted in grey.

oxygenated, the local magnetic susceptibility is reduced with respect to a resting state. This in turn brings the magnitude of the local static magnetic field and thus the local Larmor frequency closer to that of the applied field.

This also reduces local variations in Larmor frequency from site to site within the activated volume of tissue. Overall, the transverse relaxation time, T2* in an activated volume, is increased and the signal intensity, at a set observation time, is increased; there is less dephasing of the rotating nuclear magnetisation, see section 6.3. Leaving aside possible dynamic transient effects in blood oxygen supply, the BOLD hypothesis makes the unequivocal prediction of a local signal increase measured on gradient echo images. Bipolar signals, not predicted by BOLD, are now reported and their appearance has been explained by yet new peculiarities in the dynamics of the hemodynamic response. Although few doubt that fMRI signals are in some way related to stimulated brain activity there remains some considerable uncertainty about the mechanism responsible and its precise relationship to neuronal activity.

Although none of the techniques, that are available now, are capable of delivering the holy grail of non-invasive simultaneous spatial and temporal resolution, together they can provide a good deal of genuinely new information about the working brain. If no new technology appears in the next few decades then EEG/MEG will have to be combined with fMRI and PET to supplement the temporal resolution of the latter and provide localisation in the former. New technologies and developments in MRI itself are of course possible, likely even, given the progress made in the last half of this century. Progress towards a deep understanding of the human brain is certainly high on the list of priorities for the new millennium both because of the spin-off in the clinical treatment of brain disease and mental illness and because the human brain, as a facet of nature, remains one of the unsolved mysteries of the present millennium.

Questions and Problems

1. Could fMRI and EEG be recorded simultaneously to provide complimentary spatial and temporal information. Describe the technical difficulties that need to be overcome. Do the same problems apply to MEG and fMRI combinations?
2. Are there any clinical or research applications for which PET and SPECT offer a significant advantage over fMRI?
3. Which of the following are you likely to find in a small district general hospital and what role would they serve.

   Medical physicist
   Radio-chemist
   Radiographer
   Neuroradiologist
   PET
   MEG
   VEP
   Ultrasound
   CT

4. Classify SPECT, PET fMRI, EEG, MEG, optical imaging with voltage sensitive dyes, reflectance imaging and near infrared spectroscopy, into direct and indirect measures of neurophysiological activity. What implication does the direct/indirect dichotomy have for the temporal resolution of each technique?
5. A clinician wants to investigate cerebral blood flow in the foetus. Which imaging technique should he opt for and why?

Appendix A
Waves, Images and Fourier Analysis

A.1 Waves and Oscillations

Repeating or periodic phenomena are common in both nature and everyday life. The most obvious are the astronomical cycles. The progression of the days is marked by the alternating cycle of day and night, the months by the phases of the moon and the seasons by longer lasting repeating variations of daylight, temperature and the weather. On the seashore, beyond the turbulent surf, there are often unbroken lines of well-defined peaks and crests of surface water waves advancing towards the beach. In the music room, the vibrations of the piano strings, themselves periodic oscillations, induce the sound waves in the surrounding air that we hear. In physics the description and study of oscillations and wavelike behaviour, on both macroscopic and atomic scales, forms perhaps the largest part of dynamics. Similarly in signal and image processing, wave concepts play a central role. Thus we need a mathematical framework for periodic and wavelike phenomena which will allow us to describe these motions and predict their future form at different places.

In 1822 Fourier published " a theory of heat". In this work Fourier, for the first time, used periodic trigonometric functions, the sine and the cosine to solve the differential equations that arose from his calculus of heat transport in solid objects. He had deduced the differential equations, which encapsulated the physics of heat transfer, but he had no standard means to hand to solve these equations. That is to say, to find a mathematical expression which obeyed the equation, and predicted the future variation of temperature throughout the object, both in space and time. His new method of solution using the periodic trigonometric functions, not only solved the heat problem but gave birth to one of the most powerful and widely used general methods of analysis in applied mathematics, Fourier Analysis. All of modern tomographic medical imaging relies on Fourier analysis. A very large fraction of that part of electrical engineering that deals with generating, filtering and receiving signals depends on Fourier analysis.

Even the computational methods used by chemists and physicists in quantum mechanics rely on a generalisation of Fourier's method. The purpose of this appendix is to provide a gentle introduction to Fourier analysis, so that the non-mathematical reader can understand the meanings of the terms wave, period, phase, K space and Fourier Transform that are bandied about by scientists and mathematicians. We make no attempt at mathematical rigour but use analogy and illustrations to justify the main points. First of all we establish a standard mathematical description of waves and oscillations in terms of amplitude, period (or frequency) and phase. Then we add waves of different frequencies together so that we can both construct and deconstruct complicated waveforms, which may or may not repeat in time. Throughout we will use the simple illustration of surface water waves spreading on a smooth pond.

A.2 A Mathematical Description of Waves

We begin with the illustration of water waves. Consider a pond whose surface is initially unruffled. When we throw a small stone into the middle of the pond there will a splash and a local complicated disturbance of the surface. A short time later, after the transient splash has died away, we will observe a perfectly circular pattern of ridges and troughs on the surface that travel outwards at a steady rate from the splash. If we set up a vertical water level gauge on a pole in the water, at some distance from the splash, then we can record the water level variations as the waves pass by. Clearly, during the splash the water level at the measuring pole will remain constant, since the ripples have yet to arrive there. Once the first ripple reaches the pole, the water level there will continue to change up and down in an almost perfectly periodic manner, until after the entire wavegroup, created by the splash, has passed. Figure A.1 illustrates this experiment and figure A.2 shows how the height, Z of the water at the pole varies in time. Mathematically we describe this variation of Z in time by the expression

$$Z(t) = A \sin(\frac{2\pi}{T} t) = A \sin(2 \pi f t) \qquad \text{..... A.1}$$

Here A is the amplitude of the water disturbance and A is equal to1/2 of the peak to trough excursion of the water level at the pole. Sine (x) is the trigonometric function which repeats itself every 360° or 2π radians. The time interval between peaks (or troughs) of the wave is $T = 1/f$ s. The argument of the sine function is written as $2\pi t/T$. As time, t, progresses, each interval, T seconds long, corresponds to one complete revolution of the sine function in 2π radians and after each period the disturbance is repeated in a periodic fashion. We notice that the wave takes a definite time, $t=R/v$ s, to actually reach our measuring pole. Had the pole been closer to the splash then the waves will start earlier, if it is further from the splash then the waves will start later. Now suppose that we decide to repeat the experiment with a line of poles, each with its own level meter, set at intervals of a few metres. Figure A.3 illustrates the variation in time at three such poles.

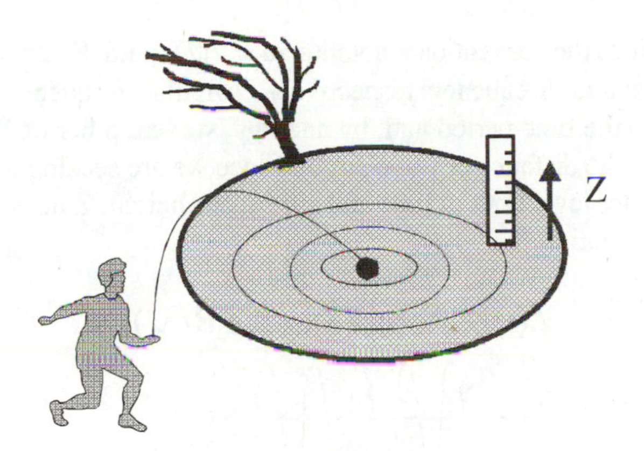

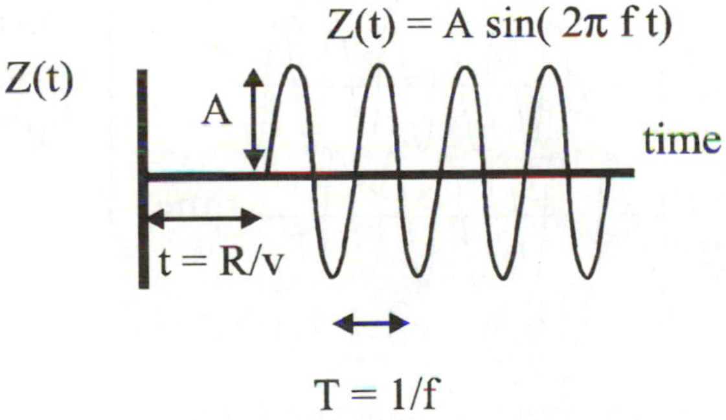

$$Z(t) = A \sin(2\pi f t)$$

Z(t)

A

time

t = R/v

T = 1/f

Figure A.1 The pond experiment.

Figure A.2 The variation in pond water level with time at a single fixed pole position. The symbol sin (x) is the standard mathematical description of the trigonometric function sine(x). The waves, travelling across the surface with velocity v, take a definite time R/v to travel from the splash to the measuring pole, a distance R away.

Clearly the oscillation will begin at closer poles earlier than at the more remote poles. Once they have all started oscillating, the variations at every pole have the same shape but shifted in time by an amount which depends on their distance from the splash. We say that these waves have different phases. In fact the complete set of waves recorded by any pole set up at a distance, R, from the splash can be described by a single mathematical expression.

$$Z(t) = A\sin\left(\frac{2\pi}{T}t - \frac{2\pi}{\lambda}R\right) = A\sin(\omega t - KR) \qquad \text{A.2}$$

Here we introduce the conventional notation $\omega = 2\pi/T$ and $K=2\pi/\lambda$ for angular velocity and spatial frequency respectively. Angular frequency is inversely proportional to the time period and, by analogy, wavenumber or K is inversely proportional to the distance or wavelength. Since we are seeking to describe the variation in water level both in time and space, our height, Z now depends both on time, t and position R

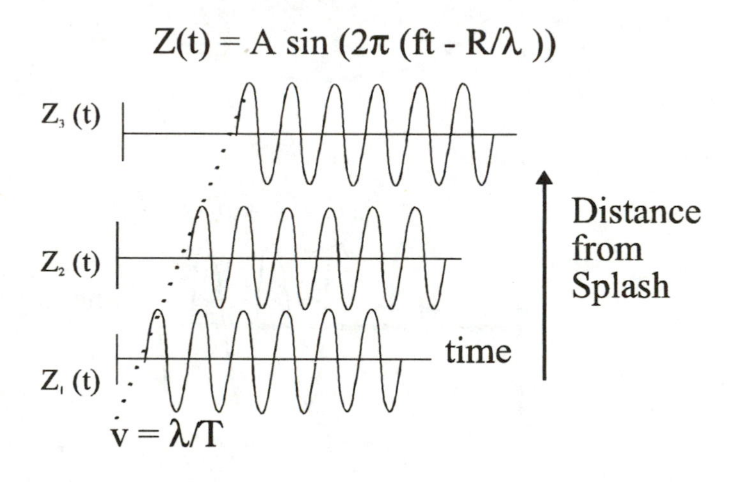

Figure A.3: The water level measured at four poles set up at equal intervals, ΔR, along a radius from the splash. The waves all travel at a velocity v and thus reach the poles at different times. The water level at all the poles oscillates with the same wave-shape, but there is a phase shift of $\phi = 2\pi\Delta R/\lambda$ between the waves, where λ is the wavelength of the wave measured across the water surface.

measured at an instant of time along the surface. In this example we are dealing with a group of travelling waves, so that the disturbance varies in both space and time. If we fix the space variable, by concentrating on a single pole, we have a wave with a definite temporal frequency, $f = 1/T$ and a phase dictated by the position of the pole that we chose. The temporal wave measured at our selected pole is described by

$$Z_R(t) = A \sin(2\pi f t - \phi) \quad \text{....A.3}$$

this is a temporal wave of frequency f and phase $\phi = 2\pi R/\lambda$. If on the other hand we freeze time, by taking a snapshot of the entire pond, then we have a picture of waves in space with a definite spatial frequency $K = 2\pi/\lambda$, and a phase set by the particular instant in time of the snapshot. The two descriptions are of course complimentary.

The majority of image processing applications of Fourier analysis deal with a picture in space that is described in terms of waves with spatial frequencies, K. In signal processing, a temporal picture is described in terms of temporal waves with frequencies f. The term K space (Fourier Space) that is used frequently in this book is mostly concerned with images. Thus in chapter 1 we describe the tomographic reconstruction process as an application of Fourier analysis purely in spatial co-ordinates. Our description of MRI in chapter 6 actually uses exactly the same concepts and mathematical techniques but appears to be more complicated. There the detected nuclear magnetic resonance signal is actually a temporal signal, described in terms of temporal frequencies, but each temporal frequency is uniquely related to a spatial co-ordinate (through the application of gradient magnetic fields) and thus time gets translated into space. MRI techniques are thus described in terms of a spatial K space description of a 2D image just like the description of CT.

A.3 Fourier Analysis

At the simplest level Fourier analysis amounts to describing any repeating wave disturbance as a sum of simpler waves like the sin (x) that we have just used. Had we looked very closely at the waveform of the pond disturbance we would have found, depending on our choice of stone and the size of the splash that it had a more complicated variation than the simple, sin (x) that we have described. In fact the splash, because it is a very localised disturbance in time, would give rise to a range of water frequencies, f_1, f_2, f_3 etc each of which would have its own wavelength. All the frequencies and corresponding wavelengths must be related by $\lambda f = v$ = constant, the speed of surface water waves. Similarly a snapshot of the pond with a group of waves occupying just part of the surface area will be described by a collection of spatial waves with a range of spatial frequencies. Fourier Analysis allows us to describe the level variations more accurately in terms of a series of sine or cosine functions with different amplitudes, frequencies and phases which, added together, would accurately mimic the observed level variations either in time or space. We will concentrate on a snapshot in which time is fixed.

The waves spreading out from the splash are circularly symmetric and thus if we describe the water level variation along one particular radius then we have a description of the entire pond. Along any radius, R, of the pond the waves will look like one of the curves, Z (R), drawn in figure A.4 corresponding to a group of waves occupying just part of the pond surface. Thus in general we will have a segment of flat surface followed by a wavegroup followed by another segment of flat surface. The real space description is Z (R) can be decomposed into a series of spatial waves with frequency K according to the recipe

$$Z\,(R) = Q_1 \sin(K_1 R) + Q_2 \sin(K_2 R) + Q_3 \sin(K_3 R) + ...etc \quad ...A.4$$

Here the coefficients Q_1, Q_2, ... describe the relative amplitudes (and relative phases) of the component waves that describe the pond ripples. In general, if we are to describe the entire pond surface, we will need a very large number of waves with spatial frequencies that are very closely spaced and which cover the range K=0 to K= K_{max}. Here zero is the DC value and K_{max} corresponds to the smallest wavelength amongst the pond ripples. The change to a continuous range of frequencies takes us from a Fourier series to a Fourier Transform whose generic form is

$$Z\,(R) = \int_{K=0}^{K_{max}} Q(K)\sin(KR)\,dK \quad ...A.5$$

Again each spatial wave, sin(KR), has an amplitude Q(K) which tells us how much of the wave sin(KR) is contained in the real space variation Z(R). (In general there will be phase variations from wave to wave and these are handled most conveniently using complex numbers. This is a detail beyond our present needs). These expressions are of no value until we have an unambiguous

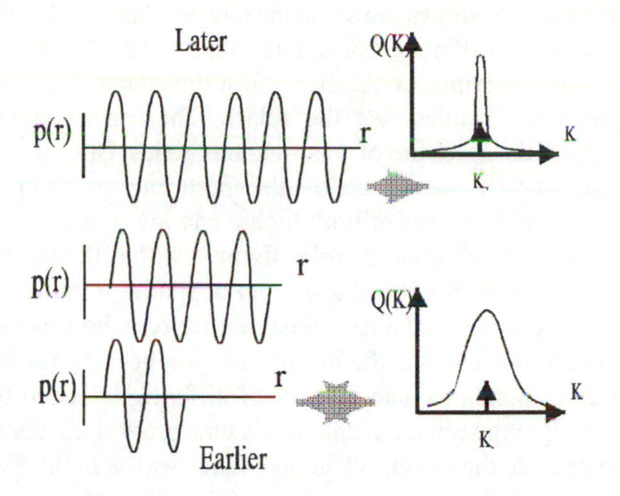

Figure A.4: The variation of pond water level along a radius to the splash taken at 3 times and the corresponding Fourier description. Although the water waves have a single frequency, short pulses of oscillation actually require a broad range of Fourier frequencies K. As the pond wave group gets longer, so the Fourier spectrum gets narrower. The centre frequency of the Fourier spectra is always the same, K_W.

scheme with which to calculate the amplitudes, Q(K). The Fourier method gives us a precise recipe for this

$$Q(K) = \int_{R=0}^{\text{Edge of the pond}} Z(R)\sin(KR)dR \qquad \ldots A.6$$

Thus the amplitude Q(K) of one wave with spatial frequency ,K is given by the amount of overlap or degree of similarity between our real space function Z(R) and the wave in question, sin(KR).

Together the expression for Z(R) and Q(K) form what is called a Fourier Transform pair. If we know Z(R) then we can decompose it into waves, if on the other hand we know Q(K) we can reconstruct Z(R) using these expressions. Two such pairs are illustrated in figure A.4. Now we have two complimentary ways of describing a snapshot of the waves on the pond. Either we can talk about a spatial map of the variation on water level amplitude with position on the pond, just like an ordinance survey map of a geographical location, or we can use the Fourier description of component wave amplitude (and phase) with spatial frequency, K. Figure A.4 shows three snapshots in time of the water level variation with radius at different times after the splash, together with the corresponding Fourier descriptions. At the earliest time most of the pond is still flat and there just a few ripples near the splash. The corresponding Fourier description consists of the spectrum of Fourier amplitudes, Q(K) plotted against spatial frequency K. In each case we see that the spectrum consists in a dominant ripple, K_w accompanied by waves of both higher and lower spatial frequencies with lower amplitudes. It will noticed, from figure A4, that the Fourier picture changes as time goes on from a broad spectrum of spatial frequencies, shortly after the splash, to a narrow one at a later time. Both spectra however are centred on the same spatial frequency, K_w, the basic wave produced by the spash. The broad spectrum corresponds to many waves of differing K added together to create the short wavegroup seen on the pond. As time progresses more and more waves are generated on the pond, all at the same wavelength. The Fourier description now has a much narrower spectrum, still centred about the dominant pond wave, K_w. Where do all the new waves with K different from K_w come from? At the earlier time most of the pond was flat and a flat line in real space requires a large number of different waves added together in antiphase, in order to represent the complete pond. As more of the pond is occupied by the waves of constant K, there is less flat line to be described and thus a smaller number of different waves are needed to represent the pond surface. This may seem rather artificial in the context of the pond waves but it illustrates a crucial issue for many areas of science and engineering. Any short pulse of a single intended frequency must necessarily involve a large number of extra frequencies. Simply switching the wave on and then, a little later, switching it off generates new fequencies. Thus in digital communication systems, where signals are transmitted as a sequence of very short pulses, both the transmitter and the receiver must be able to handle a wide range of frequencies in order to reproduce the narrow pulse shape. In section 6.4 we describe how slices are selected in MRI using a short pulse of RF waves in the presence of a gradient magnetic field. The gradient magnetic field encodes position in the patient with NMR frequency. A very short pulse of an RF wave not only excites the intended slice position but necessarily slices to either side as a result of the real spread in frequencies in the short pulse of applied RF., see figure 6.14.

2D Fourier Analysis

We have used the pond example in which waves were evident from the outset and we chose a simple one-dimensional example. Fourier analysis is not restricted to just obvious waves or periodic disturbances, nor is it restricted to just one dimension. In fact nearly all applications of Fourier analysis in medical imaging are concerned with two-dimensional or even three-dimensional objects that are not periodic. A picture, Z(X,Y) which is a function of two co-ordinates X, Y will need a two dimensional Fourier description. After a two dimensional Fourier transform of a two-dimensional surface, we arrive at a two-dimensional description in K space Q (K_x, K_y), where K_X, K_Y are spatial waves running parallel to X and Y respectively, and Q is the amplitude of the component wave at K_x, K_y.. Two-dimensional Fourier analysis uses the same idea of splitting a function into component waves, but now the component waves have a direction in both real space and K space. Figure A.5 illustrates this by comparing the real space component waves and their K space representation. The reciprocal relationship between wavelengths, L in real space and the corresponding K space length, $K = 2\pi/L$ has some important consequences in applications of Fourier analysis. Fourier amplitudes at large values of L in real space (close to the origin in K space) reflect the overall gross features of a 2D scene, whereas those at small values of L (far from the origin in K space) reflect the fine detail.

Two dimensional Fourier transforms are used extensively throughout image processing and optics. Quite apart from the purely mathematical convenience of their use they do have a very real physical significance in diffraction optics and crystallography. For example it can be shown that the variation of light amplitude in any diffraction pattern is the Fourier transform of the object that produces the diffraction. Thus if laser light is shone through a very long narrow slit onto a screen, then the pattern of diffracted light will be a series of light and dark fringes varying according to the sinc function ((sinc (x) = sin(x)/x); x being the distance along the screen measured from the centre of the pattern. In this case the pattern varies only in one direction because we used a long thin slit which corresponds to a 1 dimensional object. The sinc function diffraction pattern is the one dimensional Fourier transform of the light intensity at the slit which is sometimes referred to as a "tophat"; zero either side of the slit and constant across the slit itself. As we reduce the length of the slit to a square hole then a two dimensional diffraction pattern emerges, because now the light is now diffracted in two directions. In Figure A.6 we show a contour map of the 2 dimensional Fourier transforms of rectangular objects. These are exactly the diffraction patterns that these objects would produce if they were made into holes in an absorbing sheet.

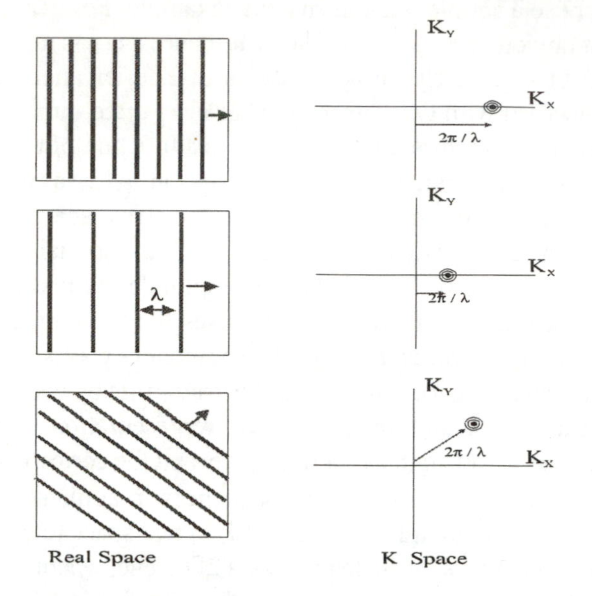

Figure A.5: The representation of component Fourier waves in real space and K space. In the left-hand panels the bars represent the peaks of plain sinusoidal waves, assumed to occupy all space, with wavelength, L. The equivalent K space contour maps are shown in the right hand panels. A single wave is here represented by a tight nest of contours centred at the co-ordinates (K_x, K_y) , the components of the wave's wavevector. The contours represent the amplitude of the wave. Note that the length scales in real space and K space are reciprocal, ie $K = 2\pi / L$.

Sampling

All tomographic imaging in medicine uses digital computers to collect the data, reconstruct the images and clean them up to remove artefacts and noise. The image data is collected as a series of discrete samples and this imposes some restrictions on the use of Fourier methods.

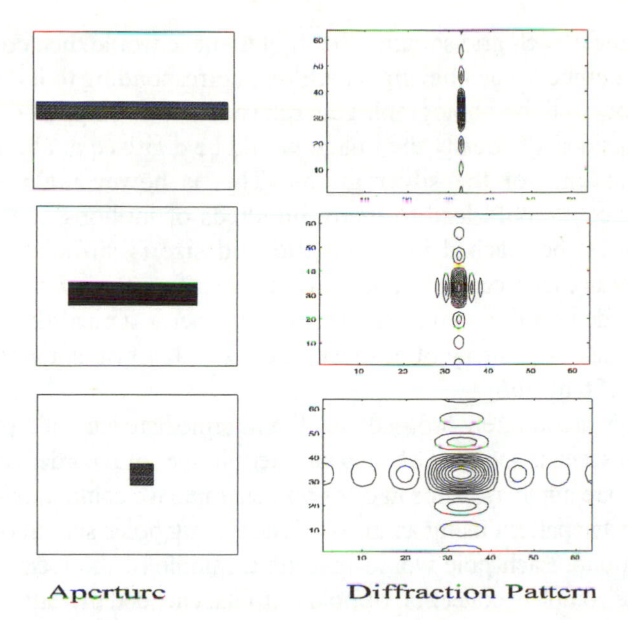

Figure A.6 : The 2D Fourier transforms (Diffraction patterns) associated with three slits or apertures. The left-hand column shows the apertures. The right hand column shows contour maps of the 2D Fourier transform computed for each aperture. These would be exactly similar to the diffraction patterns obtained by shining a laser beam through each aperture. As the slit length is reduced so the Fourier transform spreads out along the direction of the long axis of the slit.

The pond experiment will again serve as a useful illustration. A photograph of the pond waves provides such a high-resolution image of the complete scene that our eyes read that image as we would the real world. That is to say it appears to be continuous with no gaps between adjacent segments on the picture. In fact of course, under the microscope the photograph consists of a discrete array of silver grains, some of which are blackened by the exposure to light to form the image. Here the graininess is so fine that generally the human eye ignores or does not even resolve it. Computers work on numbers not the appearance of silver grains and thus any digital image must consist of an array of numbers that represent some contrast feature of the image. The photograph of the pond could be analysed by imposing a grid over the image and then counting the number of blackened grains in each grid square. The digital image would then consist of the array of these numbers, each having an address corresponding to its grid square. Clearly just how well the photograph gets relayed into the computer depends on our choice of grid size. Ideally the image would be digitised at about the same scale as the spacing of the silver grains. This is however almost entirely impractical since it would lead to many hundreds of millions of numbers. A compromise must be reached in which the grid size is sufficiently small to faithfully reproduce an acceptable amount of image information without incurring too large a burden on the computer memory. In fact a typical digital medical image will consist of an array of at most 1024 by 1024 grid squares and more typically only 256 by 256.

Image digitisation rarely proceeds via the intermediate step of a photograph. Rather the real scene is sampled using a smaller number of recorders or detectors whose outputs are numbers. Thus in our pond example we could imagine seeking to record the wave pattern using an array of measuring poles set out on a regular grid over the pond. Each pole would give us a sample of the local water level variations. The complete collection of pole outputs, encoded digitally, could then be assembled in a computer into a two dimensional picture. Each measuring pole with its connection to the computer is likely to be quite costly and so we would like to use as few as possible. The question then arises how small a number can we get away with and then, given our choice of sampling grid size, what precisely will we lose from our image. This is fertile ground for discrete Fourier analysis. It turns out that we must always use a pole spacing that is no more than 1/2 of the smallest wavelength (highest spatial frequency) that is likely to be important on the pond. In practice a spacing of about 1/5 or 1/10 of this value would used in order to deal with possible noise arising from random variations of the water level. If we chose a spacing of 1/2 wavelength then we will just be able to reconstruct the fundamental parts of the wave pattern under ideal conditions. Any feature in the pattern spanning a smaller distance than this will be lost. If a coarser grid were used then we could be in for a particularly nasty bit of trouble called aliasing. If we sample, even a simple wave, at more than 1/2 its wavelength then we will not be able to reconstruct that wave in the computer. This is called the sampling theorem and it is the undeclared first law of digital processing. Figure A.7 illustrates the problem. Aliasing is disastrous in all situations where a waveform is sampled too coarsely. The envelope of the discrete samples will represent a wave that is not actually present in the original continuous signal. In all applications of signal processing elaborate steps have to be taken to avoid the aliasing problem. The most important of these is filtering. For example if it is decided to digitise a speech signal for telephony at a rate of say 16 kHz, the average bandwidth of human speech, then the continuous signal must be heavily filtered, before digitisation, to remove all wave components above no more than 8 KHz. If this is not done then any reconstructed speech would

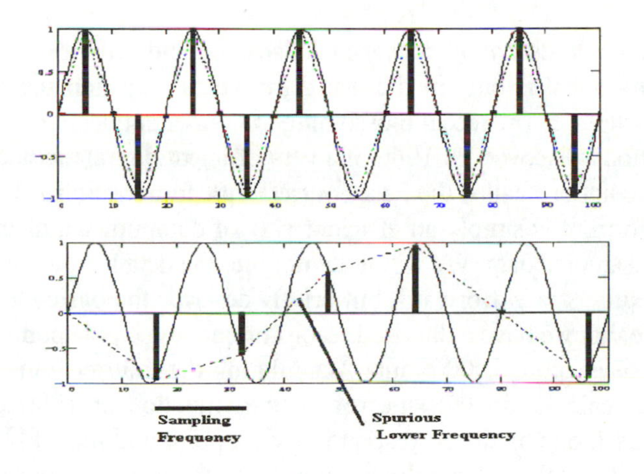

Figure A.7. The aliasing problem in digital signal processing. The first trace shows a single sine wave sampled at the rate of two samples in each period. The envelope of the samples has exactly the same frequency as the continuous wave. The lower trace shows the same wave but sampled at slightly less than one sample per period. Now the envelope of the samples represents a wave of a completely different frequency.

be heavily distorted In our pond example we would have to ensure that very small-scale mini-ripples superimposed on the main wave shapes were not shorter than the intervals between our poles.

In tomographic medical image reconstruction sampling considerations are just as important. In x-ray CT (see section 4.5) the projections are recorded by a set of x-ray detectors each of which has a finite width. This width sets an upper limit on the final spatial resolution of the image. In addition care has to taken to ensure that enough projections are collected to sample K space sufficiently finely. In MRI each projection is a time signal with temporal frequency encoding position, see section 6.4. The signal must be sampled in time sufficiently finely to resolve the highest frequency present in the signal, that is to say the sampling rate must be at least twice the highest temporal frequency. Higher resolution in MRI is obtained by using more intense gradient fields, which in turn involves higher NMR frequency modulation and thus necessitates more rapid sampling

The Fast Fourier Transform

Fourier methods have been used in science and engineering for over a hundred years but it is only in the last thirty years that their use has become nearly commonplace. The recent blossoming in Fourier applications is mainly due to the invention / discovery in 1960 of a very efficient digital method of actually doing the calculations called the Fast Fourier Transform or FFT. The FFT is not a new transform, it is simply an efficient way of computing a discrete Fourier transform of sampled data. We will not enter into the details of the method (even though it is quite straightforward) but briefly describe the basic idea.

Suppose that we have a digitised record of our single pole pond experiment, which comprises about 1000 points. Calculating the Fourier transform of this data involves calculating the sine (or cosine) function at 1000 time points, multiplying each one by the corresponding data point and then adding together the 1000 resulting terms. After this we will have obtained the Fourier amplitude at just one frequency. For each frequency the whole set of operations has to be repeated with a new sine function. It can be shown that, organised in this way, the number of calculations required to obtain the complete 1D Fourier transform of N data points will be proportional to NxN. Clearly as N increases the time taken, even on a modern computer, rises to prohibitive proportions. If we were instead interested in a 2D Fourier transform the situation is even worse. In fact, by organising the calculations in a clever way, the 1D transform can be obtained in just N log (N) operations. The reduction in computer time for 1000 points is by a factor of about 300. The several different algorithms that can perform this trick all rely on the same idea; there are very many identical intermediate multiplications, involving exactly the same factors. These only need to be calculated once, stored and reused throughout the calculation. The essential trick is to create a systematic way of identifying where these cases crop up. The invention of the Fast Fourier Transform is commonly attributed to Cooley and Tukey but, as they themselves acknowledged, Gauss used similar methods in

1805. Even though the FFT was in a sense only rediscovered in 1960 it was only then that digital computing had become sufficiently widespread for the method to have a major impact on a very wide and continually widening range of applications. Probably every digital medical image produced, now involves some use of the FFT; if not in tomographic reconstruction, then in filtering or correlation analysis.

Cleaning Up Noisy Images or Image Processing

In addition to enabling tomographic image reconstruction to take place, Fourier transforms are frequently used to modify and enhance existing pictures. Again our pond picture will illustrate the idea. Suppose that on the day of the famous pond experiment, a stiff breeze gets up after all the poles had carefully been put in place. The pond surface on that day was covered with random ripples so that instead of being a mirror the surface looked like stippled icing on a Christmas cake. What to do? Give up and try another day? A moment's reflection reveals that Fourier analysis will (with luck !) allow the experiment to go ahead. The experiment is performed as intended and then afterwards Fourier filtering methods are used to clean up the pictures. The idea is very simple. First a few pictures of the ruffled pond are taken and then the experiment goes ahead as planned. Finally a few more pond pictures are taken for good measure. Using the computer the pond pictures are Fourier analysed to obtain the spectrum of random waves created by the wind on the surface. Comparison of the before and after sets shows that, although in fine detail the wind waves vary in their shape, in Fourier space their transform $W(K_x, K_y)$ remains constant. Furthermore after Fourier analysis it is found that the splash waves are mostly at frequencies that differ from the wind waves. Although the real space picture is a mess the experimental picture in K space is a clear sum of the splash and the wind. If $W(K_x, K_y)$ is subtracted from the experiment $Q(K_x,K_y)$ and the resulting difference, Fourier inverted then a clear picture emerges of the splash waves apparently moving across a mirror-like pond. Figure A.8 illustrates the process. The rider "with luck" is important. Given that poles were already in place it would quite possible for the higher spatial frequencies of the wind waves to exceed the critical value of one half the pole spacing . Under these conditions the Fourier method would not work since then the wind waves would be aliased and spread

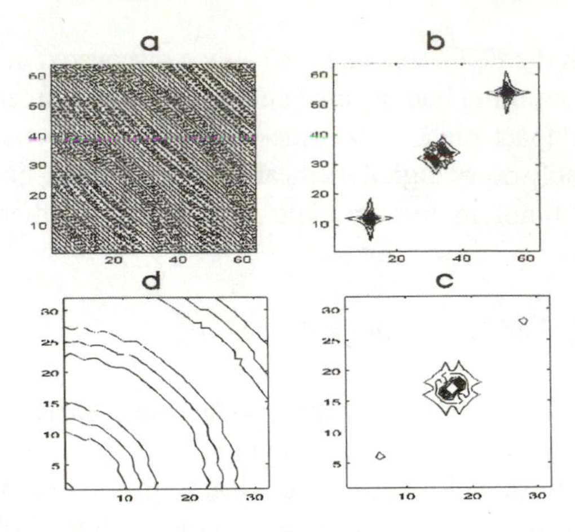

Figure A.8: Removing noise from images. One quadrant of the raw "mage" from the pond is shown in panel a. The 2D Fourier transform of this image is shown in b). Here the circular ripples are initially obscured by "noise" that comprise the shorter wavelength components(components further out in K space). A simple low pass filter applied to the K space image separates off the noise components to produce the filtered Fourier Transform shown in panel c). An inverse transform recovers a cleaned version of the original picture, shown in panel d). Here the circular ripples have a very much better signal to noise ratio. The noise has been much reduced.

throughout K space making it virtually and impossible to disentangle the required ripples from the effects of the wind.

Appendix B
Fourier Image Reconstruction

B.1 The Central Slice Theorem

In this section we will derive the exact relationship between the 1D Fourier transform of a projection and the 2D Fourier transform of the contrast function, $\mu(X,Y)$. For simplicity we will use the specific example of x-ray CT, in which $\mu(X,Y)$ is the linear x-ray attenuation coefficient for the material in the slice. In this case each element in a projection is the sum, or integral of the attenuation along a particular path taken by an x-ray beam across the slice. It will be

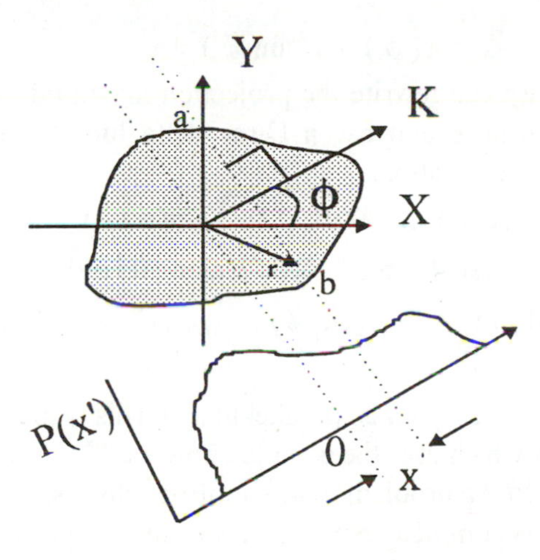

Figure B.1: The geometry of the derivation of the central slice theorem. A position within a single slice is described by the vector, r. An integral along a path from a to b passing through this point and making an angle ϕ with the Y axis of the slice produces one element of the projection $P_\phi(x')$. The profile, $F_\phi(k)$ produced by the 1D Fourier transform of the projection provides one line of data along the line K_ϕ in K space. The direction of K in K space is perpendicular to the line of sight used to create $P_\phi(x')$.

appreciated, from the discussion in chapter 1, that the final result is completely general and applies to any of the tomographic imaging schemes. The only differences are the physical nature of μ, the manner in which a projection is physically obtained and the manner in which K space is sampled. X-rays passing through a slice making an angle, ϕ, with the X axis produce the projection, $P_\phi(x')$. The magnitude of the projection, $P_\phi(x')$ at distance x' from its centre is given by the line integral through the slice from point a to point b. Thus

$$P_\phi(x') = \int_a^b \mu(X, Y, Z_{slice})dl \qquad \text{...B.1}$$

All slice points on the path a, b are represented by the relationship $\underline{\mathbf{k}} . \underline{\mathbf{r}} = x'$, where $\underline{\mathbf{k}}$ is the unit direction vector, perpendicular to the x-ray beam, which has components $k_X = \cos(\phi)$, $k_Y = \sin(\phi)$. In this case $\underline{\mathbf{k}}$ describes the direction in which the x-ray beam is translated in order to acquire the profile $P_\phi(x')$. The equation of a particular path across the slice ending at the projection position x' becomes

$$X\cos(\phi) + Y\sin(\phi) = x' \qquad \text{...B.2}$$

Using this expression we can rewrite the projection magnitude as a 2D surface integral over the entire slice and use a Dirac delta function to pick out the particular path under consideration.

$$P_\phi(x') = \int\int \mu(X, Y)\delta(X\cos(\phi) + Y\sin(\phi) - x')dXdY \qquad \text{...B.3}$$

If we take the 1D Fourier transform of $P_\phi(x')$ along its ordinate x' then we have

$$F_\phi(\underline{K}) = \int\int\int \mu(X, Y)\delta(X\cos(\phi) + Y\sin(\phi) - x')e^{-i'Kx} dXdYdx' \qquad \text{...B4}$$

where K is the spatial frequency taken parallel to x'. (This is the more usual form of a Fourier Transform, which uses the complex power $e^{-iKx} = \cos(Kx) - i\sin(Kx)$ to incorporate phase into the problem. Our simplified discussion in appendix A just used one part of this complete expression). Using the property of the Dirac delta function under the integral sign

$$\int \delta(x - a)f(x)dx = f(a) \qquad \text{...B.5}$$

we recover the 2D integral

$$F_\phi(\underline{K}) = \int\int \mu(X, Y)e^{-i(K_X X + K_Y Y))} dXdY \qquad \text{...B.6}$$

because $K\cos(\phi) = K_X$ and $K\sin(\phi) = K_Y$.

This is just a single line through the 2D Fourier transform of μ (X, Y), taken at the spatial frequency K_X, K_Y. Thus each projection, after Fourier transformation yields one line through the 2D transform. The collection of all such projections taken at successive angles φ, covering at least 180°, provides an estimate for the 2D transform throughout K space. A 2D inverse transform on this complete set of data then yields an estimate for $\mu(X,Y)$, the desired image quantity.

B.2 Filtered Backprojection

The Fourier method is rarely used in exactly this manner in practice. CT, SPECT and PET use the real space equivalent, filtered backprojection, described below. This scheme is rather more efficient in computer time than the direct Fourier method. In MRI the data arrives in the computer as a Fourier transform, and thus only a single inverse transform is required for reconstruction. Filtered backprojection is a variant of the following method originally derived by an astronomer, Radon in 1917 for the analysis of astronomical data. The derivation has 2 steps

1) We rewrite the expression for the complete 2D inverse Fourier transform of F (K) in cylindrical polar co-ordinates in K space.

2) After a little rearrangement, we obtain a sum over the polar angle, φ, of a modified version of each projection. The modification amounts to convoluting the original projection with a spatial filter function - the Radon filter. The scheme arises from the mathematical properties of Fourier transforms of convolutions.

Step 1

To begin with we write the real space contrast function μ as a Fourier sum, thus

$$\mu(X,Y) = \int\int F(K)\, e^{i(K_X X + K_Y Y)))}\, dK_X dK_Y \qquad \text{...B.7}$$

This expression uses cartesian co-ordinates K_X, K_Y to cover K space, but we would like to use cylindrical polar co-ordinates, K, φ to cover the same space. In polar coordinates in K space, the element of area becomes KdK dφ and the argument of the exponent is written as Kx'. Thus, the expression for μ becomes

$$\mu(X,Y) = \int d\phi \int K\, F(K)\, e^{iKx'}\, dK \qquad \text{...B.8}$$

This is a sum over the polar angle, φ, of 1D Fourier transforms of the product of F (K) and K. Now we use the fact that the Fourier transform of a convolution, is the product of the Fourier transforms of the individual functions

$$\int\int f(x)g(x-\tau))e^{-ikx}\,d\tau dx = \int f(x)e^{-kx}\,dx \int g(x)e^{-ikx}\,dx \qquad \text{...B.9}$$

The convolution in this case is the integral, $\int f(x).\, g(x - \tau)d\tau$, on the left hand side of this equation. Since our expression for μ involves the transform of the product, K F (K) in K space, we should be able to write this as a convolution in real space. The real space functions are the inverse transform of F (K) and the inverse the inverse transform of K. The inverse transform of F (k) is simply the single projection, $P_\phi(x')$. The inverse transform of K is not actually defined analytically since it is a linear ramp, continuing to infinity. In the practical case however there is always an upper bound, K_{max}, set by the finite sampling interval used to record the projection data. Thus we need the inverse Fourier transform of the function

$$x = |K| \;\; for\; x \le |K_{max}|$$
$$x = 0 \; for\; x > |K_{max}|$$

This is called the Radon or RAMP filter and it has the form

$$R(x) = K^2_{max}(2\, sinc\;(2\, K_{max}\, x) - sinc^2(K_{max}\, x)) \qquad \text{...B.11}$$

This is the function that is convoluted with each projection

$$\int P_\phi(x')R(x-x')dx \qquad \text{...B.12}$$

The complete 2D reconstruction then simply consists of two parts. Each projection $P_\phi(x')$ is convoluted with the radon function R (x-x'). Then results are added together.

$$\mu(X,Y) = \int d\phi \int P_\phi(x')R(x-x')dx \qquad \text{...B.13}$$

The advantages of filtered backprojection are that the filter function R (x) can be computed once, stored and used for each projection and that reconstruction can start as soon as the first projection has been measured. Thus, reconstruction and data acquisition are then interleaved in time.

One obvious outcome of convoluting a projection with the Radon filter is that the result has both positive and negative parts, as shown in the right panel in figure B.2. When all the projections, taken at different angles are summed, cells contributing no signal to the projection actually end up with no reconstructed intensity, leaving aside the effects of noise. In these cells temporary finite contributions, backprojected at one projection angle, are cancelled by equal and opposite contributions from other projection angles. This can also be seen in figure 1.4.

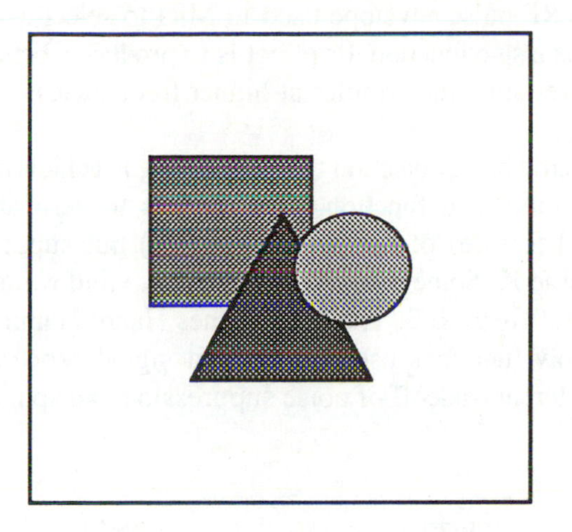

Figure B.2: a) The Radon or Ramp Filter: R (x') is the 1D Fourier transform of K for$K<K_{max}$. The function R (x) is convoluted with P (x') to yield the filtered backprojection. b) The Filtered Backprojection process. A single projection from the test slice on the left of the diagram is shown in the middle panel. The filtered projection obtained by convoluting this projection with the Ramp filter, R(x') is shown in the right panel. Notice that the filtered projection has both positive negative parts.

Practical Filters

Reconstruction from perfect, noise free, projection data is best accomplished using the ramp filter R(x') since this provides an exact result. Real data however often contains a random noise component and a ramp filter can actually accentuate the noise component in the final image. Thus in real tomographic reconstructions, the ramp filter is modified so that a reduction in image resolution is traded for an improvement in point to point fluctuations in the final image. In SPECT and PET applications, reconstruction is accomplished with filters that have names such as HAN or HAM or PARZ. Each acronym refers to a ramp filter that has been modified by one or another window function. These functions remove undesirable effects of having a sharp cut-off at K_{max} and conveniently reduce noise. In effect these modifications amount to low pass filtering, the projection data prior to reconstruction. Windowing is an example of the generic method of filtering in Fourier space that we described in Appendix A. In 6.4 we discuss the modification of the RF pulse envelope used in MRI to select a slice. There the "window function" is a sinc function. Its effect is to produce a better more square, slice profile and lower amplitude ripples at higher frequencies at the expense of a slightly wider slice.

In practical filtered backprojection the windowing is achieved by replacing K itself with more complicated function, that reduces to zero at large K. This preserves the broad features of the data (at low K) but suppresses the noise contamination at large K. Some examples of typical windowing functions are shown in left panel of figure B.3. The filter names Hann; Hamming and Parzen are standard window functions used throughout signal processing. Each one makes a slightly different tradeoff of noise suppression and spatial resolution .

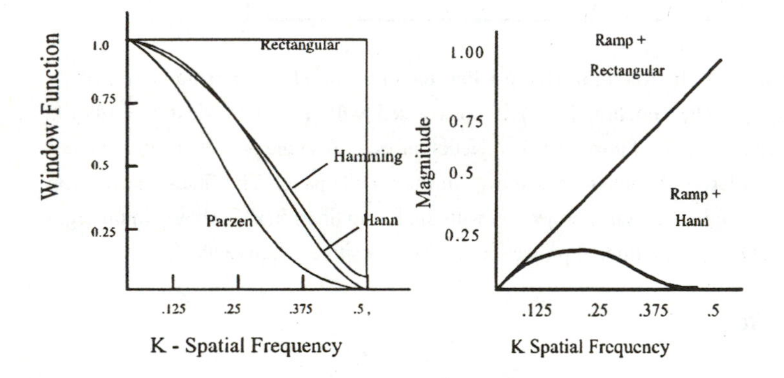

Figure B.3: a) The window functions, W(k), commonly used in filtered backprojection. b) The Fourier space filters corresponding to the combination of Radon "ramp" with rectangular and Hanning windows.

Appendix C
Mathematics of Nuclear Magnetic Resonance

C.1 Introduction

In this appendix we provide a sketch of the mathematics of NMR. NMR is fundamentally a quantum phenomenon; the nuclear magnetic moment itself does not exist within the framework of classical physics but depends entirely on a quantum explanation. Similarly the motions of the nuclear magnetic moment in the presence of an externally applied magnetic field should strictly be modelled using quantum mechanics. Curiously however probably no NMR spectrometer or MRI measuring sequence was ever designed, using the full panoply of quantum mechanics, rather a classical approximation was and probably always will be used in NMR engineering. Unfortunately this can lead to some confusion since the standard development of the theory switches back and forth between quantum and classical concepts, methods and pictures. Although somewhat less than rigorous, the hybrid account does however lead to a set of quite simple equations that can be solved to produce very accurate approximate quantitative predictions of NMR phenomena and provide a strong guide to physical intuition. An alternative rigorous wholly quantum mechanical development, on the other hand, would lead to very more complex equations and probably would not be intuitively very helpful.

We start in section 1 with a calculation of the bulk nuclear magnetisation or degree of alignment of nuclear magnetic moments in a strong static magnetic field. This relies entirely on a quantum description of quantum energy levels, split in energy, by the Zeeman effect. Once we have a nuclear magnetisation then we can treat it as though it was a single classical dynamical entity subject to classical forces. In section 2 we develop the Bloch equations that describe the precession and relaxation of the bulk nuclear magnetisation in terms of these classical forces. These phenomenological equations still form the basis of nearly all quantitative discussions in NMR and MRI

C.2 Alignment of Nuclear Moments in a Static Magnetic Field

A single nuclear magnetic moment, for example that associated with a single hydrogen nucleus, will tend to orient itself along the direction of an external magnetic field. The energy involved in this alignment process is however very small, in comparison with typical thermal energies and random exchanges of heat that tend to destroy any net magnetic polarisation. A dynamic equilibrium is established, between the two opposing tendencies resulting in slightly more moments pointing along the field than opposed to it. The task is to calculate the size of this polarisation and this requires a quantum description.

The hydrogen nucleus has an intrinsic angular momentum, $I = ½ (h/2\pi)$, and an associated magnetic moment γI, where the symbol, γ, represents the magnetogyric ratio for the proton. When placed in a magnetic field, B, the moment has two possible energies $-\gamma I B$ (along the field) and γIB (opposed to the field). This is the Zeeman effect. The restriction to just two possible energy values and thus two possible alignments is the essential quantum prediction. If there were no other forces acting then all the nuclear moments present in the field would eventually take up their lowest energy state and there would be 100% polarisation. However the difference in energy between the two states is very small, much smaller than typical thermal energies, kT. Thus, if the collection of hydrogen atoms is also in thermal equilibrium with its surroundings, at a temperature T, then there will be rapid fluctuations between the two states, with the emission and absorption of photons with energies $h\nu = 2\gamma IB$. A very much reduced static polarisation will result. In fact, at room temperature, even in an applied field of 1 Tesla, the two states are very nearly equally populated as a result of this thermal agitation. It is essential that we can calculate the precise degree of polarisation. Since there are just two orientations, the net number of moments pointing along the field is just $N^- - N^+$ and the magnetic polarisation, M is given by

$$M = \gamma\, I\ (N^- - N^+) \qquad \ldots C.1$$

Now, we invoke a general result from statistical mechanics, described originally by Boltzman in the C19, that endures the transition to quantum mechanics. If we have a system of particles in thermal equilibrium and these particles can have just two possible energies separated by an amount ΔE then the ratio of the number of particles in the two states is given by

$$\frac{N^+}{N^-} = e^{-\Delta E / kT} \quad \text{...C.2}$$

In our present case $\Delta E = \gamma IB - -\gamma IB = 2\gamma IB$. It is the size of this energy, in comparison with typical thermal energies, kT, that determines the ratio of energy level populations, and thus the degree of nuclear magnetic polarisation. If we combine our expression for M with the Boltzman factor then we obtain the result for M

$$M = N \gamma I \tanh \left(\frac{\gamma I B}{k T}\right) \quad \text{...C.3}$$

Where $N = N^- + N^+$ is the total number of protons under consideration, and tanh is the hyperbolic tangent function. We can simplify this a little more since the magnetic energy is actually very small in comparison with the thermal energy and then $\tanh(x) \sim x$, giving us

$$M = \frac{N \gamma^2 I^2 B}{kT} \quad \text{...C.4}$$

This is called Curie's law. Three important facts emerge from this equation

i) M increases linearly with the applied magnetic field, B.

ii) The nuclear magnetisation decreases linearly with increasing temperature

iii) NMR is an intrinsically weak phenomenon, since $\Delta E / kT \ll 1$

Putting the values of the physical constants into the equation, we find that the degree of polarisation at room temperature in a field of 1 T is about $1{:}10^6$. The size of the resonance signal is proportional to M and thus NMR signals are expected to be rather weak at room temperature.

Here, for convenience, we stop using the quantum picture and turn to a classical description that provides good quantitative results and provides a very useful quantitative picture of the resonance process. This switch frequently causes a conceptual problem for students. Throughout NMR and MRI there is frequent mention of nuclear spin flip angles that can take on any value. This is apparently completely at odds with the dictum that the nuclear proton spin can only have two orientations, either completely parallel or anti-parallel to the applied magnetic field. The resolution of this paradox is quite subtle and really beyond the scope of this text. A rough explanation goes like this. The up and down states are the so-called pure stationary or eigenstates of the spin in an external field. At a quantum mechanical level, it is possible to construct other temporary quantum states from these two pure states for which there is a component of the spin pointing at any angle to the applied field. In both NMR and MRI we are always dealing with very large numbers of similarly prepared spins and then the statistical average looks temporarily like a single large nuclear magnetic moment vector pointing at an angle to the external field. This is an example of what Bohr called the correspondence principle. When dealing with very large numbers of particles and a large total energy, the statistical predictions of quantum mechanics become closely similar to those of classical physics.

C.3 Classical Picture of NMR and the Rotating Frame

We will assume that we are dealing with a very large number of essentially identical proton spins, all doing nearly the same thing. We now use the term Bulk Nuclear Magnetisation, **M**, to refer to the nuclear polarisation of the collection of spins in a small region of tissue. It represents the vector sum of all the proton magnetic moments in the sample under investigation. For the most part we will treat it as though it were a single classical magnetic moment (compass needle), subject to external forces.

Classically such a magnetic moment, placed at angle to a static magnetic field, **B** experiences a torque ,**Γ** where

$$\mathbf{\Gamma} = \mathbf{M} \wedge \mathbf{B} \qquad \ldots \text{C.5}$$

The torque, acting perpendicular to both **M** and **B** (the vector product) tends to twist **M** into the direction of **B.** This is precisely what aligns a compass needle with geomagnetic North. But **M** is given by

$$\mathbf{M} = \gamma \mathbf{I} \qquad \ldots \text{C.6}$$

Where **I** stands for the vector sum of all the proton spin angular momenta in the little chosen volume. Using Newton's laws, a torque produces a rate of change of angular momentum and so

$$\Gamma = \frac{d\mathbf{I}}{dt} = \gamma \mathbf{I} \wedge \mathbf{B} \qquad \ldots \text{C.7}$$

The solution to this vector equation of motion allows us to predict the time evolution of the direction of the bulk magnetisation vector, after it has been tipped through an angle,α ,out of its equilibrium direction, and then allowed to progress, without any other external influences apart from the static field, B.

$$M_z = M_o \cos(\alpha)$$
$$M_x = M_o \sin(\alpha)\sin(\gamma B t) \qquad \ldots C.8$$
$$M_y = M_o \sin(\alpha)\cos(\gamma B t)$$

These results show that the Z component of the magnetisation remains fixed (in the initial orientation that was set by the perturbation) but the XY components undergo a simple harmonic precessional motion identical to that of a spinning top. The frequency ν, of this precession is given by

$$\nu = \frac{\gamma}{2\pi} B \qquad \ldots C.9$$

This is called the Larmor frequency. Notice that this is exactly the frequency of the photons emitted or absorbed in a spin flip transition since $\Delta E = h\nu$.

Pretty as this picture is, it isn't complete. It says that the precessional motion once started, will last forever. Intuitively we would expect the magnetic moment vector, once perturbed, to eventually return to its equilibrium state and the harmonic motion to cease. The equation is incomplete since we haven't actually included all the forces acting on **M.** This is the point at which the classical picture becomes a bit sticky and a complete quantum description would be very difficult. The magnetic moment is subjected to thermal motion, classically this is pictured as a fictitious random frictional force acting on **M**. As soon as **M** is perturbed away from its equilibrium orientation in the field, then dissipative forces (thermal motion) will tend to restore equilibrium in a time T1. In quantum terms the perturbation has altered the relative occupancy of the spin up and down states and the thermal

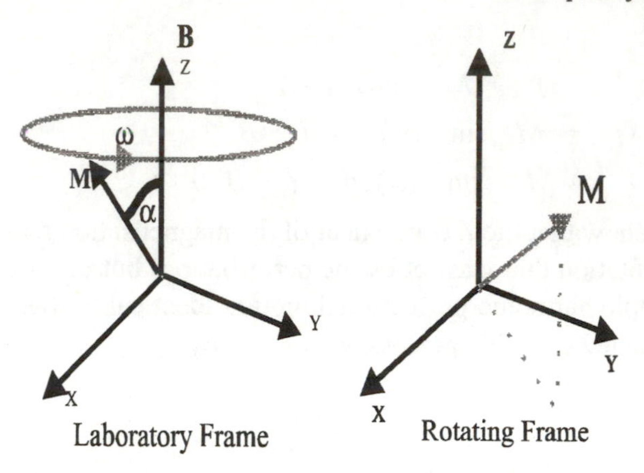

Figure C.1: A comparison of the laboratory and rotating frame pictures of magnetic precession In the laboratory frame the magnetisation precesses about the static field along Z. In a frame rotating at an angular frequency of ω, the magnetisation is static

motion induces random spin flips that serve to restore the initial equilibrium state. Classically we talk about the spin system relaxing back towards equilibrium. It is found empirically that this relaxation proceeds in such a way that the rate of change of Mz depends on the amount by which its direction deviates from that of the field. The relaxation is thus exponential in time with a characteristic time T1, the Longitudinal or Spin Lattice relaxation time. Classically this can be modelled by adding a fictitious force to the right hand side of our equation of motion for M, to mimic the effect of the thermal motion. Thus the z component obeys

$$\frac{dM_z}{dt} = \frac{M_o - M_z}{T1} \qquad \text{...C.10}$$

Where M_o is the thermal equilibrium value of the magnetisation.
The solution to this equation describes the return of the z component of nuclear magnetisation to its thermal equilibrium value, M_o

$$M_z = M_o \left(1 - e^{-t/T1}\right) \qquad \text{...C.11}$$

T1 varies with tissue type and this gives rise to the second flavour of MRI contrast

A similar treatment is given to the transverse components of magnetisation,

$M_{X,Y}$. **M** really comprises very many single spin contributions, each of which will be subjected to a quasi -static environmental field as well as the uniform externally applied field. In liquids this additional field is rather small in comparison to B (motional narrowing, see section 6.2 but it is not zero and furthermore varies from place to place within any small volume, on an atomic scale. Thus in reality each individual spin is actually precessing in a slightly different field and thus has a slightly different Larmor frequency. In consequence, if the precession begins with all spins making up M, lined up, or to be more precise, in phase, the small variations in static field causes M to break up. Some spins have slightly higher and some slightly lower Larmor frequencies and so instead of M progressing like a tightly controlled convoy, some of the components speed ahead and some lag behind. If we measure $\langle M_{XY} \rangle$, the average "length" of XY component of M, it no longer stays the same. In time, for every spin pointing in one direction there are spins pointing in the opposite direction. The spins are said to dephase in the XY plane as a result of the variation of static magnetic field from spin to spin. Experimentally the decrease in M_{XY} is approximately exponential and thus again we can write an approximate equation of motion by adding a fictitious term that accounts for the dephasing. This is the origin of the transverse relaxation time T2.

$$\frac{dM_x}{dt} = \gamma(\mathbf{M} \wedge \mathbf{B})_x + \frac{M_\infty - M_x}{T_2}$$
$$\frac{dM_y}{dt} = \gamma(\mathbf{M} \wedge \mathbf{B})_y + \frac{M_\infty - M_y}{T_2} \quad \ldots C.12$$

Since, in tissue, T2 varies with tissue type, it forms the third flavour of MRI contrast. Together the equations for **M** are called the Bloch equations, named after one of the discovers of NMR. They provide a very useful mathematical description of the nice picture of the bulk magnetisation relaxing and precessing following a perturbation. Longitudinal relaxation with a characteristic time T1 involves the return of the nuclear magnetisation to thermal equilibrium with an exchange of energy between the proton spins and the surrounding environment. Transverse relaxation, with a characteristic time T2, involves the dephasing of the constituent spins that make up the rotating magnetisation M_{XY} with no exchange of energy. This classical picture is extremely suggestive of how NMR can be manipulated using time varying magnetic fields and the Bloch equations can be used to predict what will happen. Figures C.1 and 6.3 illustrate these ideas.

Excitation using RadioFrequency Waves

Throughout both NMR and MRI, the first step in any experiment is to create the rotating magnetisation in the XY plane. This is always accomplished by using a radiofrequency field tuned to the Larmor resonance frequency, $\nu = \gamma B/2\pi$. Formally then we need to describe the motion of M under the influence of both a static field B and a time varying field

$$B_1(t) = B_1 \cos(2\pi\, \nu\, t) \qquad \ldots C.13$$

This is done formally by solving the Bloch equations, with a change in the first term, replacing B by B $+B_1$. B_1 is directed along the X or Y direction in order to twist M away from its equilibrium orientation along the static field B. parallel to z. Rather than use a brute force solution to Bloch's equation, we will use a pictorial argument. This is the rotating frame picture of magnetic resonance. We follow a development of the rotating frame given by Bleaney and Bleaney. First we have to describe how dynamics apparently changes in rotating coordinate systems.

Rotating Frames of Reference in Dynamics

We use Newton's laws to determine dynamical behaviour of moving objects on the earth. If we happen to be in the business of shooting heavy objects at high speeds from one place to another then it will quickly become apparent that, in addition to air resistance and gravity there are two other apparent forces acting on our projectile. These fictitious forces, centripetal and Coriolis arise because we are judging ballistic properties while sitting on a rotating earth. On our apparently stationary and flat earth it is easy to forget that we are in fact spinning through space with an angular velocity of 10^{-4} rads/s , the daily rotation of the earth. If we were doing our ballistics sums for projectile motions around the earth, whilst sitting on a distant planet then the earth's rotation would be obvious and its effect on the mathematics hard to overlook. In fact there is a nice mathematical trick that allows us to transfer from a stationary to a rotating frame without having to journey to a distant planet. If we consider any dynamic quantity ,**Q** we can relate its rate of change **dQ**/dt in one frame, to **DQ**/Dt the apparent rate change, in a co-ordinate frame rotating at an angular frequency ,**ω**, This can be achieved by writing

$$\frac{d\mathbf{Q}}{dt} = \frac{D\mathbf{Q}}{Dt} + \omega \wedge \mathbf{Q} \qquad \text{...C.14}$$

Returning to magnetic resonance we imagine that the time varying field ,B_1 is circularly polarised, rotating in the XY plane and that we can sit on the tip of rotating field vector and watch what is happening to the proton magnetic moment. We apply our frame transformation to the nuclear magnetisation so that

$$\frac{DM}{Dt} = \frac{dM}{dt} - \omega \wedge M \qquad \text{...C.15}$$

But , we have seen from Newton's laws (ignoring friction for the time being), that we can relate the rate of change of M in the laboratory frame, dM/dt, to the torque supplied by the static field. Thus the torque equation becomes

$$\frac{DM}{Dt} = \gamma M \wedge B - \omega \wedge M = \gamma M \wedge (B + \frac{\omega}{\gamma}) \qquad \text{...C.16}$$

Where the change in sign arises from the reversal of the order of ω and B in the vector product. This equation can be interpreted by saying that, in the rotating frame a field, $B+\omega/\gamma$ apparently supplies the torque. If we chose our rotating frame in the right way, then this field will vanish and the spin will be at rest in that rotating frame. The right choice is the resonant or Larmor frequency of the spins. If we make $\omega = -\gamma B$ then we are sitting at rest with respect to the spins. If we apply our RF field B_1 at just this frequency, with a small amplitude, then to first order the resonance frequency is not altered and, in the rotating frame, the nuclear moment only sees B_1. Generally if we supply a small circularly polarised RF field at ω then in the rotating frame the spins see a total field

$$\begin{aligned} B_z &= B + \frac{\omega}{\gamma} \\ B_x &= B_1 \end{aligned} \qquad \text{...C.17}$$

The spin then precesses about this effective total field. As the frequency approaches the Larmor frequency so the z component of the apparent field in the

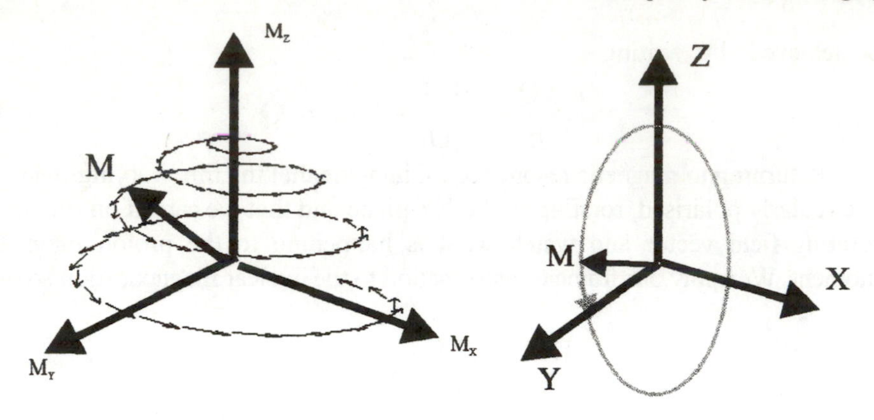

Figure C.2 : Under the influence of the Static field B and RF field B. The Nuclear magnetisation ,M, precess about Z and simultaneously tips about the X axis (the direction of B_1).In the rotating frame, M simply to rotates about X

rotating frame reduces to zero, leaving just the RF field which appears stationary, because we are rotating at its frequency. In the rotating frame the spins must precess about the B_1 field with a frequency $\omega_{precess} = \gamma B_1$, but this precession is about the X axis. Thus the application of a small RF field along the X axis results in the nuclear magnetisation being tipped away from its equilibrium orientation at a rate, $\omega_{precess}$. In general B_1 is very much smaller than B and so the spins execute many revolutions about the Z axis in the laboratory frame, in the time that they rotate about X.

BIBLIOGRAPHY

General References

Aird E.G.A (1988) *Basic Physics for Medical Imaging* Butterworth Heinemann,Oxford

Bleaney B.I. and Bleaney B. (1965) *Electricity and Magnetism* chaps 20,23: Oxford Oxfrod University Press

Eddington A. (1928) *The Nature of the physical World* ,Cambridge

Eisberg R. M. (1967) *Fundamentals of Modern Physics* New York, Wiley

Kittel C. (1965) *Introduction to Solid State Physics* 3rd edition J. Wiley

Kreel L., Thornton A. (1992) *Outline of Medical Imaging* , Butterworth Heinemann Oxford.

Sutton D (editor) (1998, Textbook of Radiology and Imaging 6th Edition Churchill Livingstone, New York

Webb S. (1988) *The Physics of Medical Imaging* . Bristol, Adam Hilger edited by S. Webb

Chapter 1 Tomography

Ambrose J. and Hounsfield G.N. (1972) *Computerised Axial Tomography* British Journal of Radiology **46**,148-149

Dümmling K., (1984) *10 years of computed Tomography- a retrospective view,* Electro Medica **52**,13-27

Hounsfield G .N. (1980) *Computed medical Imaging* (Nobel Prize Lecture) Science 210, 22-28

Radon J. (1917) *Uber die Bestimmung von Funktionen durch ihre Integralwerte langs gewisser Mannigfaltigkeiten* Leipzig Math. Phys. **69**, 262-277

Tetel'Baum S.I. (1957) *About a Method of Obtaining Volume Images with the Help of X-rays* Bull. Kiev Polytechnic Inst **22**,154-160

Chapter 2 Atomic and Nuclear Physics

Christy R.W., Pytte A (1965) *The structure of Matter : An Introduction to Modern Physics*,Benjamin, New York

Eisberg R.M.,(1967) *Fundamentals of Modern Physics*,Wiley New York

Lipson S. G., LIpson H (1969) *Optical Physics* Cambridge University Press, Cambridge

Feynman R.P. (1964) *The Feynman Lectures on Physics vols 1 and 2*, Addison Wesley, Reading, Mass

Chapter 3 Effects of Ionising Radiation and Radiation Protection

Kathren R.L., (1979) *Radiation Protection* Medical Physics Handbook 16 , Adam Hilger Bristol

Perkins A.C. (1995) *Nuclear Medicine: Science and Safety* ,John Lubbery ,London ,1995

Turner J.E. (1995) *Atoms, Radiation and Radiation Protection* , second edition Wiley Interscience, New York

Valetin J., Webb G.A.M (1993) *Medical uses of radiation : Retaining the Benefit but recognising the harm* , Swedish Rad. Prot. Inst. , and NRPB

Wall B. F., Hart D. (1991) *The potential for dose reduction in diagnostic radiology*, NRPB

ICRP (1991) *1990 Recommendations of the International Commission on Radiological Protection* ICRP pub 26 Annals of the ICRP **21** ,1-3

Hughes J.S., Shaw K.B.,O'Riordan M.C., (1989) *Radiation Exposure of the UK population -1988 Review* NRPB -R227 , HMSO London

Shrimpton P.C., Wall B.F., Jones D. G., Fisher E.S., Hillier M.C.,Kendall G.M., Harrison R.(1986) *A National Survey of doses to Patients Undergoing a Selection of Routine X-ray Examinations in English Hospitals*, NRPB -R200, HMSO ,London

UNSCEAR 1988 . *Sources,Effects and Risks of Ionising Radiation* ,United Nations Scientific Committee on the effects of Atomic Radiation. 1988, Report to the General Assembly, UN sales Publ. E.82.IX.8, United Nations ,New York

Chapter 4 X-rays

Barrett H.H., Swindell W. (1981) *Radiological Imaging Vols 1,2,* Academic Press London

Chesney D.N,Chesney M.O.,(1984) *X-ray Eqiupment for Student Radiographers* ,Blackwell Scientific,Oxford

Dobbs H.J., Husband J.E.,(1985) *The role of CT in the staging and radiotherapy planning of prostatic tumours,* Br.J. Radiol. **58**,429-536

Hounsfield G.N.,(1973) *Computerised transverse axial scanning (tomography) Part 1: Description of the system* Br.J. Radiol.,**46**,1016-1022

Mees C.E.K., James T.H. (1966) *The theory of the photographic process,* 3rd ed. Macmillan New York

Mistretta C.A.,(1979) *X-ray Image intensifiers : The Physics of Medical Imaging* ed A.G. Haus, AAPM New York

Chapter 5 Nuclear Medicine

Ansell and Rotblatt(1948)

Anger H. O. (1957) *A New instrument for mapping gamma-ray emitters* Biology and Medicine Quarterly Report UCRL-36553 January 1957,38

Belcher E.H., Vetter H. (1971) *Radioisotopes in Medical Diagnosis* Butterworths London

Bailey D.L., Jones T., Spinks T.J., Gilardi M.C., Townsend D.W. (1991) *Noise equivalent count measurements in a neuro-PET scanner with retractable septa* , IEEE Trans. Med. Imaging **10**,256-260

Maisey M.N., Britton E., Gilday D., eds (1982) *Clinical Nuclear Medicine* , Chapman & Hall London

Phelps M.E. (1986) *Positron Emission Tomography: principles and quantitation in Positron Emission Tomography and Autoradiography* ed Phelps M.E.,Mazziotta J.C., Schelbert,Raven, New York.

Sharp P.F., Gemell H.G., Smith F.W., (1998) *Practical Nuclear Medicine* ,Oxford Medical Publications,Oxford

Wrenn F.R. Good M.L., Handler P. (1951) *The use of positron emitting radioisotopes in nuclear medicine imaging,* Science **113**,525-527

Watson J.D.G. Myers R.S., Frackowiak R.S.J. Hajnal J.V.,Woods R.P., Mazziotta J.C., Shipp S. Zeki S. (1993) *Area V5 of the Human Brain:: Evidence from a combined study using Positron Emission Tomography and Magnetic resonance Imaging* , Cerebral Cortex, Mar/Apr1993 **1047**, 79-94

Chapter 6 NMR and MRI

Abragam A. (1983) *Principles of Nuclear Magnetism* Oxford: Oxford University Press

Bloch F. (1946) *Nuclear Induction* Physical Review **70**,460-474

Bloembergen N., Purcell E.M. Pound R.V. (1948) *Relaxation Effects in Nuclear Magnetic Resonance absorption* , Physical Review 73,679- 712

Fukushima E. (1989 *NMR in Biomedicine: The Physical Basis* Key Papers in Physics American Institute of Physics New York , AIP Edited by E.Fukushima

Hahn E. L (1950) *Spin Echoes*, Physical Review **80**,580-594

Lauterbur P.C. (1973) . *Image Formation by Induced local interactions : examples employing nuclear magnetic resonance*

Mansfield P. (1977) *Multi-Planar image formation using NMR spin echoes.* J.,Phys. **C10**:L55-58

Matwiyoff N.A. (1990) *Magnetic Resonance Workbook* ; New York, Raven Press

Morris P.G (1986) *Nuclear Magnetic Resonance Imaging in Medicine an Biology* Oxford , Oxford University Press

Purcell E.M., Torrey H.C.,Pound R.V.(1946) *Resonance absorption by nuclearmagnetic moments in a solid* Phys. Rev. **69**,37-38

Rinck P.A.(Editor) (1993) *Magnetic resonance in Medicine –The basic textbook of the European magnetic resonance forum* , Blackwell Scientific Oxford

Taylor R. (1988) *Nuclear Magnetic Resonance Imaging* Phys. Med. Biol. 983) Principles of Nuclear Magnetism

Turner R. (1992) *Magnetic Resonance Imaging of Brain function* American Journal of Physiologic Imaging 3,136-145

Vlaardingerbroek M.T.,den Boer J.A. (1996) *Magnetic Resonance Imaging* Springer Berlin

Chapter 7 Ultrasound

Fish P. (1990) *Physics and Instrumentation of Diagnostic Medical Ultrasound* Wiley Chichester

Foley W.D. (1991) *Colour Doppler Flow Imaging* Andover Medical publications, Boston

Taylor K. (editor) (1990) *Duplex Doppler Ultrasound*, Churchill Livingstone New York

Wells P.N.T. ed (1977) *Ultrasonics in Clinical Diagnosis*, Churchill Livingstone, Edinburgh

Woodcock J.P. (1979) *Ultrasonics*, Medical Physics Handbook ,Adam Hliger Bristol

INDEX

1

A

B

C

D

E

G

H

I

K

L

M

N

O

P

Q

R

S

T

U

V

W

X

Z